PRINCIPLES OF DATA FUSION AUTOMATION

For a complete listing of the *Artech House Radar Library*,
turn to the back of this book

PRINCIPLES OF DATA FUSION AUTOMATION

Richard T. Antony

Artech House
Boston • London

Library of Congress Cataloging-in-Publication Data

Antony, Richard T.
Principles of data fusion automation / Richard T. Antony
p. cm.
Includes bibliographical references and index.
ISBN 0-89006-760-0 (alk. paper)
1. Computer algorithms. 2. Database management. 3. Multisensor data fusion.
4. Decision support systems. I. Title.
QA76.9.A43A58 1995
006.3—dc20 95–15178
CIP

British Library Cataloguing in Publication Data

Antony, Richard T.
Principles of Data Fusion Automation
I. Title
629.89263

ISBN 0–89006–760–0

685 Canton Street
Norwood, MA 02062

International Standard Book Number: 0–89006–760–0
Library of Congress Catalog Card Number: 95–15178

10 9 8 7 6 5 4 3 2 1

This book is dedicated to my grandfather for being a superb role model;
my father, who encouraged me to write;
my mother, who encouraged me to become an engineer;
my children, Christopher and Allison, who with few complaints logged off the computer each evening so Dad could work on his project;
and my dear wife, Sandra, who has always helped me follow my dreams.

Contents

Preface

In its most general sense, data fusion is the process of combining evidence to support intelligence generation. Numerous real-world problems benefit from the combination of similar and dissimilar source information. Biological systems develop situation awareness through the analysis of sensory-derived information. Robotic vehicle control systems attempt to emulate the biological situation development process by combining the outputs from multiple on-board sensors. Earth resource studies increasingly rely on the correlation of satellite-based multispectral imagery with map-based information products. Future airport safety systems might one day fuse ground-based three-dimensional (3D) Doppler radar data with on-board range-finding sensors to identify potentially dangerous flight profiles.

Because of the central role intelligence plays in Command and Control (C^2), the Department of Defense (DoD) maintains a considerable interest in data fusion. Due to the combined effects of (1) the shortages of trained analysts, (2) the increasing volume of sensor-derived information, (3) the increased speed and lethality of weapon systems, and (4) the time-critical nature of military decisions, data fusion has been deemed a DoD *critical technology*. Because of both the author's Army-domain background and the metaphorical similarity between biological and tactical situation awareness, the tactical Army domain provides an excellent backdrop for investigating the general principles of data fusion automation. The principles that are developed extend readily to both air and naval applications as well as to a wide range of non-DoD problems.

Although biological systems routinely develop real-time situation understanding by fusing mixed-resolution, noisy, and ambiguous sensory information, the development of algorithms capable of generating a coherent and comprehensive situation awareness based on incomplete, uncertain, and possibly erroneous remotely sensed information presents a considerable challenge. Less than a decade ago, many tactical data fusion system developers sought the *Golden Algorithm*. With hindsight, the search for the "perfect" (but elusive) algorithm was much like the search for the pot of gold at the end of the rainbow. After extensive research and development efforts, both the difficulty and the diversity of the data fusion task are more widely appreciated.

Ideally, a large-scale data fusion system should be designed using a top-down, requirements-driven approach. System-level requirements would drive data-source information requirements, which in turn drive the development of sensor systems that provide the required information. In practice, data fusion systems have generally been developed in a rather bottom-up fashion, driven largely by the capability of available sensor systems, current doctrine, logistics considerations, computational performance limitations, and fiscal constraints.

The inherent difficulty of expectation-based machine reasoning, the information-limited nature of sensor data, and the highly constrained computational resources of fielded systems has led algorithm developers to explore a wide range of candidate information-processing disciplines—from conventional, well-established numerical-based algorithms to symbolic-based reasoning paradigms; from fractal-based modeling to the newest connectionist approaches. With such a wide diversity of technical approaches and no over-arching theory to guide paradigm selection, technical approaches tend to be driven by individual and institutional backgrounds or the latest, highly-touted "hot technology."

Although sophisticated detection, tracking, estimation, and pattern classification algorithms do exist, data fusion algorithm development has generally not kept pace with continually expanding system requirements and user expectations. This shortfall can be attributed, at least in part, to the lack of a formal engineering-level theory of data fusion and the rather fragmented nature of the research and development community.

From an information analysis perspective, data fusion automation requires the development of algorithms that are both *effective* and *practical.* To be effective, algorithms that address real-world problems must be both *robust* and *context sensitive.* Due to limited computational resources and the time-critical nature of decision making, to be operationally practical, the *algorithms,* as well as all required *database support,* must be highly efficient.

This book establishes a highly intuitive, top-down, system-level view of the data fusion process that reveals key principles associated with the development of both effective and practical fusion algorithms. Because technical jargon is avoided, mathematical formalisms are kept to a minimum, and numerous motivational examples are offered, readers with little more than a general technical background should be able to easily follow the material. The top-down character of the analysis tends to break down barriers between the traditional monolithic paradigms and the intuitive nature of the treatment yields insights that supports the development of considerably more powerful problem-solving approaches than are traditionally employed.

In attempting to identify "underlying (or general) principles" of data fusion, we will make numerous generalizations. Problem-solving paradigms will be "forced" into taxonomies, generic processing models will be formulated, and reasoning approaches will be categorized. Although such generalizations can require significant assumptions and simplifications, they can also permit deep insights into inherently complex issues. The author hopes that the reader will find that the benefits of this approach outweigh its shortcomings. As appropriate, simple metaphors are used to foster a better understanding of the material.

Due to the wide range of potentially relevant classes of external input (human observations, output from electromagnetic, acoustic, seismic, and imaging sensors) and the wide range of applicable analytical approaches (time-series analysis, image understanding, rule-based reasoning, simulated annealing, graph-based search, statistical decision theory), this book does not attempt to provide a comprehensive treatment of individual technical approaches. However, a list of recommended references is offered in the bibliography as a starting point for further study.

The book is organized into three parts. Part I describes a simple biologically motivated data fusion process model that reveals a number of important algorithm and database issues. The model attempts to provide a high enough level-of-abstraction view of the fusion process so as to avoid the traditional dilemma associated with the direct comparison of competing techniques (for example, tradeoffs between Bayes and Dempster–Shafer evidence accumulation methods) yet a low enough level of abstraction to offer meaningful insights into both the strengths and limitations of the wide range of problem-solving techniques. Although not a formal information-theoretic-based approach, the model provides a conceptual unification among a wide range of apparently disparate information-processing disciplines.

Part II discusses database-related issues that are relevant to the data fusion problem. Traditionally, there has been little intersection between the *data fusion* and the *database* research and development communities. This book attempts to demonstrate that there exist critical relationships between these two disciplines. In the near-term, the insights offered in Part I should help algorithm developers avoid some of the pitfalls of conventional algorithmic approaches; while in the longer term, the construction of database kernels, such as the one outlined in Part II, can facilitate the development of algorithms that better capture the character and sophistication of human problem solvers.

Part III brings together the algorithm principles established in Part I and the database principles established in Part II. Taken together, these principles represent critical requirements for effective machine-based reasoning. A number of robust, context-sensitive, and efficient data fusion algorithms are presented that demonstrate the practical advantages of effective database management system support.

This book would not have been possible without the Department of the Army's 6.1 Basic Research and 6.2 Exploratory Development programs and the enthusiastic support of my research over the past 12 years by a succession of first- and second-level managers, initially at the former Harry Diamond Laboratories (Charles "Harvey" Edwards, Richard Johnson, and Philip Emmerman) and later at the Army Communications-Electronic Command's Intelligence and Electronic Warfare Directorate (IEWD) (George "Bill" Mitchell, Calvin Eanes, Richard Poisel, James Mulligan, and Russell "David" Grubb). In addition, I am indebted to Professors Harry Van Trees and Clayton Stewart for their invitation to participate in the visiting Research Fellow program offered by the Center of Excellence in C3I at the George Mason University and their encouragement to begin the preparation of this manuscript.

I would like to acknowledge the credit due the members of the Joint Directors of Laboratories Data Fusion Group (DFG). My membership on this active and highly professional working group over the years helped me focus my research efforts and permitted me to become associated with many of the prominent members of the data fusion R&D community, including Edward Waltz, James Llinas, David Hall, Franklin White, and Otto Kessler. The impetus to write this book was spawned, in part, by the DFG's charter to enhance the intellectual basis of data fusion and foster stronger ties among technologists, system developers, and the user community.

The following individuals reviewed all or parts of this document and provided numerous constructive ideas: Clayton Stewart, Dennis Beude, Hanan Samet, and Richard Poisel. Additionally, I would like to thank two fellow researchers, Douglas Chubb and Terence Cronin, for the many stimulating technical discussions we have had over the years. Terry provided me with numerous references from his extensive personal math library. Christopher Bogart offered a number of suggestions regarding the book's organization. Several technical discussions regarding the shortcomings of conventional database systems with Michael Brichetti reinforced my interest in spatial data structures. Galen Jaskowiak provided an editorial review of a number of the chapters. Special thanks go to Richard Thompson, the senior graphics artist at IEWD who professionally prepared all of the figures and patiently supported me through several revisions. Finally, the numerous helpful suggestions and recommendations provided by the anonymous reviewer are gratefully acknowledged.

Introduction

Part I provides a broad perspective view of the tactical data fusion problem, identifies significant limitations in conventional algorithmic approaches, and recommends enhancements that can lead to the development of more powerful automated fusion algorithms. Chapter 1 presents an overview of the breadth, scope, and complexity of the data fusion problem. A widely referenced functional-level fusion model is discussed and justified using several natural metaphors.

In Chapter 2, algorithm development is shown to involve a two-step process. In step 1, a problem-solving paradigm is selected; in step 2, an algorithm is developed and refined. While much has been written about the development of specific classes of algorithms, the paradigm selection process has received considerably less attention. Because of this book's emphasis on establishing general principles of algorithm development, Chapter 2 focuses on paradigm selection issues rather than on the details associated with specific algorithms. An abstract fusion process model is developed, leading to a taxonomy of canonical problem-solving forms. Each canonical form exhibits well-defined characteristics that can be used to distinguish classes of algorithms. For a given fusion task, the selection of the proper canonical problem-solving form is shown to enhance both the quality of the fusion product and efficiency of the fusion process. While most conventional fusion algorithms are shown to be associated with the lower order canonical forms, human analysts routinely solve comparable problems using significantly higher order canonical form approaches.

Chapter 3 provides a number of illustrative examples intended to help clarify the fusion process model and explore the advantages and disadvantages of the various canonical problem-solving forms. In Chapter 4, individual problem-solving paradigms are mapped into the canonical problem-solving form taxonomy. Chapter 5 summarizes Part I by reviewing the relationships between algorithm characteristics and real-world problem-solving requirements. Fundamental limitations of the lower order problem-solving paradigms are shown to be significant impediments to the successful automation of complex fusion tasks.

An often stated goal of tactical data fusion is the production and maintenance of a "common picture of the battlefield." Thus, at least conceptually, there must exist a common database that can be accessed by all echelons within a spatially distributed network of analysis nodes. Since both the inputs to and the outputs from the fusion process can be stored, searched, and potentially even manipulated within the system database, the fusion process model proposed in Part I reveals critical relationships between the database organization and problem-solving paradigms. For instance, while simple, highly combinatoric fusion algorithms possess relatively high computational demands, they generally impose relatively small demands on database management systems. On the other hand, more computationally efficient, highly robust, and context-sensitive fusion algorithms require sophisticated database search and templating of massive domain knowledge bases.

Because many of the higher order canonical form algorithms proposed in Part I can severely stress existing database management systems, Part II addresses database design issues associated with the support of the higher order canonical form algorithms. Chapter 6 identifies database requirements that are derived from the proposed fusion process model. Conventional database systems are shown to provide relatively inefficient support to highly robust and context-sensitive fusion algorithms. To overcome such limitations, Chapter 7 establishes a top-down database design methodology based on the data fusion system requirements presented in Chapter 6. Chapter 8 outlines the architecture of a powerful database kernel that both supports the development of higher order canonical form algorithms as well as ensures highly efficient execution of those algorithms. Chapter 9 presents a number of database kernel extensions, including enhanced query languages, query optimization, and higher level-of-abstraction database shells. Chapter 10 summarizes the database issues developed in Part II.

Part III brings together the algorithm issues developed in Part I with the database issues developed in Part II. Chapter 11 summarizes the important relationships between algorithms and database services. Chapter 12 discusses advantages of the proposed database in terms of its support to the underlying data fusion system requirements for spatial, temporal, and hierarchical reasoning. Chapter 13 outlines a number of sophisticated higher order canonical form algorithms that demonstrate the strong relationships that exist between database representation/organization and robust, context-sensitive, and efficient machine reasoning. Chapter 14 offers a number of concluding remarks, recommends approaches for evolving more sophisticated reasoning paradigms, and suggests directions for future research and development.

PART I

Algorithm Issues

Data fusion enhances situation awareness and decision making through the combination of information obtained from one or more sources. Part I explores:

1. The inherent difficulties of combining evidence;
2. The limitations of traditional data fusion algorithms;
3. Effective methods of overcoming many of these limitations.

CHAPTER 1

Introduction and Perspectives

1.1 COMMAND AND CONTROL METAPHOR

The game of chess provides a natural metaphor for the Command and Control (C^2) process. Both chess players and battlefield commanders require a clear picture of the playing field to properly evaluate their own options as well as those of their opponent. In both chess and C^2, opposing players command numerous individual resources (pieces or units) that possess specific characteristics and capabilities. Both the resources and the strategies employed potentially vary over time. Groups of chess pieces are analogous to higher level organizations on the battlefield. The chess board itself represents domain constraints to movement that are analogous to terrain, weather, logistics, and other characteristics of the military problem. Player-specific strategies are analogous to tactics, while legal moves are analogous to doctrine. In both domains, the overall high-level objective of an opponent may be known but specific tactics and subgoals are typically unknown.

In spite of a chess player's complete:

(1) knowledge of the location of all the opponent's forces (pieces on the chess board),
(2) knowledge of all domain constraints (the chess board),
(3) knowledge of all the opponent's doctrine (legal moves),
(4) control over all his own assets,

chess is a highly challenging game. Although the two problems are metaphorically similar, the tactical commander typically possesses a considerably less accurate and less complete view of the playing field than a chess master. If a chess player was provided incomplete and inaccurate knowledge of the location and disposition of the opponent's pieces, the classically well-structured game of chess would begin to resemble the popular children's game of "Battleship" where players must effectively guess the location of an adversary's forces. Because individual sensors provide only limited observables, coverage, resolution, and accuracy, the analysis of individual sensor reports tend to lead to ambiguous and highly local interpretations of observables. The domain constraints for the tactical problem are generally considerably more complex than the well-structured playing field in chess. In addition, doctrinal knowledge in the tactical domain is both more difficult to use and less reliable that its counterpart in chess. Therefore, whether performed manually or supported by automation, the development of a coherent and comprehensive tactical situation awareness tends to be a highly challenging problem.

The data fusion discipline seeks to provide a holistic approach to information analysis that addresses the inherent complexities of the situation awareness problem. Despite a more global perspective and the use of all the available sensor-derived information, the nondeterministic nature of the domain, and the largely expectation-based character of the reasoning process virtually assures that there will exist some degree of uncertainty in the fusion product. This uncertainty can be minimized by employing an analysis process that is both robust and context-sensitive. Robust algorithms require appropriate modeling of the phenomenon under observation, while reasoning in context can provide effective constraints on expectation-based reasoning.

Although an individual consumer of fusion product may have very restricted information requirements, a data fusion system must maintain a rather global perspective to ensure that the significance of individual information fragments are recognized and properly utilized. The importance of analyzing individual pieces of data within a global framework is easily demonstrated by analogy. For example, although at the outset of a criminal investigation a single strand of hair at a crime scene might appear insignificant, it may ultimately prove to be the key piece of evidence that discriminates among several suspects. Similarly, although the small wedge-shaped rock at the top of a great archway is often dwarfed by the larger stones, without this keystone the arch could not stand. Thus, it is not just the information content (information measure) of a piece of data but also its relationship to other pieces of information that can be vital.

In general, the higher the level of command, the more abstract the required fusion product. For instance, at a *company* organization level, high-accuracy targeting information might be required; while at the *division* organization level, a coarser resolution representation might be more desirable. Because both the domain entities and the fusion products possess a multiple level-of-abstraction character, data fusion inherently requires a holistic, multiple level-of-abstraction approach to information combination.

1.2 SENSORY FUSION—BIOLOGICAL MOTIVATIONS

Even the most primitive animals sense their environment, develop *situation awareness,* and react to the acquired information. This situation awareness supports the survival of the species by assisting in the acquisition of food and in successfully avoiding being eaten by another animal. For instance, a barn owl fuses visual and auditory information to accurately locate mice under very low light conditions, while a fly avoids danger by responding to threatening visual and turbulence cues. Both the sensing and processing aspects of biological systems have evolved to support the development of more effective situation understanding. Color vision, for instance, is thought to have been a natural adaptation that permitted apes to locate ripe fruit among vegetation. In a similar manner, the development of artificial situation awareness systems requires the development of appropriate sensor systems and the ability to effectively fuse the acquired information in a manner that supports effective decision making.

Biological situation awareness can rely on a single highly sophisticated sense or on multiple, often less sophisticated senses. Examples of both models abound. The hawk has a highly acute visual search and tracking capability, and the dog has an excellent sense of smell. Sexual attraction may primarily depend on sight (plumage), smell (pheromones), or sound (mating call).

In biological systems that depend on multiple senses, the fusion process can involve either cueing between the senses or the true combination of information provided by independent senses. As an example of the former, the sound of a siren cues the visual sense to the presence and the general direction of an important event. As an example of the latter, at a considerable distance, an airborne glider, propeller-driven aircraft, helicopter, and jet may be difficult to distinguish. However, the characteristic sound of each craft (or lack of sound, in the case of the glider) coupled with the visually sensed information (approximate size and speed) might permit high-confidence identification.

In many cases, multiple sensory fusion is critical to a successful decision. Food that looks appetizing may be extremely salty (taste required), spoiled (smell required), or stale (touch required). At the other extreme, the fusion of multiple sensory inputs may be unnecessary if the senses provide redundant information that effectively overspecify the decision process. For example, bacon frying in a pan need not be seen, smelled, *and* tasted to be identified, because each of these senses could independently support the decision process.

Multiple sensory fusion need not involve only the conventional five senses of sight, sound, smell, taste, and touch. For instance, somatic signals, such as acidity of the stomach plus visual or olfactory cues, can trigger acute hunger pains. Vision coupled with inner-ear balance information and muscle feedback signals are fused to support motor control. In addition to its highly sensitive sense of smell, which can detect faint traces of blood, and hearing, which can detect the sound of a struggling fish up to one mile, a shark possesses receptors in its head that measure close-range electrical fields given off by animals.

Although a single sense might be easily deceived, multiple sensory fusion can reduce the likelihood of a decision error. Hunting and fishing, in fact, rely on the successful deception of multiple senses. Duck hunters use both visual decoys and duck calls to attempt to simultaneously provide deceptive visual and auditory information. Because a deer can sense danger based on the senses of smell, sound, *or* sight, to deceive the deer's threat awareness system, a shrewd hunter must mask his scent (or stay downwind), make little or no noise, and remain motionless if the deer looks in his direction.

As will be observed throughout Part I, multiple sensory fusion in biological systems provides a powerful metaphor for the development of multiple sensor data fusion systems. As with any metaphor, the study of a simpler or better understood phenomenon can often provide important insight into a more complex or less familiar one.[1]

1.3 THE FUNCTIONAL FUSION MODEL

A high-level functional view of the data fusion process is shown in Figure 1.1. The model was developed by the Data Fusion Group of the Joint Directors of Laboratories (JDL) Technical Panel for C3I.[2] The model both defines the scope of the data fusion domain and establishes a common framework that facilitates dialogue between researchers, technologists, developers, and users. The model has evolved over the years in an attempt to better characterize the fusion process [1].

There is a large class of sensing systems that can provide input to the fusion process. Hall [2] presents a tabular listing of various sensor classes, indicating both the basic observables and the information that is typically derivable from those observables. Since most sensors generate information relative to individual objects (for example, aircraft, tanks, and weapon launch), *object refinement* (level 1) typically represents the first stage of the analysis process. Level-1 fusion is associated with:

1. Object detection;
2. The association of new detections with known objects (observed previously by either the same or a different sensor);

1. Although a detailed understanding of fusion in biological systems is still in its infancy, the general utility and effectiveness of multiple sensory fusion are well understood.

2. The JDL Data Fusion Group (DFG) consists of representatives from DoD laboratories, other government agencies, and associate members from private industry and universities. The group's charter includes fostering the exchange of information and the development of a data fusion community infrastructure. The Joint Services Data Fusion Symposia held since 1987 are planned and sponsored by the DFG.

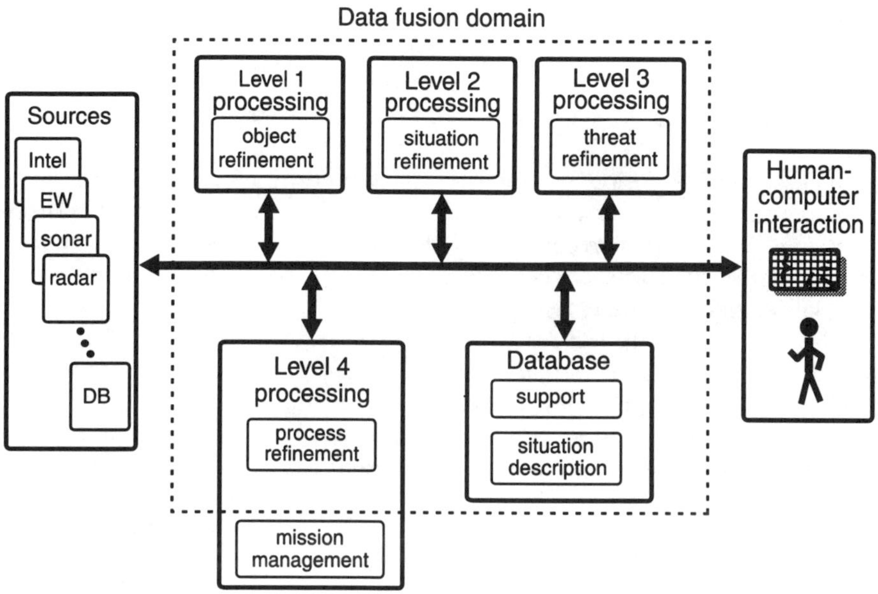

Figure 1.1 High-level functional model of the data fusion process.

3. Refinement of object attributes (such as position, velocity, and status);
4. Object typing, classification, and identification.

Level-1 fusion benefits from:

1. The use of heterogeneous sensors that provide at least partially independent sensor-derived information;
2. The employment of spatially distributed sensors to enhance the coverage, perspective, and reliability of the measurements;
3. The application of non-sensor-derived domain knowledge that can provide important problem-solving constraints.

Much of the early development oflevel-1 fusion algorithms focused on statistical-based approaches to detection, association, track estimation, classification, and identification.

In recent years, there has been an increase in automating the higher level fusion functions. Level-2 fusion (*situation refinement*) effectively extends and enhances the completeness, consistency, and level of abstraction of the situation description produced by level 1 to:

1. Compensate for the information-deficient and error-prone measurement space;
2. Resolve ambiguity in the level-1 products;
3. Develop higher level interpretations of the sensor-derived information.

Whereas level-1 fusion products are generated by reasoning based, for the most part, on sensor observables, the dynamic inputs to level-2 fusion processes come primarily from the current *situation description.* Level 2 performs a higher level analysis of level-1 fusion products based on organizational, logistical, doctrinal, and tactical knowledge of military forces.

The first stage of level-2 fusion performs *situation abstraction,* which includes both *situation generalization* and *situation specialization. Situation generalization* represents the bottom-up aggregation or abstraction of information for the purpose of generating situation awareness with respect to higher level-of-abstraction entities or organizations. For instance, given a priori knowledge that *corn fields* are typically associated with general *farming operations,* the recognition of an individual corn field from multispectral imagery indirectly provides evidence for a higher level-of-abstraction entity (a farm). When inadequate or inconsistent information exists to confirm such a generalization, the sensor network may be tasked (via information requests to level 4) to provide information that may either substantiate or refute the generalization.

Situation specialization represents top-down reasoning for the purpose of deducing or inferring subordinate elements or entities. An observed or inferred farming operation in a certain geographic region, for instance, suggests the existence of possibly unobserved, or unobservable, lower level-of-abstraction elements (such as barns, ponds, tractors, and stables). Evidence of a stable may suggest the existence of certain, as yet unobserved, residents of that building (horses). Once again, sensor tasking may be required in order to provide evidence to confirm or refute the existence of an inferred but unobservable element.

Taken together, *situation generalization* and *situation specialization* permit the development of the structural, organizational, and functional relationships among domain elements. *Situation abstraction* attempts to fill in missing information and to develop a more complete and integrated situation representation than is possible using reasoning based strictly on direct observables. Employed opportunistically, both situation generalization and situation specialization tend to drive the overall fusion system toward a consistent, comprehensive, and multiple level-of-abstraction situation description.

Situation Assessment

Situation assessment provides a higher level interpretation of the evolving dynamic situation description in terms of the relationship among perceived events and the status of entities and organizations. Situation assessment relies on expectation-based reasoning that depends on nodal, structural, and functional analysis. Although significant variations in the precise definition of the process are found in the literature [3], its general objectives are reasonably well defined.In this book, situation assessment is considered to be *threat-neutral.*

Threat Refinement

Threat refinement (level 3) identifies potential enemy intent and friendly force vulnerabilities based on the analysis of the current disposition of both friendly and enemy forces.[3] In effect, threat assessment represents a specialization of the current situation assessment. Since the situation is inherently dynamic, threat assessment must deal with both potentially near-term and long-term implications of the current situation. For instance, the identification of "an enemy missile battery moving toward City A" represents a high-level situation assessment product. Although the near-term threat might be minimal, in the long-term, two friendly divisions could be within range of the fully operational battery.

Because a potential threat can exist in either the presence or absence of a friendly force action, threat assessment represents a natural boundary between the data fusion domain and the balance of the command and control process. Thus, *Static threat assessment* resides on the fusion side of the notional boundary between data fusion and the remainder of the C^2 domain, while *dynamic threat assessment,* which is based on war gaming and engagement modeling, and *mission planning,* which seeks to both maximize friendly force effectiveness and minimize their losses, falls outside the scope of the fusion domain.

Process Refinement

Level-4 fusion seeks to optimize the overall fusion system performance with respect to a set of functional objectives by supporting global control of both the collection and analysis resources. In general, there exist a hierarchy of collection requirements extending from general surveillance to highly directed collection of detailed, high-resolution, high-accuracy measurements with respect to specific targets. Different classes of sensors lend themselves to different roles within the overall information requirements.When not controlled or guided by such a global strategy, sensors act as "vacuum cleaners," collecting, along with vital information, totally redundant, unnecessary, and unwanted information. The analysis of massive amounts of irrelevant data can severely burden both manual and automated fusion systems. In addition, level 4 fusion plays an important role in assessing the quality of the data to be fused.

---

3. The literature is somewhat inconsistent in its treatment of the term "enemy intent." However, since the principles to be developed do not depend on whether *intent* is established as a level-2 or level-3 product, we need not be concerned with such inconsistencies. Because we assume that level-2 analysis is "threat-neutral," *intent* will be treated as a level-3 product.

The importance of monitoring the quality of the information to be fused has been observed experimentally in predetection fusion systems [4].

Based on recognized information deficiencies and potentially available collection assets, level 4 generates prioritized information requirements that are sent to the collection manager. In addition, level 4 provides top-down control of the three lower fusion levels to help focus the overall reasoning process. The current fusion system objective, for example, might be to generate high-confidence and high-resolution locations of critical communication nodes and specific high-valued targets, while at the same time maintaining moderate-resolution wide-area surveillance of multiple areas of interest. Thus, level 4 seeks to optimally focus the information collection process and ensure the effective use of the acquired information.

Figure 1.2 provides a multiple level-of-abstraction view of the situation awareness process.

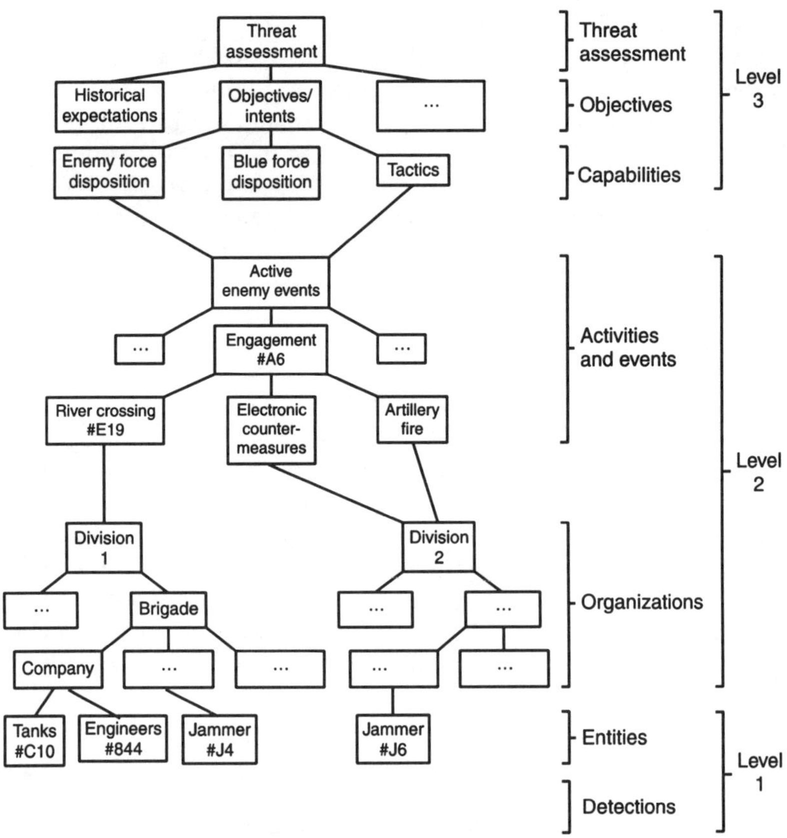

Figure 1.2 Multiple level-of-abstraction view of a dynamic tactical situation.

The following example demonstrates many of the key concepts that have been introduced so far. In particular, the example illustrates:

1. The role of the four-level functional fusion model;
2. The benefit of combining the output from multiple, heterogeneous sensors;
3. The advantage of employing non-sensor-derived domain information.

Consider a highly simplified tactical scenario involving the movement of an enemy truck convoy to resupply an engineer battalion engaged in constructing a temporary bridge across a tactically important river. The terrain in the vicinity is rather hilly, and the trucks are following a secondary road. Three "friendly" sensors—a Doppler moving target indicator (MTI) radar, a forward-looking infrared radar (FLIR), and an electronic support (ES) system—are observing the region of interest. An MTI radar can detect moving targets in a high-clutter environment; a FLIR produces an image of the thermal properties within its field of view; and an ES system measures parameters of electromagnetic systems, such as radar. Thus, the individual sensors measure relatively independent target attributes. For the purposes of this example, the sensors are assumed to be operating under conditions that favor both their individual performance and the effective combination of their output (specifically, high signal-to-noise ratio, adequate field-of-view, continuous coverage, and full registration among their outputs).

Level-1 fusion can potentially overcome a number of the inherent problems associated with traditional single-sensor processing:

Problem 1: *The MTI radar resolution cell size and target revisit rate are too low to permit reliable track association.*

Solution: Rather than a strictly statistical track association model, a road-following behavioral model could be utilized to help focus the expectation-based reasoning process. Even though inadequate sensor resolution, registration errors, and other deficiencies might not permit a radar target to be unambiguously associated with a roadway, a target can be inferred to be on a road if (1) the resolution regions associated with a sequence of detections intersect the road, and (2) the estimated vehicle speed is consistent with a road-traveling vehicle.

Problem 2: *With a conventional tracker, terrain masking and nonradial target motion tend to lead to highly fragmented target tracks.*

Solution: Given knowledge of both the underlying road network and the sections of the roads that are masked from the radar due to terrain, electronic attack (EA), or weather and sections of the road that are tangential to the radar, target tracks can be "tentatively" extrapolated along the road through such masked regions.[4] If subsequent detections are consistent with the extrapolated tracks, the track projection operation is confirmed. Similarly, because of the highly predictive nature of a road-following trajectory model, occasional missed detections due to purely statistical phenomenon (such as low signal-to-noise ratio) can also be tolerated. Thus, context-sensitive reasoning has the potential to significantly reduce track fragmentation.

Problem 3: *Because a Doppler radar can only detect moving targets, most of the relatively stationary equipment at the bridging location is undetectable.*

Solution: The FLIR system can detect targets based on their thermal properties, while an ES sensor can detect and possibly classify vehicles based on their distinctive electronic emissions. In addition, if the equipment was observed by the radar as it moved to its current location, the existence of multiple stationary targets can be inferred, based on the convergence of the various target tracks.

Problem 4: *Both the FLIR and ES systems have limited locational accuracy.*

Solution: The use of sophisticated signal-processing techniques, correlation between the FLIR and the ES, and/or correlation with the relatively high-range and azimuth resolution radar return (for moving targets) can greatly improve target location accuracy.

Problem 5: *The trucks have no unique electronic signature that can be used to support target classification.*

Solution: The observed speed, spacing, and road association of the trucks coupled with indications from both the FLIR and the MTI radar backscatter can potentially overcome this deficiency.

Although analysis based on the measurements from any single sensor may be inadequate, by properly combining the measurements from all three sensors, the generation of a coherent, comprehensive, and logically-consistent level-1 situation awareness product is conceptually feasible.

In level 2, the individual targets and events generated in level 1 are abstracted to produce a higher level-of-abstraction situation understanding. Targets observed near the river might be characteristic of elements of an engineer battalion and, because of their presence near a river and on the opposite side from the friendly forces, a bridge-building mission might be inferred. Based on the extent and the timelines of the various operations, trucks following each other at regular intervals might also be inferred to be either a supply convoy for the engineer battalion or part of the assembly operation in support of an enemy force river crossing. Regions where a large number of targets move into, but not out of, could be a staging area.

---

4. MTI radars can only detect moving targets that possess a radial component of motion with respect to the radar.

When targets do begin to move out of this region, the next stage of the enemy plan might be unfolding. Thus, level-2 analysis:

1. Reduces the impact of incomplete, ambiguous and erroneous level-1 products;
2. Hypothesizes completely unobserved, unobservable, or higher level-of-abstraction elements;
3. Identifies missing information and supporting information requirements.

This latter information becomes input to level 4.

In level 3, the threat posed by the hostile forces is developed by taking into account, for example, the overall disposition of friendly and enemy forces, apparent enemy strategies and objectives, friendly force vulnerabilities, range of observed or suspected weapons systems, and minimum travel time to contact. Since there can be multiple explanations for the observed bridge construction operation, the potential threat depends on the answer to questions such as:

1. Does the bridging operation offer an adversary a significant tactical advantage?
2. Is the bridging operation merely a prudent measure to ensure that a successful retreat is possible?
3. Does the bridge construction represent a form of deception?

Only if the bridging operation offers a significant tactical advantage to the opposing force does a significant threat exist. Since the higher level-of-abstraction situation awareness products possess considerable uncertainty, reducing the ambiguity of the threat assessment product requires a comprehensive evaluation of the available level-1 and level-2 products in the broadest possible context.

Lacking full knowledge of an adversary's intent, the maintenance of multiple competing situation and threat assessment hypotheses is inevitable. Competing hypotheses represent different interpretations of the available data, based on separate contexts, constraints, or reasoning models. The moving targets in the above example, for instance, might be either supply vehicles or combat vehicles. Support for valid hypotheses can be expected to increase over time, while support for false hypotheses or discontinued activities will tend to decay. For example, a relatively large number of vehicles are observed and the bridge has already been completed, combat vehicles might be the most likely hypothesis.

Figure 1.3 offers a representative expansion of each of the first four fusion levels into multiple subprocesses at progressively greater levels of detail. In level 1, when a new target is detected, its existence is verified, its location and trajectory refined, and its probable classification developed. In level 2, the target's association with a particular military unit is established, the association between the unit and the current enemy organization is hypothesized, and a description of the overall situation is developed. In level 3, the potential threat to the friendly forces is evaluated.

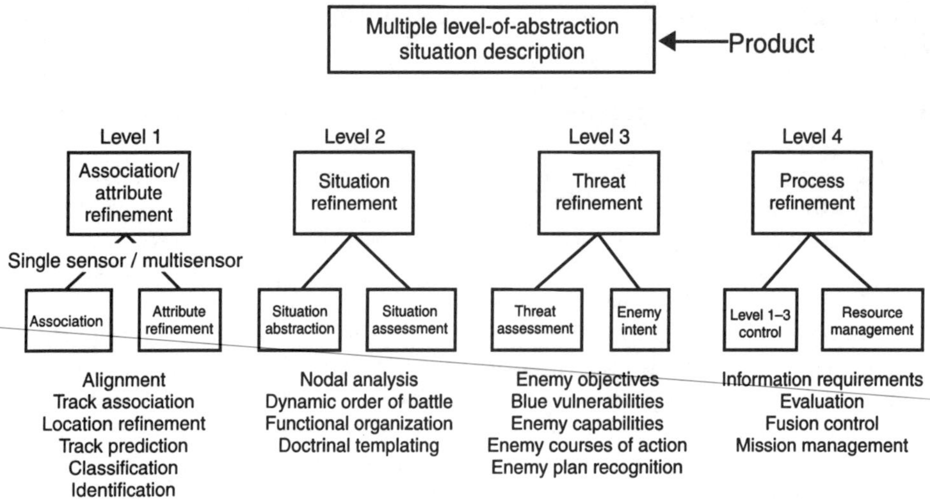

Figure 1.3 Representative functionality associated with the four fusion levels.

A number of the underlying characteristics of the tactical data fusion problem are summarized in Table 1.1. The overall situation awareness task consists of numerous dependent and independent multiple level-of-abstraction functional processes. Due to the three-dimensional nature of the world, spatial reasoning underlies the information combination process. Since individual sensors might observe the same target at different times and have very different processing and reporting delays, and because input from the situation description is asynchronous with respect to the sensor acquisition process, temporal reasoning is an underlying requirement of the fusion process. The problem domain of interest inherently possesses a hierarchical organization. The dynamic, nondeterministic character of the domain, the limited information content of sensor reports, and the nature of expectation-based reasoning ensure that there will exist a degree of uncertainty in the fusion products. Finally, real time performance is mandated by the time-critical nature of the tactical environment. All of the these characteristics will be explored in considerable detail in the following chapters.

1.4 FUSION SYSTEM ARCHITECTURES

The objective of the tactical data fusion process is to generate a "coherent intelligence picture." From a situation awareness perspective, sensor requirements should be dictated by the information product requirements of the user. However, rather than being the driver of the process, modern data fusion systems have generally been driven by the development and deployment of sensor systems. Therefore, data

Table 1.1
Characteristics of the Tactical Data Fusion Problem

Fusion System Character	*Inherent Problem-Solving Requirements*
Multiple level-of-abstraction features, organizations, and processes	Hierarchical reasoning
3-D character of the physical world	Spatial reasoning
Dynamic, evolving situation	Temporal reasoning
Uncertain data; uncertain decisions	Use all relevant information sources: sensor, domain knowledge, and domain constraints; maintain multiple hypotheses
Real-time requirements	Distributed processing; task decomposition; efficient algorithms; efficient data representation

fusion systems are effectively information-processing systems that are "built on top" of a data acquisition system. Given a specific set of sensors, a fusion system (whether fully manual or partially automated) must be optimized to support an appropriate range of functions, such as weaponeering, critical node analysis, enemy situation assessment, threat assessment, interdiction, and deception planning, based on specific user requirements (for example, accuracy, response time, and areas of interest). Level-4 fusion attempts to optimize the overall system performance with respect to multiple conflicting system objectives and constraints.

The architecture of a fusion system can range from highly *centralized* (or *monolithic*) to highly *distributed. Distributed* data fusion systems vary considerably with respect to the actual distribution of data, processing, and control. The appropriate distribution of all three elements can permit highly concurrent processing. *Centralized* data fusion systems offer the benefit that all data (and thus the maximum available information) exist at one location; whereas in *distributed* fusion systems, information can potentially be "lost" as a result of local processing and analysis. On the other hand, *spatially centralized* data fusion systems tend to have extensive and possibly unrealizable communications and data-processing requirements. Distributed data fusion is treated in considerable detail in [5]. Systems have been proposed that employ a hybrid approach, incorporating both centralized fusion for similar source sensors and distributed fusion nodes for dissimilar source sensors [6].

In addition, centralized systems are highly vulnerable to failure because of the critical nature of the central node in such an architecture. From a tactical perspective, because it represents a physically large, easily recognized critical element in the intelligence preparation and dissemination processes, a centralized system is highly vulnerable to attack. Thus, for multiple and somewhat independent reasons, the current trend is toward more spatially distributed data fusion systems. However, in many cases, the selected architecture is constrained by the nature of a particular sensor suite and their associated individual processors as well as the available communication bandwidths between nodes.

The blackboard (BB) problem-solving paradigm [7] provides a very natural, highly distributed architecture for large-scale data fusion system applications. The BB paradigm metaphorically models the collaboration among a group of human problem solvers seated around a chalkboard (the BB) that maintains the current state of the overall problem and a description of all existing subproblems. Each participating problem solver can possess inherently different knowledge, problem-solving styles, and experience. As their expertise permits, participants autonomously (and asynchronously) attempt either to solve subproblems or further decompose existing subproblems. A moderator maintains some degree of centralized control over the problem-solving session that can range from direct control over the evaluation of all participant's recommendations, solutions, and chalkboard updates (*dictatorial*) to no intervention, permitting individuals to independently update the state of the chalkboard (*anarchy*).

The BB problem-solving *metaphor* provides a conceptual framework for communication and result sharing among a group of specialized knowledge sources (KS) that cooperatively and/or competitively interact, under some degree of centralized control, with the evolving problem solution state maintained by the BB.[5] The BB paradigm is really a specialized form of object-oriented (OO) reasoning, which will be discussed in Part II. In general, *objects* possess arbitrary characteristics and degrees of autonomy. Objects can either directly or indirectly interact. Indirect interaction occurs when individual objects autonomously modify the overall problem solution state held by the BB. Modifications include posting locally generated solutions, partial solutions, hypotheses, and unsolved problems. From this viewpoint, the BB itself is merely a specialized object that maintains the problem solution state.

As shown in Figure 1.4, the functional-level fusion model of Figure 1.1 can be readily mapped into a BB-based problem-solving paradigm. The four functional process levels can be implemented as hierarchical sets of task-decomposed KSs. The interconnections between nodes represent KS communication paths. The dynamic situation description is maintained by the BB. Individual fusion processes can operate synchronously, be triggered by other processes, or operate fully asynchronously, responding autonomously to the state of the BB. Thus, control of individual fusion processes (the KSs) can range from highly distributed to highly centralized. Table 1.2 summarizes important characteristics of the BB paradigm that are relevant to the data fusion process. The organization of the BB is an important element of the paradigm.

5. Although KSs are often production rule–based, in general, this need not be the case.

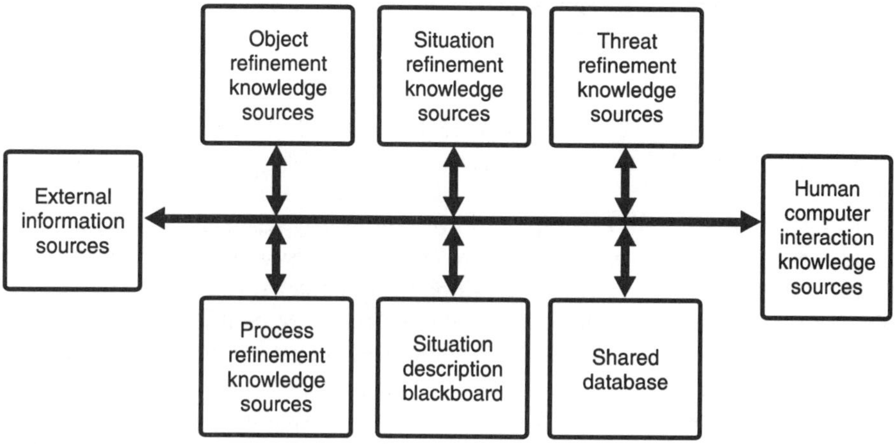

Figure 1.4 Natural mapping of the functional level data fusion model shown in Figure 1.1 to a blackboard-based implementation architecture.

Table 1.2
Blackboard Properties Relevant to Data Fusion

Fusion System Character	*Inherent Problem-Solving Requirements*	*Relevant BB Paradigm Properties*
3-D character of physical world	Spatial reasoning	Efficient spatial/semantic database organization
Dynamic evolving situation	Temporal reasoning	Dynamic uncertainty management
Multiple level-of-abstration entities and organizations	Hierarchical reasoning	Multiple level-of-abstraction KSs, BB partitioning
Uncertain data, uncertain reasoning knowledge	Use of a relevant info sources, maintain multiple hypotheses	Context-sensitive KSs, truth maintenance
Real-time requirements	Task decomposition, distributed processing, efficient top-down/global control, efficient database access	Local plus hierarchical KS control, inherent distributed processing potential, effective BB/DB representation

In Part II, a database organization that facilitates the implementation of a BB-based fusion system is recommended.

1.5 JUSTIFICATIONS AND ILLUSTRATIONS OF THE FUNCTIONAL FUSION MODEL

HEARSAY [7] was a ground-breaking effort to build a speaker-independent, connected-speech understanding system. The input to the system was a digitized signal that represented the utterance of a complete sentence from a particular task domain. The goal of the program was to generate the most likely syntactically correct sentence that was spoken. With a 1,011 word vocabulary, HEARSAY achieved 87% accuracy for speaker-dependent, continuous speech recognition.[6]

The process of understanding speech was based on a hierarchical reasoning approach supported by a BB-based system architecture. An utterance was first segmented and classified; syllables, and then words were identified. Since all possible pairwise word sequences in the task domain were known, constraints were applied to help eliminate errors in interpretation. Next, phrases were identified. Finally, the overall grammar of each candidate sentence was evaluated to identify the most likely interpretation of the overall utterance.

Both the HEARSAY problem-solving paradigm and its processing architecture readily map into the four-level data fusion model just discussed. Raw speech signals are analogous to raw sensor output. In HEARSAY, speech phonemes were extracted using predominantly signal-processing techniques. Therefore, the first stage of processing is analogous to level-1 fusion. At progressively higher levels of abstraction, grammatical knowledge is used to impose constraints on the prediction of words that may precede or follow other words. Alternative hypotheses for individual words, phrases, and sentences (analogous to level-2 fusion) were generated and meaning (analogous to level-3 fusion) was developed. In this way, the input information and all higher level products were interpreted within a progressively more global context.

In a data fusion system, raw sensor data is typically processed using predominantly signal-processing techniques, while more symbolic reasoning approaches are utilized to identify targets, units, and organizations and to attempt to assess intent and threat. In both the speech understanding and tactical data fusion applications, the data-driven process of collection and bottom-up processing is augmented by top-down, multiple level-of-abstraction reasoning. In both systems, by virtue of their higher level knowledge and more global perspective, the higher level processes util-

6. For speaker-dependent applications, contemporary speech recognition systems have achieved significantly better performance than HEARSAY. However, human performance is at least an order of magnitude better than the best speech recognition systems [8].

ize lower level products as well as provide a degree of control over those processes. Given both the system requirements and constraints, level-4 fusion seeks to optimally reduce the entropy of the fusion products through top-down control of both information collection and analysis process.

A further justification of the functional fusion model is offered in Figure 1.5. First, *input information classes* are associated with the three levels of the model. In addition, the answers to the questions *where, when, what, why, how,* and *so what* are associated with various levels of the model. Finally, the *information product classes* are associated with each level of the functional fusion model. Although the boundaries between the levels are not necessarily crisp, the multiple level-of-abstraction nature of the inputs, the fusion process, and its products are readily apparent.

1.6 DATA FUSION METAPHORS

Because tactical situation understanding requires the development and maintenance of an adequate multiple level-of-abstraction "picture" of a dynamic situation, the data fusion process can be compared to assembling a complex jigsaw puzzle for which no picture of the completed scene exists. Assembling a jigsaw puzzle contain-

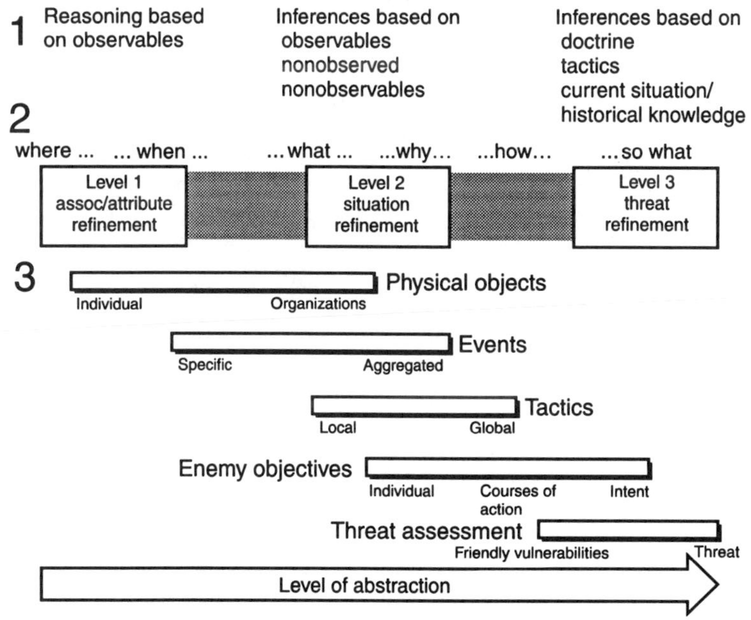

Figure 1.5 Three interpretations of the first three levels of the functional fusion model.

ing hundreds or thousands of pieces requires painstakingly placing individual pieces until either an adequate characterization of the picture has been developed or the reconstruction is complete.

Consider the puzzle shown in Figure 1.6 depicting a landscape scene that includes many aggregate objects (forest, meadow), each composed of numerous more primitive objects (trees, ground cover). Individual puzzle pieces might represent very low level-of-abstraction information (a fragment of bark on a single tree trunk, providing only color and partial texture). In a tactical data fusion problem, a specific puzzle piece is analogous to an estimated parameter derived from a single sensor measurement; a primitive object might be a single vehicle, while an aggregate object might be a company- or division-level force.

Because very little context exists at the beginning of the puzzle construction process, problem solving depends predominately on very general expectation-based reasoning. Since multiple interpretations of individual puzzle pieces are possible (a blue puzzle piece could represent either sky or water), without an established context, the placement of pieces tends to be quite difficult. This problem is closely analogous to performing level-1 fusion before a comprehensive situation description exists (system startup). However, as more level-1 products are developed (trees, grass, cliffs) and higher level-of-abstraction fragments (*analogous to level-2 products*) of the picture emerge (forest, meadows, mountains), the understanding of the current situation establishes a partial context that supports subsequent analysis. As the scene begins to take shape, the expectation-based reasoning process is more con-

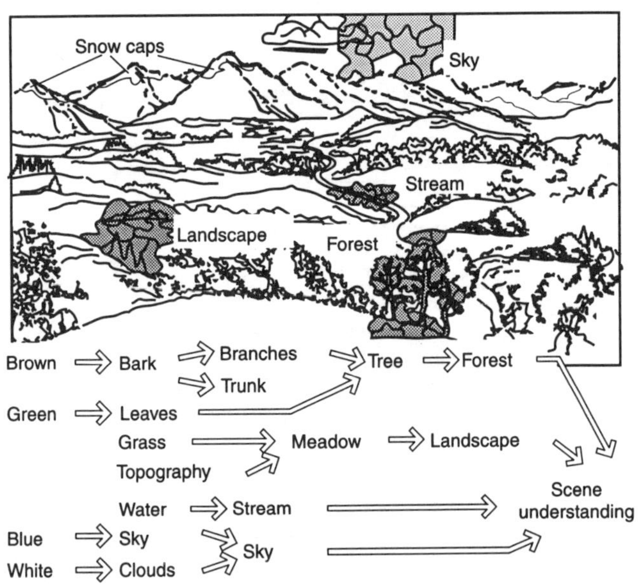

Figure 1.6 Data fusion puzzle-solving metaphor.

strained, making the placement of subsequent pieces easier. In addition, these constraints support a goal-driven search for specific missing pieces (*analogous to level-4 collection requests*). At some point, even though many puzzle pieces are yet unplaced (*undetected information*) and perhaps many puzzle pieces are actually missing (*information not collectible by the available sensor suite*), a coherent, multiple level-of-abstraction understanding of the overall scene begins to emerge.

Throughout the four levels of both the data fusion and the puzzle assembly process, the use of knowledge that is *not* explicitly contained within the sensor-derived information or the puzzle piece can provide a significant contribution to the fusion product. For the example shown in Figure 1.6, a high-level interpretation of the partially completed puzzle might be: "The scene appears to be a summer photograph containing a *western* U.S. mountain range." In this case, the specialization of the mountain range (*western*) was possible because the scene contained deciduous trees in full leaf (which implies nonwinter) in the foreground in addition to snow-capped mountain peaks in the distance. For summer snow cover to exist, the mountain peaks must be higher than eastern U.S. mountains. Thus, the use of domain context knowledge permits a more "complete" interpretation of the scene.

The process of solving a complex puzzle or developing a reliable automated situation awareness inherently requires *both* bottom-up (*deductive-based*) and top-down (*goal-driven*) reasoning. Identifying and fitting one information fragment often supports the identification and fitting of other related fragments. In the case of puzzle assembly, because there exist complex relationships among the primitive, composite, aggregate, and organizational entities in the scene, fitting a single piece not only reduces the overall entropy of the puzzle but can make adding several other pieces considerably easier. Just as duplicate or erroneous puzzle pieces would complicate the puzzle-assembly process, so does redundant or irrelevant sensor-derived information burden the situation understanding process.

In both the puzzle-assembly problem as well as the tactical situation understanding problem, it is difficult to define when an *adequate* representation of the situation has been achieved. In the case of the puzzle metaphor, although the general landscape characteristics might be evident, missing puzzle pieces could depict denizens of the woodland community that can be hypothesized but for which no evidence yet exists. In addition, individual puzzle pieces can contain partial or ambiguous information. The presence of a section of a log wall in the evolving scene, for instance, suggests the possibility that a log cabin exists even though the pieces that contain the balance of the cabin have not yet been discovered.

If many pieces of the puzzle are missing, puzzle assembly and subsequent analysis can be significantly more difficult. A simple example illustrates that, likewise, inadequate sensor-derived information cannot necessarily be overcome by any form of analysis. Figure 1.7(a) shows a cluster of azimuth and elevation measurements that are associated with two separate targets. Since the clusters have significant spatial overlap, no algorithm may be capable of reliably separating the measurements that are due to the individual targets. However, with the addition of a third information dimension (*range*), two distinct, easily separable clusters are readily apparent, as observed in Figure 1.7(b). Whereas even an extremely sophisticated algorithm may have been unable to discriminate between the two targets in the former case, an extremely simple algorithm can distinguish the targets in the latter case. Thus, although the use of non-sensor-derived information and sophisticated domain-sensitive reasoning can enhance problem solving, such knowledge may not be able to compensate for inadequate sensor-derived information.

1.7 CONTEXT-SENSITIVE REASONING

As observed in the puzzle assembly metaphor just discussed, human reasoning can be highly sensitive to domain context. Because of the importance of domain context to the development of effective data fusion algorithms, a simple analogy between biological fusion systems and artificial fusion systems is offered. Consider the information combination and decision-making task of a pedestrian who is attempting to cross a four-lane highway at a busy intersection. Once the light turns red, the pedestrian must decide when it is safe to step into the intersection. Although the problem appears rather trivial, the required data analysis task can be quite complex to emulate.

To maintain a metaphorical similarity with the comparable artificial data fusion application, the primary observables are assumed to be the *location* and *speed* of the several closest vehicles traveling toward the intersection. Although it *is* possible to formulate a decision based solely on this information set, early parental training and previous experience suggests that the rich context within which these vehicles are embedded should be considered. This contextual knowledge includes the time of day, the apparent condition of the road surface, and the current weather condi-

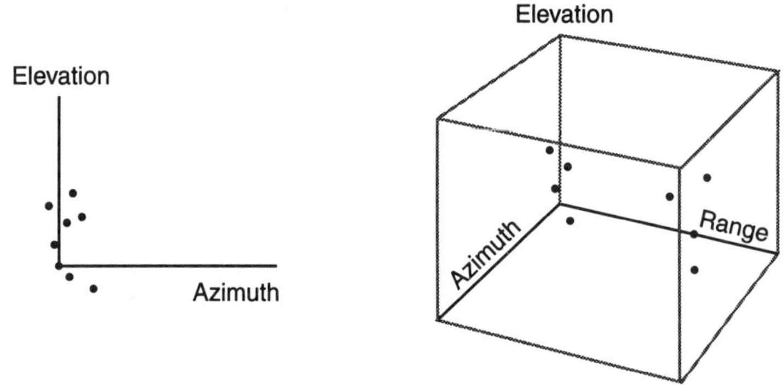

Figure 1.7 (a) Two-dimensional measurements from two targets, and (b) three-dimensional measurements for the same two targets.

tions. Each of these conditions might be relevant to an assessment of whether, once the light turns red, the various cars are likely to stop before reaching the cross walk or whether one or more cars are likely to "run the light." For instance, if the road is slick, it is after dark, and the car nearest the intersection is a shiny red sports car, extra caution might be warranted.

Although the required stopping distance for the cars may be predictable based on their location and speed, few pedestrians would want to rely on such mathematical calculations. The shortcomings of such an algorithm, of course, is that many relevant constraints (besides nominal braking rate) have been effectively ignored in such a calculation. While a seasoned pedestrian can rapidly combine diverse sets of observations, facts, and heuristics before performing a task as routine as crossing a road, the development of algorithms that emulate such performance is nontrivial, in part, because the process can rely on the availability of extensive domain knowledge. For many artificial data fusion applications, such a requirement entails the storage, search, and maintenance of potentially large databases.

While a human's visual system facilitates the collection of both "target" information and much of the domain information that establishes the problem-solving context, tactical sensor systems, for the most part, provide only remotely sensed "target-related" information. Thus, in artificial data fusion applications, the domain context generally must be supplied by a priori knowledge bases, including geographic and cultural feature maps, typical target behavior models, consistency constraints, and logistics requirements.[7]

The complexity of both the puzzle-assembly and the road-crossing decision problems illustrate the potential difficulty of fusing data from diverse information sources. In both cases, effective data fusion required the combination of both explicit observables and domain context knowledge. As we will see, rather than relying on rigid mathematical transformations, effective data fusion algorithms can require the use of extensive domain knowledge and much more sophisticated information combination approaches.

Because prototype data fusion systems are generally developed using relatively small databases, critical performance problems can develop when such systems are interfaced with large-scale domain databases. Since database search space size and computational requirements can scale exponentially with data set size, such performance problems are not readily overcome by either higher performance hardware or concurrent implementations. As if to add "insult to injury," much of the non-sensor-derived domain knowledge that must be available "on demand" to such

---

7. Much of the domain knowledge that is relevant to the tactical data fusion problem was actually derived by a sensor of some description (for example, terrain elevation generated from stereo imagery, feature data generated by multispectral satellite mapping systems, or weather sensors). Since some of these sensors provide dynamic information, the distinguishing characteristic between the *primary* sensor-derived information and context knowledge is that the principal purpose of the latter is to provide constraints on the interpretation of the former.

reasoning systems will be used infrequently, if at all. Thus, while potentially valuable, the maintenance of large-scale domain knowledge bases can entail a significant cost, in terms of both overhead and performance. The representation and organization of such information so as to help mitigate both of these problems is the subject of Part II of this book.

1.8 DATA FUSION LITERATURE

Waltz and Llinas [3] provide an overview of the tactical data fusion problem domain, present a wide range of issues related to level-1 to level-3 fusion, outline a number of specific system approaches, and discuss performance evaluation criterion. Hall [2] discusses many of the principal techniques used to build fusion algorithms, organized within a functional-level taxonomy developed by the author. Taken together, these two books offer an excellent introduction to the data fusion problem and process, and they provide an extensive list of references to the myriad of technical issues associated with the problem domain. More recently, Manyika and Durrant-Whyte [5] present a formal information-theoretic approach to data fusion, with a focus on decentralized processing and analysis. When probabilistic models of actions and information update are appropriate, the Bayesian-based framework permits the calculation of the expected utility of the sensing and fusion process. Decision making based on a strict information-theoretic approach is complicated by the required rigor and the associated computational difficulties associated with finding optimal actions. The present book seeks to build a conceptual bridge between more formal, quantitative approaches (axiomatic-based paradigms) and less formal, more qualitative approaches (artificial intelligence-based paradigms) to data fusion.

Two major annual national fusion symposia offer a valuable forum for technologists, developers, and users.[8] Other conferences focus on fusion subdisciplines (for example, Combat Identification System (CIS) Conference and Automatic Target Recognition (ATR) Conference) or specific problem-solving approaches (for example, Expert Systems, neural networks, Bayesian inference, and fuzzy-based reasoning). A bibliography is found at the end of the book for those interested in a more in-depth treatment of many of the subjects covered.

References

[1] White, F., O. Kessler, and D. Noble, "Data Fusion Model," *Proc. 1993 Joint Services Data Fusion Symp.*, Vol. 1, Laurel, MD, June 14–18, 1993, pp. 49–74.

[2] Hall, D.L., *Mathematical Techniques in Multisensor Data Fusion*, Norwood, MA: Artech House, 1992.

8. The Joint Directors of Laboratories' Joint Services Data Fusion Symposium and the SPIE National Sensor Fusion Conference.

[3] Waltz, E., and J. Llinas, *Multisensor Data Fusion*, Norwood, MA: Artech House, 1990.

[4] Kadar, I., et al., "Robust Centralized Predetection Fusion for Enhanced Target Detection," *Proc. Signal Processing, Sensor Fusion and Target Recognition* III, I. Kadar and V. Libby (eds.), *SPIE*, Vol. 2232, April 1994, pp. 37–55.

[5] Manyika, J. and H. Durrant-Whyte, *Data Fusion and Sensor Management: A Decentralized Information-Theoretic Approach*, Englewood Cliffs, NJ: Prentice Hall, 1994.

[6] Kadar, I., et al., "Pre-Detection Fusion for Enhanced Surveillance," *Substance Identification Analytics*, J.L. Flanagan et al. (eds.), Proc. Europto Series, Innsbruck, Austria, October 4–8, 1993, *SPIE*, Vol. 2093, 1993, pp. 512–523.

[7] Jagannathan, V., R. Dodhiawala, and L.S. Baum (eds.), *Blackboard Architectures and Applications*, San Diego, CA: Academic Press, 1989.

[8] Erman, L.D., et al., "The Hearsay II Speech Understanding System: Integrating Knowledge to Resolve Uncertainty," *Computing Surveys*, Vol. 12, No. 2, June 1980.

CHAPTER 2

Data Fusion Process Model

2.1 INTRODUCTION

Chapter 1 presented an overview of the tactical data fusion problem domain, discussed a functional-level fusion model, and explored several metaphors. This discussion provided insight into the inherent difficulty of generating a comprehensive situation understanding by combining disparate, fragmented, and potentially unreliable information. This chapter discusses a biologically motivated fusion process model that leads to a taxonomy of canonical problem-solving forms that permits a categorization of conventional fusion paradigms and promotes a top-down view of the algorithm development process.

Data fusion automation inherently involves a two-stage process. First, a problem-solving paradigm must be selected from among a wide range of candidate approaches. Second, a specific algorithm is developed based on that paradigm and tailored to achieve some required level of performance. For example, in stage 1, a developer might select a neural network paradigm; and in stage 2, the network type (multistage perceptron), the number of stages (3), the number of nodes per stage (20, 12, 3, respectively), and the training algorithm (backward error propagation) must be determined. Since a considerable effort may be required to settle on the final network parameters and then train the network, the vast majority of the algorithm development effort is associated with stage 2.

In this chapter, we focus of the first stage of the algorithm development process. Individual algorithm development efforts rarely expend significant resources evaluating alternative paradigms. In some cases, the paradigm selection is "by default" ("the contract called for an Expert System solution," "our corporation specializes in the design of second-order Kalman filters"). Perhaps the underlying reason for the lack of interest in the paradigm selection process is that there exists no widely accepted, formal selection methodology. Lacking a formal theory, paradigm selection tends to be rather subjective. However, unless the selected problem-solving paradigm adequately supports the underlying problem requirements, the resulting algorithm may fail to achieve acceptable performance regardless of the level of effort applied. One of the objectives of this book is to demonstrate the importance of selecting an appropriate problem-solving paradigm.

A number of attempts have been made to formalize the paradigm selection process. In terms of data fusion applications, Hall and Linn [1] developed a taxonomy of appropriate level-1 problem-solving approaches and Beude [2] identified suitable level-2 and level-3 problem-solving paradigms. Although these taxonomies offer a coarse guide to paradigm selection, neither reference proposes a general criteria for paradigm selection from within the class of applicable approaches.

Because there does not exist a widely accepted, unified theory that permits a formal comparison of paradigms, empirical comparisons provide an obvious alternative. A number of metrics have been developed and investigated for evaluating individual algorithms, for example, [3]. *Measures of performance* (MOP), for example, use parametrics such as detection probability, error rate, and target classification accuracy to compare the *performance* of competing approaches. *Measures of effectiveness* (MOE), on the other hand, attempt to capture the *effectiveness* of an algorithm in an operational environment using such measures as accuracy of information or target nomination rate. Even with carefully planned empirical studies, there is no guarantee that either the qualitative or the quantitative advantages of individual paradigms will be revealed. Generalizations based on such empirical metrics are even more suspect.

Thus, paradigm selection often relies on professional judgment, experience, and anecdotal evidence. Unfortunately, considerable disagreement exists among technologists as to the relative strengths and weaknesses of numerous candidate paradigms. For instance, claims and counterclaims about the potential of neural networks to generalize from examples are offered. Disagreements persist regarding whether Dempster-Shafer methods are more or less general than Bayesian methods. Fuzzy rule-based systems are viewed by some to be more general than conventional rule-based systems, while others argue they are subsets of expert systems. Likewise, there exist varying opinions on the relative merits of Bayesian inference networks versus more traditional statistical pattern recognition approaches.

The principal difficulty of comparing individual problem-solving paradigms is the obvious "apples and oranges" nature of such comparisons. A more detailed look at this familiar metaphor reveals important insights into the difficulty of comparing paradigms. In the case of actually comparing "apples" and "oranges," their *similarities* (both are fruits, both are similar in size and shape, both grow on trees, both produce seeds) tend to be greater than their *differences* (different flavors, pH, texture). Obviously, the smaller the differences between two entities (be they fruits or problem-solving paradigms), the harder they are to distinguish; conversely, the larger the differences between two entities, the more easily they can be distinguished. Thus, "apples" and "oranges" are distinguishable by the differences between their least significant properties.

Consider the advantages of the development of a *complete* taxonomy that exhaustively represents all meaningful categories. In a complete taxonomy of food classes, the class "fruit" summarizes the properties that are common to both apples and oranges. Such a taxonomy would clearly distinguish the food groups "fruits" and "grains," for example. Because taxonomies readily support *interclass* comparisons at all levels of abstraction, nonsibling leaf nodes (nodes that do not share the same parent) in the taxonomy are easily contrasted.

On the other hand, taxonomies do not support *intraclass* comparisons. However, by ensuring that the taxonomy is *complete*, the *differences* between leaf nodes within a single class tend to be relatively insignificant. For example, in the case of the class "fruit," a more complete taxonomy would include both "citrus" and "noncitrus" nodes. With this modification, an important criterion has been established for distinguishing "apples" from "oranges."

This discussion provides a simple view of both the potential strengths and weaknesses of a taxonomical organization of problem-solving paradigms; such organizations effectively discriminate among entities that are rather dissimilar but do not distinguish entities that are very similar. However, as long as leaf-node entities are more similar than different, the value of distinguishing them tends to be relatively small. Thus, the development of a "complete" problem-solving form taxonomy can permit an investigation of the fundamental properties of the various paradigms, which, in turn, can yield important insights into the paradigm selection process.

The taxonomy-based development presented in this chapter seeks to overcome the following three shortcomings:

1. The lack of a formal theory for comparing paradigms;
2. The inherent shortcomings of purely empirical approaches to paradigm evaluation;
3. The disadvantages of highly local, single-level-of-abstraction paradigm comparisons.

The next section establishes a highly intuitive model of the data fusion process. Section 2.3 presents biologically motivated extensions to this model that lead to the identification of 15 classes of fusion problems (Section 2.4) and a taxonomy of 16 canonical problem-solving forms (Section 2.5). Section 2.6 discusses the relationships between the 15 fusion classes and 16 canonical problem-solving forms. Section 2.7 provides a general discussion of constraint-based reasoning, and Section 2.8 offers a number of concluding remarks.

2.2 FUSION PROCESS MODEL

The goal of data fusion, as well as most other forms of data processing, is to turn *data* into useful *information.* In perhaps the simplest view of the data fusion process, all the required information is assumed to be present within a set of sensor-derived measurements. Thus, the role of data fusion is the extraction of information embedded in a data set ("separating the *wheat* from the *chaff*") by some form of filtering or transformation. As shown in Figure 2.1, such a fusion algorithm, $F_1(O, T, K)$, can be characterized as a function of:

1. The observables, O;
2. The current single level-of-abstraction tactical situation description, T (target track files, target locations);
3. A set of a priori declarative knowledge, K (distribution functions, templates, constraint sets, or filters).

The output of the fusion process provides updates to the situation representation and is fed back to the reasoning knowledge base. Thus, as sensor-derived information becomes available, this information and the relevant situation description information are transformed by the function, F_1. The output of this process is fed back to update the situation representation as well as to support the refinement of the reasoning knowledge base.

Signal processing, statistical hypothesis testing, target localization based on the intersection of two independently derived error ellipses, and target identification based on the correlation of an image with a set of rigid templates are simple exam-

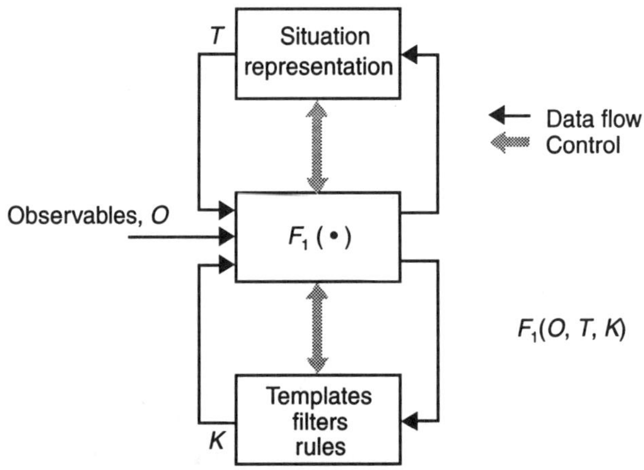

Figure 2.1 Simple, single level-of-abstraction fusion model.

ples of such a fusion model. In the case of the latter example, O is the input image, K is a set of templates, and the situation description T is updated as a result of a successful template match.

In general, this "information extraction" view of the fusion process makes a number of unstated, simplifying assumptions. Examples of such assumptions include:

1. A relatively small number of distinguishable targets;
2. Adequate sensor update rates;
3. A homogeneous sensor suite;
4. Adequate information content in the sensor observables;
5. Relatively high-resolution sensors;
6. High-reliability sensors;
7. Full sensor coverage of the entire area of interest;
8. Stationary Gaussian random processes;
9. A linear process model.

These assumptions tend to greatly simplify the data analysis process. For example, rigid templating approaches perform best when a set of observables closely matches a single template and is relatively uncorrelated with the balance of the templates. Track association algorithms perform well on radar return from closely-spaced detections of a single moving target. In general, the combination of similar features is more straightforward than the combination of disparate features. When the sensor data possesses adequate information content, high-confidence analysis is possible. High signal-to-noise ratios tend to enhance signal detection. High-resolution sensors reduce ambiguity and uncertainty with respect to feature measurements (location, frequency). High-reliability sensors maximize sensor availability. Adequate sensor coverage provides a more complete view of the areas of interest. Many forms of statistical-based reasoning are simplified when signal interference is modeled by a Gaussian random process. Finally, when superposition holds, relatively simple analytical processing models may be sufficient.

Conventional algorithmic techniques such as statistical decision theory, matched filtering, classical pattern recognition, parameter estimation, and rigid templating generally prove satisfactory when the aformentioned assumptions are valid. Typical applications where such assumptions are realistic, include:

1. Track assignment in low target-density environments;
2. Tracking ballistic target, which obey well-established physical laws;
3. Classification of entities based on radar types.

However, many data fusion tasks exhibit one or more of the following complexities:

1. A large number of target and nontarget entities (at a particular sensor resolution, garbage trucks may be indistinguishable from armored personnel carriers);
2. Within-class variations of individual targets (hatch open versus hatch closed);
3. Very low data rates can exacerbate track association problems;
4. Multiple sensor classes often generate disparate numeric and symbolic observables that can be difficult to combine;
5. Inadequate sensor coverage of areas of interest due to terrain, interference, and weather;
6. Inadequate set of sensor observables;
7. Inadequate sensor resolution;
8. Registration and measurement errors;
9. Inadequate a priori statistical knowledge (unknown prior and conditional probabilities, multimodal density functions, or non-Gaussian and nonstationary statistics);
10. Delays in processing and communication of measurements and/or analysis products;
11. High level-of-abstraction information requirements (not merely platform location and identification);
12. Complex propagation phenomenon (multipath, diffraction, atmospheric attenuation);
13. Purposely deceptive behavior.

When any of these complexities exist, the sensor-derived information may be incomplete, ambiguous, erroneous, and difficult to combine and/or abstract. In such cases, a fusion model based strictly on the composition among (1) the observables, (2) a single level-of-abstraction situation description, and (3) a set of rigid templates or filters tends to be fundamentally inadequate.

Rather than simply "extracting" information from sensor-derived data, the data fusion process model must be extended to include the combination, consolidation, organization, and abstraction of information in order to (ideally) maximize both the total information content and the utility of the resultant product. The information content of the fusion product can be enhanced in at least four ways.

1. New sensors can be developed to provide additional measurables; existing sensors can be improved to provide better resolution, accuracy, sensitivity, and reliability.
2. Additional similar sensors can be employed to improve the coverage and/or confidence in the domain observables.
3. The use of dissimilar sensors can increase the dimensionality of the observation space by measuring at least partially independent target attributes (a

radar may offer excellent range and azimuth resolution, while an electronic intelligence (ELINT) sensor can support target identification).
4. Domain knowledge can be applied to provide real-world constraints on the decision process.

While the first three recommendations effectively *increase* the information content of the observables, the incorporation of domain knowledge can *reduce* the effective dimensionality of the decision process by providing problem-relevant constraints. Thus, the objective of applying domain constraints is to reduce the *dimensionality* of the decision process to that of the available sensor-derived information.

Figure 2.2 depicts the logical enhancement of Figure 2.1 to include relevant non-sensor-derived domain knowledge *D*. While the observables *O* might be considered *explicit* domain knowledge (because they are explicitly provided by the sensors), the domain knowledge *D* will be considered *implicit* domain knowledge. Although human analysts routinely use both explicit and implicit domain knowledge to perform fusion tasks, automated approaches typically rely on the former.

As an example of the utility of such implicit domain knowledge, consider the extrapolation of the track of a ground-based vehicle that has been observed moving along the relatively straight line path shown in Figure 2.3. Although the target is actually a wheeled vehicle traveling along a road with a hairpin curve just beyond the

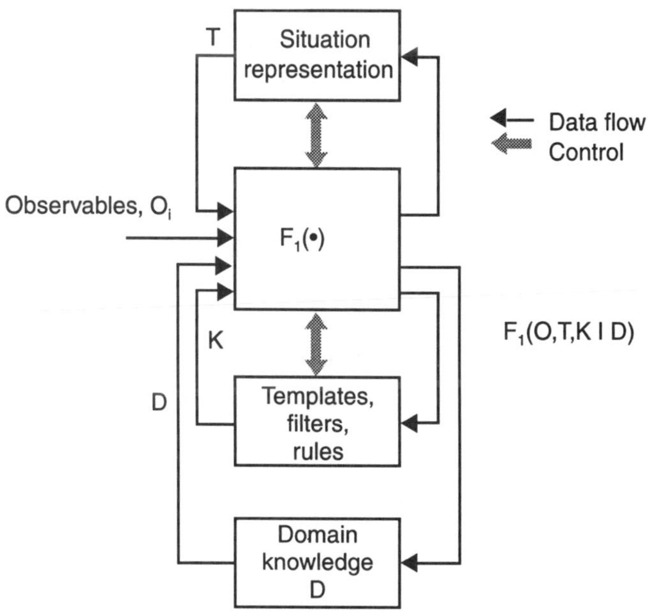

Figure 2.2 Single level-of-abstraction fusion model supporting context-sensitive reasoning.

last detection point, a purely statistical-based tracker may attempt to extend the track through the hill (which explains the hairpin curve) and into the lake on the other side.[1]

Target tracking of high-speed ballistic projectiles or relatively low-speed aircraft and naval vessels using statistical-based algorithms has been highly successful. However, many classes of problems do not lend themselves to such solutions. For instance, even civilian aircraft can maneuver rapidly enough so that "a very few of the existing schemes among the many in the literature (even adaptive ones with maneuver detection) can track such a target with good accuracy" [4]. Attempting to adapt statistical-based algorithms to the tracking of ground-based targets tends to be even more problematic. For example, terrain and cultural features tend to have a significant influence on ground-target movement. As a result, algorithm performance

Figure 2.3 Road-following target tracking model.

1. In general, a statistical tracker performs both track assignment and a track extension based on a statistically based motion model. The basic principles of the first order Kalman filter tracking model are outlined in Chapter 4.

tends to be unsatisfactory when relevant domain constraints are ignored. In much the same way that a system of equations with too many unknowns does not produce a unique solution, the "missing" domain knowledge in Figure 2.3 effectively led to an "underdamped" tracking solution.

The road-following target-tracking problem illustrates the critical role of paradigm selection to algorithm development. Rather than the failure of a statistically based tracker, the prior example demonstrates its misapplication. In applying a strictly statistical approach to the ground target-tracking problem, an assumption was made (perhaps unwittingly) that the deterministic domain constraints were either irrelevant or at least insignificant. However, in this particular application, the domain constraints proved to be considerably stronger than the relatively weak constraints on platform motion provided by a statistical tracking model. Thus, the primary objective of paradigm selection must be to ensure that the requirements of the fusion task match the inherent capability of the selected approach.

Because, as depicted in Figure 1.1, the situation description M is the composite product of the first three levels of the fusion process (individual object descriptions, composite object descriptions, and threat assessment), the fusion process model must be further generalized. Figure 2.4 depicts a generalization of the fusion process model depicted in Figure 2.2 that supports the combination of observables O, the a priori reasoning knowledge K, and the multiple level-of-abstraction/multiple perspective situation description M. All three classes of knowledge are potentially sensitive to (or conditioned on) the implicit domain knowledge D. Clearly, the models in Figures 2.1 and 2.2 represent special cases of the model depicted in Figure 2.4. The level-4 control loop has been included in Figure 2.4 to illustrate its relationship to both the fusion process and the sensor-tasking element.

Figure 2.5 depicts the essential elements of the data fusion domain and illustrates the high-level interaction between the various elements of the fusion process. In this view, the reasoning knowledge K, the domain knowledge D, and the current situation description M are resident within the system database.

2.3 BIOLOGICALLY MOTIVATED FUSION PROCESS MODEL EXTENSIONS

This section proposes a number of extensions to the fusion process model just presented that leads to the identification of 15 different classes of fusion problems and the development of a true taxonomy of problem-solving paradigms. As discussed in Chapter 1, the fusion of multiple sensor data to enhance situation understanding in a tactical environment is metaphorically similar to multiple sensory fusion to enhance situation understanding in biological systems. In this chapter, we attempt to exploit this similarity by developing a biologically motivated model of the data fusion process that yields important insights into both algorithm and database issues.

Important objectives of this effort are to:

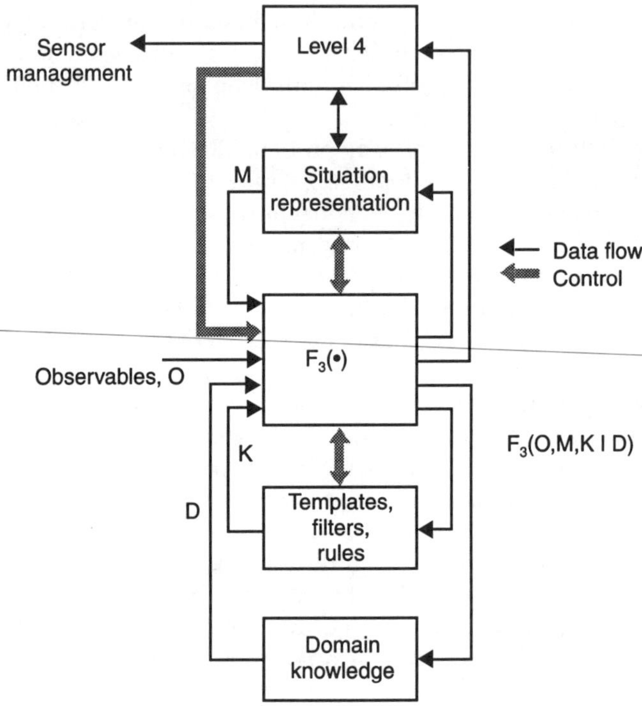

Figure 2.4 Multiple level-of-abstraction fusion model supporting context-sensitive reasoning.

1. Offer a unified framework for identifying both the similarities and differences of disparate data-processing approaches;
2. Help identify shortcomings of conventional automated fusion algorithms;
3. Reveal the critical role of the system database to both algorithm development and algorithm performance;
4. Support the establishment of a formal set of database requirements;
5. Identify opportunities for developing more powerful problem-solving approaches.

2.3.1 Short-, Medium-, and Long-Term Knowledge

The biological data fusion metaphor relies on the association of the primary elements of the fusion process with *short-term*, *medium-term*, and *long-term memory*.[2] Short-

2. In humans, relatively recent memories appear to be stored for a period of time in the hippocampus in a midprocessing state between short-term and long-term memory. This observation helps explain why a concussion can cause the victim to permanently lose the memory of all events occurring from a few minutes to a few days before the accident.

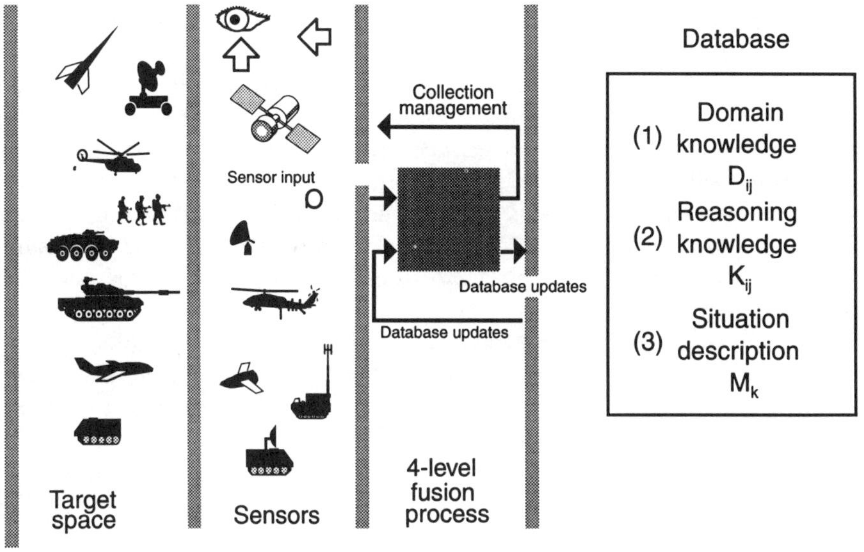

Figure 2.5 Relationship between the three classes of knowledge and the data fusion domain.

term memory is associated with highly transient short-term knowledge, medium-term memory maintains dynamic but somewhat less transient medium-term knowledge, and long-term memory maintains relatively static long-term knowledge. Thus, just as short-, medium-, and long-term memory are associated with the durability of the information in biological systems, short-, medium-, and long-term knowledge relate to the durability of the information in an artificial data fusion system.

Sensor data represents a fusion system's short-term knowledge, while a fusion system's relatively static factual and procedural knowledge base represents long-term knowledge. Since the goal of both biological and artificial situation awareness systems is the development and maintenance of the *current relevant perception* of the environment, the dynamic *situation description* represents the fusion system's metaphorical medium-term memory. In both biological and tactical data fusion systems, *current* emphasizes the character of the dynamically changing scene under observation as well as the time-evolving interaction among a network of distributed fusion processes. Due to memory limitations and the critical role of medium-term memory in system survivability, only *relevant* states can be maintained by either biological or tactical situation awareness systems. *Perception* indicates the inherent uncertainties in the fusion process due to the information-limited measurements, the nondeterministic nature of the domain, and the uncertainties in the reasoning process. Thus, in both biological and tactical data fusion systems, there can be a disparity between perception and reality.

In general, biological situation awareness requires the fusion of short-term, medium-term, and long-term knowledge. As depicted in Figure 2.6, the elements of the general fusion process model shown in Figure 2.4 can be readily associated with these three classes of knowledge. The sensor observables O represent short-term declarative knowledge; the situation description M represents medium-term declarative knowledge; K represents static declarative knowledge; D can include both static and dynamic non-sensor-derived declarative domain (context) knowledge; and F_3 represents long-term procedural reasoning knowledge. Therefore, just as in the biological analog, the tactical data fusion process inherently requires the interaction among short, medium, and long-term declarative knowledge as well as long-term procedural knowledge. Figure 2.7 summarizes this high-level knowledge taxonomy.

Medium-term knowledge is acquired and forgotten quickly, while long-term declarative and procedural knowledge are learned and forgotten more slowly. As suggested by Figure 2.6, except for the timescale, learning of long-term knowledge and update of the situation description are analogous operations.

2.3.2 Context-Sensitive Knowledge

In general, knowledge can be either *context-sensitive* or *context-insensitive.* In this book, context is treated as a conditional aspect of knowledge. In statistics, for instance, a process or function is explicitly constrained due to inherent dependencies among variables or functions (for example, $f(x_1, x_2 \mid x_3 = a)$). In a similar fashion, context represents the implicit dependencies and conditioning that exist due to the state of the current situation representation and constraints imposed by the domain and/or the environment.

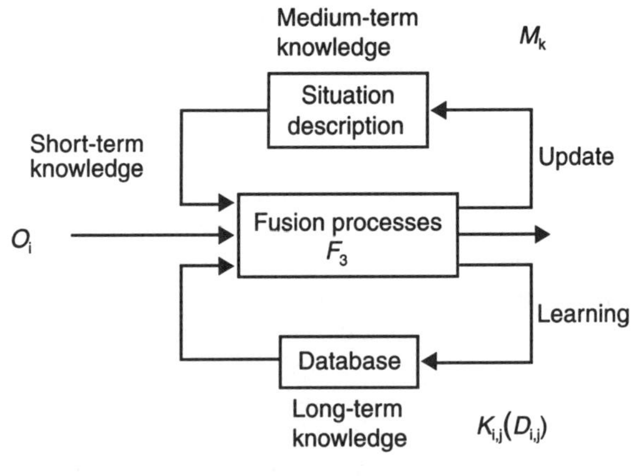

Figure 2.6 Biologically motivated metaphor for the data fusion process.

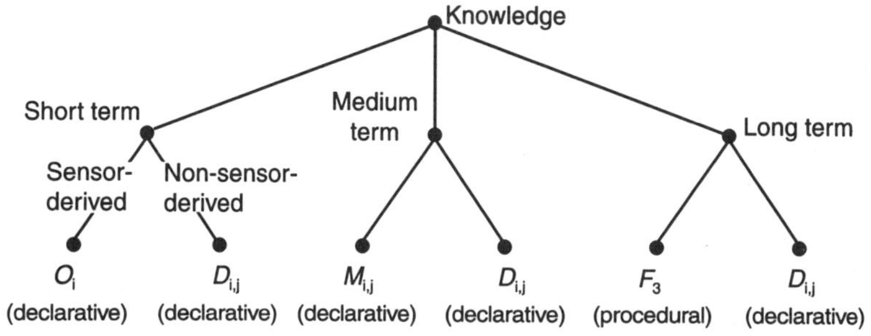

Figure 2.7 Taxonomy of short-, medium-, and long term knowledge.

Context-sensitive knowledge implicitly depends on either the current situation awareness product, the output of nonprimary sensors (atmospheric, environmental sensors), or non-sensor-derived domain knowledge. For instance, a rule such as "prefer track updates that are on or near a road" implicitly requires the determination of whether a road exists in the vicinity of the target track. *Context-insensitive* knowledge, on the other hand, is information that possesses no intrinsic conditioning. For instance, "vehicles of class *X* hold 2,000 gallons of fuel" is factual knowledge that is independent of the current state of the fusion process, any environmental condition, or the slope of a particular hillside.

There exist two primary classes of *context-sensitive knowledge. Context-specific* knowledge applies only for a specific, fixed context; a change in context state can invalidate such knowledge. *Context-general* knowledge is specialized to the current context before it is applied. Similarly, there exist two primary classes of context-insensitive knowledge. *Generic knowledge* is valid, independent of the domain context. Cultural and geographical feature data, physical laws, and equipment specifications represent generic knowledge. Finally, *context-free* knowledge simply ignores any possible context sensitivity. Fixed mobility resistance maps and statistical-based tracking models often represent context-free knowledge. Figure 2.8 illustrates the taxonomy of knowledge with respect to its context sensitivity.

Short-term knowledge is generally dynamic, perishable, and highly context-sensitive. Due to its inherent context sensitivity, short-term knowledge requires interpretation within context. Medium-term knowledge is less perishable and learned and forgotten at a slower rate than short-term knowledge. Medium-term knowledge includes the relatively dynamic, context-sensitive situation description at all levels of abstraction, such as target track files (level 1), the dynamic Order-of-Battle (level 2), and the current threat assessment (level 3). Due to its inherent context sensitivity, medium-term knowledge also requires interpretation within context.

Long-term knowledge is relatively nonperishable information that may or may not be context-sensitive. As just discussed, *context-insensitive* long-term knowledge is either generic knowledge, such as terrain/elevation, soil type, vegetation, waterways,

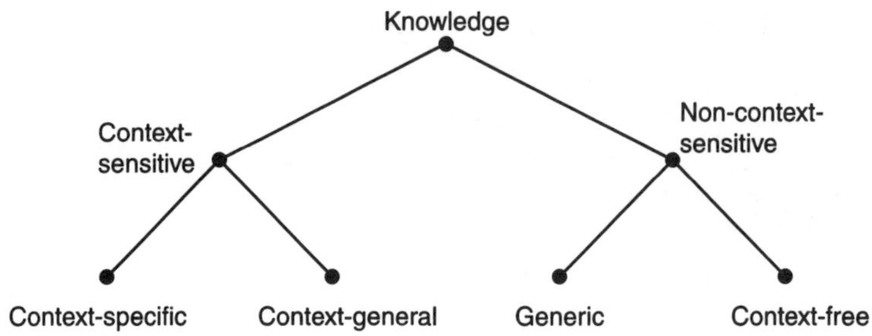

Figure 2.8 Taxonomy of knowledge context sensitivity.

cultural features, system performance characteristics, and coefficients of fixed-parameter signal filters, or context-free knowledge that simply ignores the domain context. *Context-sensitive* long-term knowledge is specialized knowledge, such as enemy Tables-of-Equipment, context-conditioned rules, doctrinal knowledge, and special-purpose two-dimensional map overlays (mobility maps, field-of-view maps). Thus, the specialization of long-term knowledge can either be embedded (*context-specific*) or conditionally dependent on dynamic or static domain knowledge (*context-general*).

Attempts to overcome obvious limitations of context-free algorithms have generally relied on context-embedded algorithms. The use of algorithms that are *conditioned* on relevant domain knowledge, on the other hand, may significantly improve algorithm effectiveness and generality. As will be discussed in Chapter 7, the separate maintenance of each of the four classes' context-sensitive knowledge can potentially enhance both algorithm extensibility and database maintainability.

2.3.3 Procedural Knowledge

Although data fusion algorithms can use various combinations of short-term, medium-term, and long-term declarative knowledge, all algorithms require some form of procedural knowledge. A taxonomy of procedural knowledge is established next as a simple means of characterizing algorithms. Procedural knowledge will be treated as long-term *declarative knowledge* with associated *control knowledge*. Long-term declarative knowledge, in turn, can be considered to be either *specific* or *general*. *Specific* declarative knowledge represents fixed (static) facts, transformations, or templates, such as filter transfer functions, decision trees, sets of explicit relations, object attributes, exemplars, and rigid templates.

While specific long-term declarative knowledge represents a set of fixed attributes, *general* declarative knowledge can characterize not only the value of individual attributes but the relationships among the attributes as well. Production rule condition sets, parametric models, and semantic constraint sets are examples of general long-term declarative knowledge. *Specific* long-term declarative knowledge fundamentally supports only non-model-based reasoning, while *general* long-term declarative knowledge supports more model-based reasoning approaches.

Fusion algorithms that rely on *specific* long-term declarative knowledge are most natural when:

1. The decision process has relatively few degrees of freedom (attributes, parameters, dimensions);
2. The decision process attributes are relatively independent (no complex inter-dependencies among attributes);
3. All relevant reasoning knowledge is static.

Thus, specific long-term declarative knowledge is natural for static problems characterized by relatively small state spaces and/or closed world domain constraints.[3]

For problems that inherently possess multiple, dependent attributes, reliance on specific long-term declarative knowledge can lead to inherently nonrobust algorithms. While such algorithms may perform reasonably well for highly constrained problems, their performance tends to fall off rapidly as the problem is generalized. In addition, computational and/or search requirements tend to increase combinatorially with the problem size.

General long-term declarative knowledge is most appropriate when one or more of the following hold.

1. The decision process has a relatively large number of degrees of freedom.
2. The relationships among attributes are significant (attribute dependency).
3. The reasoning process is temporally sensitive.

Algorithms that employ general declarative knowledge tend to be more robust than those that employ specific declarative knowledge.

Two broad classes of control knowledge can be defined: *rigid* and *flexible. Rigid* control knowledge is appropriate for simple, routine tasks that are static, inflexible, and relatively context-insensitive. The calculation of the correlation coefficient between an input data set and a set of stored exemplar patterns is an example of a simple rigid control strategy. *Flexible* control knowledge, on the other hand, supports multiple hypothesis, opportunistic, mixed-initiative strategies as well as context-sensitive algorithm control. In addition to the flexibility of the control knowledge, the control element can be categorized as either *single level of abstraction* or *multiple level of abstraction.*[4] The former represents relatively local control, while the latter sup-

3. Many of the classic AI problems (blocks world, the eight puzzle, and checkers meet these requirements) [5].

4. For consistency, the declarative element of procedural knowledge could be similarly treated as either single or multiple level-of-abstraction. However, problem-solving characteristics appear to be adequately covered by associating such characteristics with just the control element.

ports more global control. Therefore, algorithm control can range from rigid, single level-of-abstraction, to flexible, multiple level-of-abstraction. The latter supports more global problem solutions as well as more efficient performance than does strictly rigid, or even flexible, but single level-of-abstraction control.

Based on all the possible combinations of the two classes of declarative knowledge and the four classes of control knowledge, eight classes of procedural knowledge can be defined:

- Rigid, single level-of-abstraction, specific procedural knowledge;
- Rigid, multiple level-of-abstraction, specific procedural knowledge;
- Flexible, single level-of-abstraction, specific procedural knowledge;
- Flexible, multiple level-of-abstraction, specific procedural knowledge;
- Rigid, single level-of-abstraction, general procedural knowledge;
- Rigid, multiple level-of-abstraction, general procedural knowledge;
- Flexible, single level-of-abstraction, general procedural knowledge;
- Flexible, multiple level-of-abstraction, general procedural knowledge.

Figure 2.9 depicts the biologically motivated taxonomy of both declarative and procedural knowledge. Although numerous problem-solving taxonomies exist in the literature, for example, [1,6,7], what distinguishes the taxonomy offered in this chapter from many others is the motivation behind its development. In spite of the top-down character of all taxonomies, most are actually developed in a *bottom-up* fashion by first "clustering" techniques and then grouping the clusters into more general categories. The taxonomy proposed in this chapter, on the other hand, is based on a *top-down* specification of algorithm characteristics. Thus, rather than an ad hoc organizational framework for existing techniques, the proposed taxonomy provides a "capability-based" classification scheme for problem-solving paradigms.

In general, algorithms that support the general fusion model depicted in Figure 2.4 depend directly on:

1. Sensor-derived (short-term declarative knowledge) O;
2. The fusion process products (medium-term declarative knowledge) M;
3. The static declarative knowledge base (long-term declarative knowledge) K;
4. Indirectly on the a priori domain knowledge base D.

Thus, a fusion algorithm can be represented as an abstract function of short-, medium-, and long-term declarative knowledge, conditioned on the non-sensor-derived domain knowledge base, that is,

$$F_3(O,M,K \mid D) \tag{2.1}$$

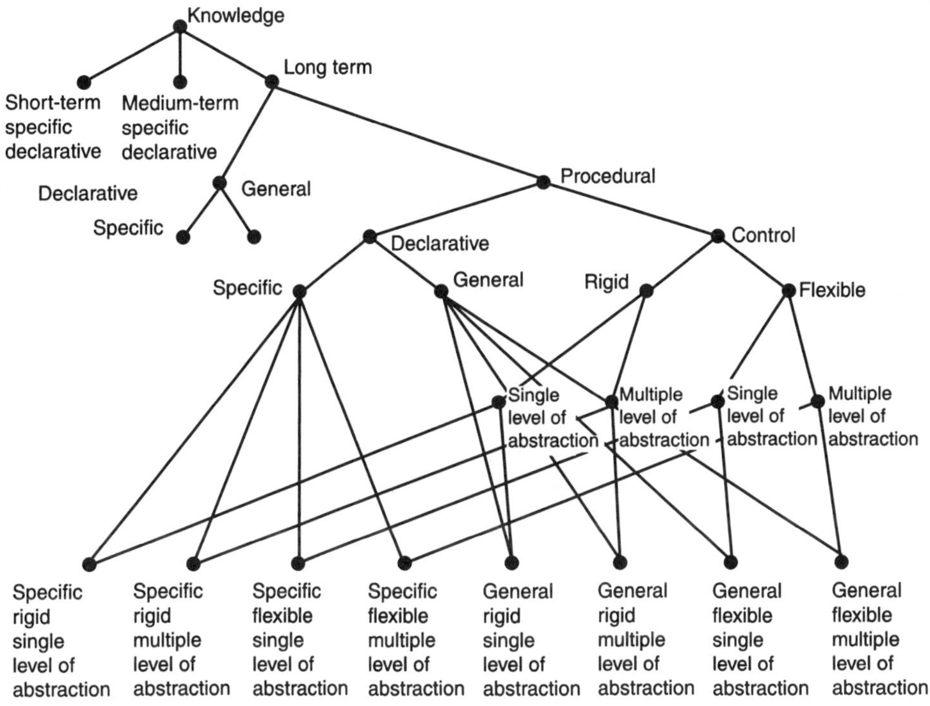

Figure 2.9 Biologically motivated taxonomy of knowledge.

As stated earlier, while both short-term and medium-term declarative knowledge are context-specific, long-term declarative knowledge may or may not be context-sensitive. Thus, K conceptually consists of two separate components

$$K = K'(\textit{context–sensitive}) + K''(\textit{context–insensitive}) \qquad 2.2$$

The advantages of an abstract data fusion algorithm formulation, such as the one given by (2.1) and (2.2), include the following.

1. All relevant classes of knowledge are explicitly depicted.
2. The context-sensitive nature of fusion algorithms is clearly shown.
3. The highly dynamic character of data fusion is apparent (since F_3 depends on short-term and medium-term knowledge).
4. The static declarative knowledge base can be logically separated into context-sensitive and context-insensitive components.

Fusion Classes

Rather than the functional view of (2.1), the data fusion process can be viewed as the general composition among short-term declarative, medium-term declarative, long-term declarative, and long-term procedural knowledge. As a direct result of this characterization, 15 unique data fusion classes can be identified as illustrated by Table 2.1, which represents all possible combinations of one, two, three, and four classes of knowledge. By associating individual algorithms with a fusion class, a simple algorithm characterization can be developed. For instance, only fusion classes that utilize medium-term knowledge can be sensitive to the current evolving situation, while only fusion classes that depend on long-term declarative knowledge can be sensitive to static domain constraints.

Table 2.1
Fusion Classes

	Declarative Knowledge Class			
Fusion Class	*Short-Term Knowledge*	*Medium-Term Knowledge*	*Long-Term Knowledge*	*Procedural Knowledge*
1	•			
2		•		
3			•	
4	•	•		
5		•	•	
6	•		•	
7	•	•	•	
8				•
9	•			•
10		•		•
11			•	•
12	•	•		•
13		•	•	•
14	•		•	•
15	•	•	•	•

Because all four classes of knowledge may be resident within the system database, rather than being a peripheral issue, the system database must be considered to be a critical element of a data fusion system. An important objective of Part II of this book is to establish guidelines for the development of database management systems that support the full range of data fusion algorithm requirements.

Implicit and Explicit Procedural Knowledge

The abstract algorithm, $F_3(\bullet)$, defined in Section 2.1, can be implemented using any of the previously defined eight classes of procedural knowledge. When the procedural element of an algorithm is associated with a rigid composition operation, such as cross-correlation and intersection, it is considered to be *implicit* or *default* knowledge. Because implicit procedural knowledge possesses both a default declarative element and default control, it is always associated with the lowest order class of procedural knowledge (specific declarative knowledge and rigid, single level-of-abstraction control knowledge). Thus, because fusion classes 1–7 do not employ explicit procedural knowledge, these classes of fusion algorithms tend to represent rather simple, rigid problem-solving approaches. On the other hand, algorithms that employ *explicit* procedural knowledge can range from rigid, single level-of-abstraction, non-model-based algorithms to flexible, multiple level-of-abstraction, model-based algorithms. As will be illustrated in Section 2.4 and re-emphasized in Section 2.5, fusion classes that employ explicit long-term procedural knowledge can potentially support robust algorithms that are sensitive to both the static and dynamic domain contexts.

2.4 DATA FUSION CLASS EXAMPLES

In this section, an example of each of the 15 fusion classes is offered to demonstrate the structural diversity of fusion algorithms. Table 2.2 summarizes the attributes of the four generic classes of knowledge. As mentioned in the last section, the fusion class taxonomy identifies whether the algorithm has an implicit or explicit procedural element as well as categorizes the algorithm with respect to its use of short-, medium-, and long-term declarative knowledge.

Class 1: (short-term declarative knowledge) • (short-term declarative knowledge)

The cross-correlation of two time-varying signals or the generation of a constant false alarm rate detection threshold are examples of the composition of two short-term knowledge sources. In the latter case, for instance, spectral, spatial, or temporal averaging may be used to develop a dynamic estimate of signal interference (noise, clutter, chaff) that controls a system parameter, such as the system sensitivity

Table 2.2
Attributes of the Four Classes of Knowledge

Knowledge Classes	*Attributes* Association	Data Class	+ Learning/ Forgetting Rate
Short-term declarative	Sensor output	Dynamic	Fast
Medium-term declarative	Situation description	Dynamic	Moderate
Long-term declarative	A priori declarative knowledge	Static	Slow
Procedural	Reasoning knowledge	Static	Slow

and/or the decision threshold. The product of the process (an adaptive decision threshold) represents medium-term knowledge.

Class 2: (medium-term declarative knowledge) • (medium-term declarative knowledge)

The correlation among multiple elements of the current situation description (medium-term declarative knowledge) is a typical example of class 2 fusion. For example, the attributes of two objects that were acquired by different classes of sensors may be merged into a single object, or two track fragments may be joined based on spatial/temporal proximity and attribute similarity. Again, the composition product normally represents medium-term knowledge.

Class 3: (long-term declarative knowledge) • (long-term declarative knowledge)

In general, class 3 fusion represents abstractions, generalizations, updates, and modifications of the long-term knowledge base. Many Intelligence Preparation of the Battlefield (IPB) products, such as mobility and river-crossing maps, Avenues of Approach (AOA) maps, and fields-of-view (FOV) maps represent long-term knowledge products. These products are derived by performing set operations among strictly long-term declarative knowledge bases, such as cultural feature, elevation, foliage, soil type, and drainage maps. The product of the composition process represents long-term declarative knowledge.

Class 4: (short-term declarative knowledge) • (medium-term declarative knowledge)

The correlation between a data sample and a set of data clusters is a typical example of the composition of short-term knowledge and medium-term knowledge. If a positive correlation exists, the data sample is added to the cluster; otherwise, a new cluster is formed. Thus, the composition product (current cluster set) represents medium-term knowledge.

Class 5: (medium-term declarative knowledge) • (long-term declarative knowledge)

The composition of the current situation description (medium-term declarative knowledge) with a rigid doctrinal template (long-term declarative knowledge) is a typical example of class 5 fusion. A mobility map might be developed by intersecting the current weather conditions map (medium-term declarative knowledge) with elevation and soil-type maps (long-term declarative knowledge). Network reconstruction requires the correlation of emitter locations (medium-term declarative knowledge) with templates of likely networks (long-term declarative knowledge). Many non-model-based situation generalization, situation specialization, situation assessment, and threat assessment operations involve class 5 fusion operations. The fusion product normally represents medium-term declarative knowledge.

Class 6: (short-term declarative knowledge) • (long-term declarative knowledge)

The composition between a signal and a template is a common example of class 6. Object classification is often performed by correlating a data sample (short-term declarative knowledge) against a set of object templates (long-term declarative knowledge). Conventional linear signal filtering requires the convolution of a signal (short-term declarative knowledge) with a previously defined filter transfer function (long-term declarative knowledge). A database query for all natural features (terrain, elevation, waterways, soil type, surface roughness) and cultural features (roads, railroads, towns) in the vicinity of a new target report is yet another example of the composition of short-term declarative knowledge (the query window) and long-term declarative knowledge (natural and cultural features). The composition product normally represents medium-term knowledge.

Class 7: (short-term declarative knowledge) • (medium-term declarative knowledge) • (long-term declarative knowledge)

Many level-1 fusion operations require the composition among short-term, medium-term, and long-term declarative knowledge. For example, in a conventional Kalman filter tracker, sensor data (short-term knowledge) is passed through a Kalman filter (long-term declarative knowledge) to evaluate the potential that the detection may be associated with all candidate track files (medium-term declarative knowledge). The product of class 7 fusion operations generally represents medium-term knowledge.

Class 8: (procedural knowledge) • (procedural knowledge)

Class 8 fusion represents operations that generate new procedural knowledge from existing procedural knowledge. Procedural knowledge can be extended by the addition of new rules, procedures, or constraints. General machine learning is an example of class 8 fusion.[5]

Class 9: (short-term declarative knowledge) • (procedural knowledge)

Procedural-based algorithms that directly manipulate short-term declarative knowledge represent class 9 fusion approaches. Model-based object-detection paradigms, pattern classification, and object recognition are typical examples. The product of a class 9 fusion algorithm generally represents medium-term knowledge.

Class 10: (medium-term declarative knowledge) • (procedural knowledge)

Class 10 fusion represents the procedural-based manipulation of the current situation description. Thus, rather than simple composition of medium-term knowledge, more general algorithms are supported. Data abstraction operations such as model-based doctrinal templating and network reconstruction are typical examples. Again, the product generally represents medium-term declarative knowledge.

Class 11: (long-term declarative knowledge) • (procedural knowledge)

Class 11 fusion represents the manipulation or transformation of static declarative information products. A typical example of this fusion class is the generation of a context-insensitive mobility resistance map based on slope, soil type, vegetation, and nominal seasonal weather conditions (long-term declarative knowledge) and a set of heuristic rules-of-thumb (long-term procedural knowledge).

---

5. Typical neural net learning is more closely associated with class 5 because the learning process generally involves minimizing the mean square decision error for a set of labeled input data. In this case, the data is short-term declarative knowledge, the weight matrix is long-term declarative knowledge, and output is medium-term knowledge.

Class 12: (short-term declarative knowledge) • (medium-term declarative knowledge) • (procedural knowledge)

Model-based target classification is an example of the composition between sensor data (short-term knowledge), previously developed target classifications (medium-term knowledge), and a target model (explicit procedural knowledge). The classifier product updates the current situation awareness (medium-term knowledge).

Class 13: (medium-term declarative knowledge) • (long-term declarative knowledge) • (procedural knowledge)

Model-based recognition and identification of military units using the logistical and doctrinal constraints associated with such units is a typical example. For instance, an algorithm (explicit procedural knowledge) might evaluate the relative relationships of subordinate elements (medium-term declarative knowledge) of a unit with respect to the local maneuver barriers, cultural features, and relevant line-of-sight barriers (long-term declarative knowledge). The product of the process is generally medium-term knowledge.

Class 14: (short-term declarative knowledge) • (long-term declarative knowledge) • (procedural knowledge)

Unlike the class 9 target classification algorithm previously discussed that was insensitive to static domain knowledge, in a class 14 algorithm, sensor-derived data (short-term knowledge) can be composed with a context-sensitive target model (procedural knowledge) that *is* dependent on non-sensor-derived domain knowledge (long-term declarative knowledge). The product of the composition process would be medium-term declarative knowledge.

Class 15: (short-term declarative knowledge) • (medium-term declarative knowledge) • (long-term declarative knowledge) • (procedural knowledge)

The domain-sensitive target-tracking algorithm discussed at the beginning of this chapter represents a simple example of a class 15 fusion algorithm. In that example, a sensor detection (short-term declarative knowledge) was intersected with the topographic and cultural feature data (long-term declarative knowledge) to evaluate road association. If the target error ellipse intersects a road, candidate track files (medium-term declarative knowledge) are tentatively extended based on a road-following target model (explicit procedural knowledge). The product of the process generally represents medium-term declarative knowledge.

Thus, a wide range of apparently unrelated problem-solving classes can be viewed as well-defined forms of data fusion. Table 2.3 provides a summary of the 15 fusion classes. The next section explores the underlying character of such algorithms by investigating the character of their associated procedural elements.

Table 2.3
Representative Examples of the 15 Classes of Data Fusion

	Declarative Knowledge				
Class #	*Short Term*	*Medium Term*	*Long Term*	*Explicit Procedural*	*Product*
1	Signals				Correlation of two signals
2		Track fragments			Merging of two track fragments
3			Geographic features		Static IPB overlay preparation
4	Data sample	Cluster set			Clustering
5		Situation description	Rigid doctrinal template		Doctrinal templating
6	Signal		Linear filter		Linear filtering
7	Detection	Track files	Kalman filter		Statistical tracker
8				Rules	Learning new rules
9	Imagery			Model-based pattern recognizer	Model-based object detection
10		Known emitter locations		Network construction rules	Network reconstruction
11			Geographic features	Rule base	Knowledge-based mobility map generation
12	Sensor data	Track files		Model-based classifier	Sequential classification
13		Situation description	Domain constraints		Doctrinal templating
14	Sensor data		Domain constraints	Context-sensitive target model	Model-based target classification
15	Target detection	Track files	Domain constraints	Context-sensitive tracking model	Smart tracking algorithm

2.5 CANONICAL PROBLEM-SOLVING FORMS

All algorithms possess a procedural element that can range from simple and rigid to very complex and flexible. Because the correlation of two time-series signals involves a well-established, rigid mathematical operation, a correlation algorithm is an example of the former. On the other hand, since it requires multiple-staged, nonstandard manipulation operations, the road-tracking algorithm discussed in Section 2.2 is an example of the latter. Although both examples rely on a "correlation" operation, the first example employs the "default" notion of correlation, while the "correlation" between a target and the road network involves less well-defined (non-default) operations. As a direct consequence, the procedural element in the road-tracking algorithm must be explicitly defined.

For algorithms that employ relatively rigid and well-established manipulation operations, the procedural knowledge will be referred to as *implicit*. On the other hand, when nondefault manipulation is required, the procedural knowledge will be referred to as *explicit*. As observed in Table 2.1, since they do not employ *explicit* procedural knowledge, fusion classes 1–7 employ *implicit* procedural knowledge.

In Section 2.3.3, eight classes of procedural knowledge were defined based on two classes of long-term declarative knowledge and four classes of control knowledge. Because *implicit* procedural knowledge tends to be rigid and well defined, its declarative element is always *specific* and its control element is always *rigid* and *single level-of-abstraction*. Thus, the procedural knowledge associated with fusion classes 1–7 is always specific, rigid, and single level-of-abstraction. Fusion classes 8–15, on the other hand, can employ any of the eight classes of procedural knowledge.

Algorithms that rely on specific declarative knowledge tend to treat problem attributes as independent, while those that rely on general declarative knowledge can reason with dependent problem attributes. As a result, algorithms that depend on specific declarative knowledge generally employ *non-model-based* reasoning, while those that depend on general declarative knowledge can support *model-based* reasoning. As a result of this association, eight basic algorithmic forms can be defined:

- *Rigid, single level-of-abstraction, non-model-based algorithms* use specific declarative knowledge and rigid, single level-of-abstraction control knowledge;
- *Rigid, multiple level-of-abstraction, non-model-based algorithms* use specific declarative knowledge and rigid, multiple level-of-abstraction control knowledge;
- *Flexible, single level-of-abstraction, non-model-based algorithms* use specific declarative knowledge and flexible, single level-of-abstraction control knowledge;
- *Flexible, multiple level-of-abstraction, non-model-based algorithms* use specific declarative knowledge and flexible, multiple level-of-abstraction control knowledge;
- *Rigid, single level-of-abstraction, model-based algorithms* use general declarative knowledge and rigid, single level-of-abstraction control knowledge;
- *Rigid, multiple level-of-abstraction, model-based algorithms* use general declarative knowledge and rigid, multiple level-of-abstraction control knowledge;

- *Flexible, single level-of-abstraction, model-based algorithms* use general declarative knowledge and flexible, single level-of-abstraction control knowledge; and
- *Flexible, multiple level-of-abstraction, model-based algorithms* use general declarative knowledge and flexible, multiple level-of-abstraction control knowledge.

In addition to the class of procedural knowledge, problem-solving paradigms can be distinguished as being either predominately *generation-based* or predominately *hypothesis-based.* The former paradigm represents data-driven problem-solving approaches; the latter represents goal-driven approaches. Generation-based approaches that use only specific declarative knowledge represent *specific transformations,* while generation-based approaches that use general declarative knowledge represent *model-based transformations.* Hypothesis-based approaches that use only specific declarative knowledge will be characterized as *explicit templating* approaches, while hypothesis-based approaches that use general declarative knowledge can be characterized as *model-based constraint satisfaction* approaches.

Therefore, given eight algorithmic forms and two basic problem-solving paradigms, 16 canonical problem-solving forms can be defined as identified in Table 2.4. In order to differentiate between fusion classes and canonical algorithm forms, *fusion classes* will be denoted by Arabic numbers and *canonical problem-solving forms* will be denoted by Roman numerals. The odd-numbered canonical problem-solving forms are generation-based, while the even-numbered forms are hypothesis-based.

Table 2.4
Definition of the 16 Canonical Problem-Solving Forms

	Procedural Knowledge						
	Declarative		*Control*				
Canonical Form #	*Specific*	*General*	*Rigid*	*Flexible*	*Single Level of Abstraction*	*Multiple Levels of Abstraction*	*Gen/Hyp*
I	•		•		•		Gen
II	•		•		•		Hyp
III	•		•			•	Gen
IV	•		•			•	Hyp
V	•			•	•		Gen
VI	•			•	•		Hyp
VII	•			•		•	Gen

	Procedural Knowledge						
	Declarative		*Control*				
Canonical Form #	*Specific*	*General*	*Rigid*	*Flexible*	*Single Level of Abstraction*	*Multiple Levels of Abstraction*	*Gen/Hyp*
VIII	•			•		•	Hyp
IX		•	•		•		Gen
X		•	•		•		Hyp
XI		•	•			•	Gen
XII		•	•			•	Hyp
XIII		•		•	•		Gen
XIV		•		•	•		Hyp
XV		•		•		•	Gen
XVI		•		•		•	Hyp

The proposed canonical problem-solving form taxonomy can potentially be extended by identifying various forms of specific and general declarative knowledge as well as forms of rigid and flexible control. For example, general declarative knowledge can be decomposed into simple parametric-, semantic- and statistical-based models, for instance. Statistical models, in turn, can be decomposed into various forms of parameter estimation. Although conceptually straightforward, the further extension of the taxonomy is problematic because some problem-solving techniques will not fall neatly into a single category (recall the "apples and oranges" discussion in Section 2.1). Thus, an informal "grouping" of problem-solving techniques may be a more realistic objective than to attempt to generate a proper extension of the taxonomy. Because of the inherent difficulty of contrasting rather similar approaches and our focus on general algorithm principles, we sidestep these difficulties and focus on the important characteristics of the 16 primary canonical forms.

In the next section, the relationships between fusion classes and canonical forms are discussed. In the following two chapters, the characteristics of each of the 16 canonical problem-solving forms are explored in considerable detail.

2.6 RELATIONSHIP BETWEEN THE FUNCTIONAL-LEVEL AND PROCESS-LEVEL FUSION MODELS

This section outlines the relationship between the *functional-level fusion model* shown in Figure 1.1 and the *fusion class* and *canonical problem-solving form* taxonomies just defined. Each of the three representations will be seen to offer important characterizations that support data fusion algorithm development. The *functional-level fusion model* provides a top-down, organizational framework that supports functional-level requirements and the evaluation of architectural issues. The *fusion class taxonomy* provides insight into the structural characteristics of algorithms (the types of inputs, the basic nature of the procedural element (*implicit* or *explicit*)). The *canonical problem-solving form taxonomy* provides insight into the potential robustness, context sensitivity, and efficiency of classes of algorithms.

This chapter has focused on the development of a process-level model that can provide insight into the comparison of classes of algorithms. As discussed at the beginning of this chapter, algorithm development inherently involves a two-stage process. The first stage requires an assessment of the problem-solving requirements, leading to the identification of a suitable class of paradigms. The road-following target-tracking example discussed earlier can be used to illustrate this process.

First, the problem-solving requirements of the task must be identified. The key information requirements for a road-following vehicle-tracking algorithm are:

1. The new target location;
2. The track history for all candidate tracks;
3. The knowledge of the existing road network;
4. A road-following track-extension procedure.

The first two elements represent *short-term* and *medium-term declarative knowledge*, respectively. The third element represents *specific long-term declarative knowledge*, and the fourth element represents *explicit procedural knowledge*. Because this task requires the composition among short-, medium-, and long-term declarative knowledge and explicit procedural knowledge, the tracking algorithm represents a fusion class 15 task (see Table 2.1).

At this point, the structural character of the algorithm has been specified. Since target tracking is fundamentally a data-driven task, a generation-based algorithm is most appropriate. Further algorithm refinement requires a detailed investigation of its intended functionality. Because the algorithm employs explicit procedural knowledge, any of the eight classes of procedural knowledge can be used to develop the application. In perhaps the simplest approach, if the error ellipse associated with a new target detection intersects the road database, the target could be declared to be on the road. Next, the distance between the new detection and all known road-following targets would be computed; the new detection would then be associated with the closest such target track. Such an algorithm employs *specific declarative knowledge* and *rigid, single level-of-abstraction control*.

A more flexible control strategy would be required if the target error ellipse was *near* but did not actually intersect a road.[6] If the target is determined to be near a road and the terrain in the vicinity of the target does not support off-road mobility, the moving target could be inferred to be on the road. Although the procedural knowledge is still rather *specific*, a more *flexible control* element is required.

By introducing behavioral models, target tracks can be evaluated with respect to models associated with (1) constant velocity motion along the roadway, (2) vehicle-convoy-following, and (3) off-the-road rendezvous at a staging area. Each model could be independently applied and the track extended only if one of the models adequately explained the observations. Although it relies on *general declarative knowledge*, the algorithm employs a rather *rigid control* strategy (evaluate-and-test).

By employing both *general long-term declarative knowledge* and a *flexible control* element, sophisticated vehicle class–dependent behavior models (separate models for armored personnel carriers, fuel trucks, and jeeps) could be utilized. For example, if the target is determined to be part of a convoy, the road is unsurfaced, rain is falling at 1" per hour, and the time is 3:00 a.m., the behavior model would preclude track extensions that appear to violate relevant mobility constraints.

Although each of the problem-solving approaches just discussed employed a single level-of-abstraction control element, all four algorithms could potentially benefit from the use of multiple level-of-abstraction control. For example, candidate tracks could be located using top-down multiple resolution spatial search of all targets that are within 2 km of the new detection. Tracks that pass this coarse filter would then be tested against the road database to permit a further pruning of the candidate track files.

An analysis of both the fusion class taxonomy and canonical problem-solving form taxonomy reveals a number of important observations. Since level-1 and level-4 fusion are closely associated with short-term declarative knowledge, fusion classes 1, 4, 6, 7, 9, 12, 14, and 15 are most closely associated with those two levels in the functional fusion model. Because level-2 and level-3 fusion processes may not explicitly employ short-term knowledge (direct observables), these two levels are most closely associated with remaining fusion classes.

As already mentioned, all 15 classes of fusion algorithms require some form of procedural knowledge. Fusion classes 1–7 depend on implicit or implied procedural knowledge. Implicit procedural knowledge represents simple default processes, such as convolution, correlation, or set operations. Thus, implicit procedural knowledge always possesses specific declarative knowledge and rigid, single level-of-abstraction control knowledge. For the fusion class 1 example offered in Table 2.3, the *correlation* integral is an implicit form of long-term procedural knowledge. Similarly, the fusion class 6 example involves the composition of a signal (short-term knowledge) and

6. Because of registration errors, finite sensor resolution, and other uncertainties, ambiguity is an underlying element of many data fusion tasks.

the finite impulse response function of a filter (long-term declarative knowledge). The composition function (*convolution*) is an implicit form of procedural knowledge.

Fusion classes that involve only implicit procedural knowledge represent simple canonical form I and II approaches to problem solving (conventional signal-processing operations that require straightforward filtering, correlation, and convolution operations). Fusion classes that do not depend on either short-term or medium-term knowledge (classes 3, 8, 11) do not directly support the development of medium-term fusion products. Therefore, only fusion classes that depend on explicit procedural knowledge and either short or medium-term knowledge (classes 9, 10, and 12–15) support the development of sophisticated data fusion algorithms. Further, only those fusion classes that employ long-term declarative knowledge support context-sensitive reasoning. Therefore, only fusion classes 14 and 15 support context-sensitive level-1 fusion, and only fusion classes 13 and 15 support context-sensitive level-2 and level-3 data fusion. Because each of these fusion classes employ explicit procedural knowledge, such fusion algorithms can use any of the 16 canonical problem-solving forms.

Figure 2.10 summarizes the general relationships between the functional fusion model, the 15 fusion classes, and the 16 canonical problem-solving forms.

In summary, the first stage of the algorithm development process involves the three following primary steps.

1. A functional-level analysis of the algorithm requirements is performed using the fusion model shown in Figure 1.1.
2. The *fusion class* is determined by identifying the classes of relevant knowledge that are to be exploited.
3. Based on the required algorithm robustness and context sensitivity, an appropriate *canonical problem-solving form* is selected.

If a class 6 fusion algorithm is determined to be the most appropriate non-model-based algorithm, for example, a selection must be made between a generation-based approach (canonical form I) and a hypothesis-based approach (canonical form II). On the other hand, if the algorithm requires explicit declarative knowledge (fusion classes 9–15), a selection must be made as to the most appropriate of the 16 canonical problem-solving forms.

The canonical form selection process can have a significant impact on the effectiveness of the algorithm. For example, consider the development of an algorithm that is to provide optimal detection of a known signal in Gaussian random noise. The optimal detector is a matched filter (specific long-term declarative knowledge) through which the input waveform (short-term knowledge) is passed. When the desired signal arrives, the output of the matched filter will be maximized and is expected to cross a predetermined threshold. Both the transformation and the thresholding process imply rigid, single level-of-abstraction control. Because it

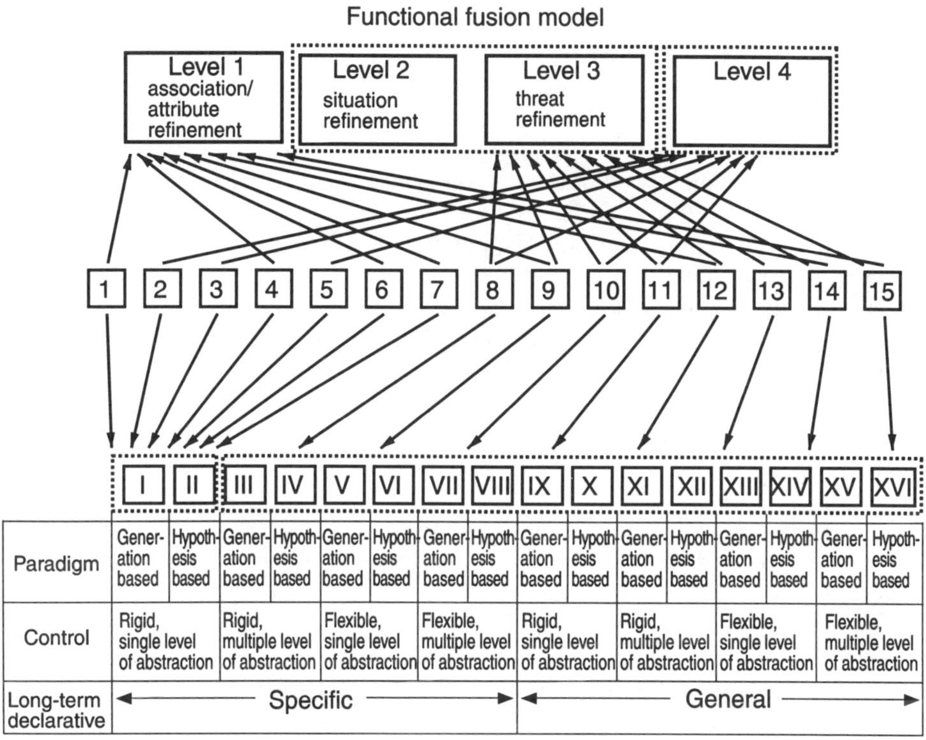

Figure 2.10 Relationships between the four-level functional fusion model and the fusion process model.

provides a natural match between problem requirements and the paradigm characteristics, a canonical form I approach is entirely adequate for this problem.

On the other hand, more complex and highly context-sensitive problems can require the use of more powerful problem-solving forms. As will become apparent by the end of the next chapter, data fusion automation is often pursued using problem-solving approaches that are fundamentally mismatched to the underlying problem requirements. Thus, the *problem requirements must dictate the problem-solving approach.* The balance of Part I focuses on the investigation of this simple yet critical algorithm development principle.

2.7 CONSISTENCY ANALYSIS

Consistency analysis is a conceptually simple and yet powerful reasoning paradigm for exploiting problem domain constraints [8]. In essence, observations and hypotheses are analyzed to evaluate their consistency with expectation-based constraints. In the target-tracking example just discussed, the *consistency* between the observed target behavior and a road-following behavioral model was the primary evaluation criterion. Viewed from this perspective, consistency analysis is a form of *templating*.[7] When an observation or hypothesis is *not* consistent with such constraints, its likelihood or plausibility is diminished. Consistency analysis-based reasoning supports the use of both single level-of-abstraction and multiple level-of-abstraction constraint sets.

Figure 2.11 depicts a number of general consistency criteria classes that are appropriate for a tactical situation awareness system. In such a paradigm, both short-term (observables) and medium-term knowledge (object, composite objects, hypotheses) are compared to relevant physical, logical, logistical, organizational, tactical, and doctrinal constraints. The general consistency criteria shown in Figure 2.11 can be further decomposed as illustrated in Figures 2.12 and 2.13.

Consistency analysis supports the use of powerful evaluation criteria. For example, the principle of the *conservation of resources* has the following three implications:

1. A physical entity can only be at one place at a time.
2. Without attrition or resupply, the amount of nonexpendable equipment remains the same over time.
3. Expendable equipment and materials require resupply.

Consider a simple application of consistency analysis to a tactical data fusion problem. Suppose a particular tank battalion that was *known* with high confidence to be at a particular location at time t_1 is *suspected* of being 10 km to the north at

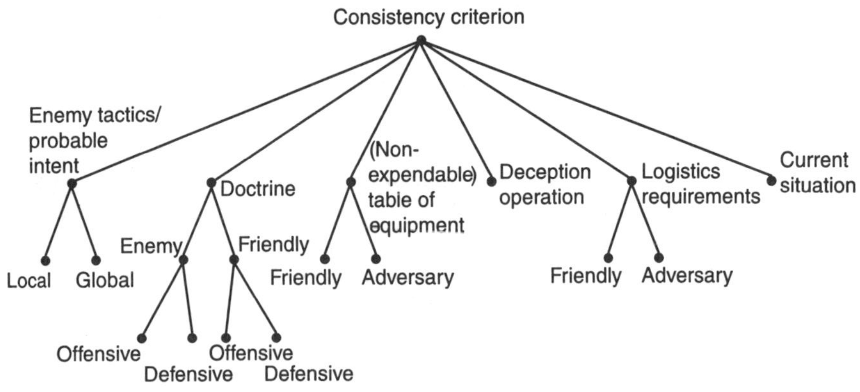

Figure 2.11 Consistency analysis view of certainty management.

7. Although templating is discussed throughout this book, Section 13.4 provides the most comprehensive treatment.

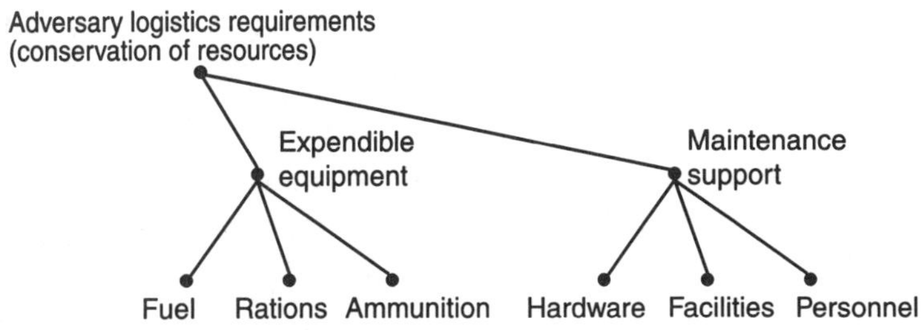

Figure 2.12 Extension of consistency analysis view of "adversary logistics requirements" in Figure 2.11.

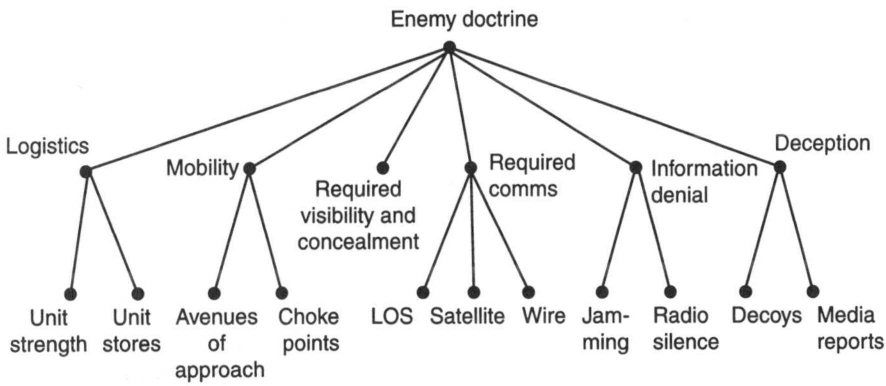

Figure 2.13 Extension of the consistency analysis view of "doctrine" in Figure 2.11.

time t_2. The hypothesis (tank battalion 10 km to the north) can be evaluated with respect to all relevant consistency criteria. For example, although the hypothesis may be consistent (or at least not inconsistent) with (1) *enemy tactics/probable intent,* (2) *doctrine,* and (3) *logistics* considerations, given the intervening terrain, the maximum travel time may be inconsistent with the *vehicle mobility capability.* On the other hand, although the new location may be consistent with *maneuver capability,* the hypothesis may be inconsistent with respect to the unit's required *logistics* support (at its present location, no reliable fuel delivery capability exists). Finally, although the apparent new location of the unit may be inconsistent with the *enemy tactics* as well as inconsistent with either *enemy offensive* or *defensive doctrine,* the hypothesis may be highly consistent with a possible *deception operation.*

Consistency analysis not only offers a powerful expectation-based reasoning methodology for confirming or denying individual hypotheses, but it can support the abstraction of lower level hypotheses as well as support the assessment of complex interrelationships among hypotheses at multiple levels-of-abstraction. Thus, the interpretation of sensor reports "in full context" implies an interpretation with respect to (1) the current overall situation description; (2) environmental constraints; and (3) physical, doctrinal, tactical, and logistical constraints.

As discussed previously, dependence on *context-specific* knowledge does not support algorithm extensibility to new domains or to changes in an existing domain. Traditional overland route-planning algorithms, for example, are typically based on nominal mobility parameters that are not dynamically sensitive to changing environmental conditions. Although the algorithm may perform adequately under nominal conditions (dry weather, daylight, mid-summer), the recommended route plans may degrade rapidly when the true context deviates significantly from the nominal assumptions.

If, on the other hand, the algorithm is implicitly sensitive to the current context conditions, the path development algorithm can readily be extended to accommodate new classes of targets, new environmental conditions, and new logistical constraints that are typically sensitive to the local soil conditions, current weather conditions, time of day. By separating context-sensitive and non-context-sensitive knowledge, the former can readily be modified with little or no effect on the latter. Therefore, the separation of context-sensitive and context-insensitive knowledge can potentially enhance algorithm extensibility.

2.8 CONCLUDING REMARKS

The four-level fusion model depicted as Figure 1.1 provides a top-down functional-level view of the data fusion process. Because there exists no widely embraced formal approach to algorithm selection and because the development of meaningful empirical comparisons of competing approaches is problematic, few objective paradigm selection guidelines exist. As a result, paradigm selection has tended to be a highly subjective process that often reflects the background or formal training of the developer.

This chapter developed a taxonomy-based characterization of problem-solving paradigms. Chapter 3 presents numerous examples that both illustrate and differentiate the 16 canonical problem-solving forms. In Chapter 4, conventional technical approaches will be organized within this taxonomy. Although many of the problem-solving paradigm "taxonomies" that have been proposed in the literature have been developed by clustering similar techniques, the proposed taxonomy provides a mutually exclusive and exhaustive decomposition of algorithm classes that can be used to characterize the strength and weaknesses of classes of paradigms. As a direct result, the taxonomy provides direct support to requirement-based paradigm selection.

By understanding the relationship between the classes of functional operations required (fusion classes) and classes of problem-solving methodologies (canonical problem-solving forms), the appropriateness of the various paradigms can be evaluated. For problems that demand fundamentally more robust, context-sensitive, and efficient approaches than are offered by traditional algorithms, more sophisticated approaches are required. The fusion process model and associated canonical problem-solving form taxonomy developed in this chapter will be used in subsequent chapters to illustrate the development of classes of more powerful algorithms.

By categorizing a wide range of conventional technical approaches with respect to the canonical form problem-solving form taxonomy, Chapter 4 permits the inherent capabilities of such approaches to be compared and contrasted. Chapter 5 summarizes the characteristics of various canonical problem-solving forms and the general algorithm principles developed in Part I.

References

[1] Hall, D.L., and R.J. Linn, "A Taxonomy of Algorithms for Multi-Sensor Data Fusion," *Proc. 1990 Joint Service Data Fusion Symp.*, Laurel, MD, May 1990, pp. 593–610.

[2] Beude, D.M., "Extension of the Data Fusion Process Model Based on Concepts of Automated Reasoning," paper prepared for Arvin/Calspan, March 1991.

[3] Sweet, R., *Proc. Command and Control Evaluation Workshop*, MORS C2 MOE Workshop, Military Operations Research Society, January 1985.

[4] Bar-Shalom, Y., *Multitarget-Multisensor Tracking: Applications and Advances*, Bar-Shalom, Y. (ed.), Norwood, MA: Artech House, 1992, p. xiii.

[5] Barr, A., and E.A. Feigenbaum, *Handbook of Artificial Intelligence*, Stanford, CA: Stanford University Computer Science Department, 1980.

[6] Shapiro, S.C., and D. Eckroth (eds.), *Encyclopedia of Artificial Intelligence*, New York: Wiley, 1987.

[7] Hafner, A.N., "Identification of Decision Related Information Elements: A Taxonometric Model of Information Schemas," *Proc. 1987 Tri-Service Data Fusion Symp.*, Laurel, MD, June 9–11, 1987, pp. 556–564.

[8] Clark, J.J., and A.L. Yuille, *Data Fusion for Sensory Information Processing Systems*, Boston, MA: Kluwer Academic Publishers, 1990.

CHAPTER 3

Examples of the 16 Canonical Problem-Solving Forms

In Chapter 2, a biologically motivated fusion process model was developed that led to the identification of 16 canonical problem-solving forms. This chapter outlines a series of progressively more sophisticated algorithms for several data fusion applications intended to demonstrate both the strengths and limitations of the various canonical form problem-solving approaches. Although the general characteristics of a number of algorithms are discussed, a comprehensive treatment of individual technical approaches is beyond the scope of this book.

As stated in the introduction, this book will identify foundational principles of both algorithm development and database design. Readers interested in a more detailed treatment of specific algorithms may want to begin their investigation with [1] or [2]. The bibliography at the end of the book provides references for the particular applications addressed in this chapter. Chapter 4 provides a brief description of a wide range of technical problem-solving paradigms that have been used to construct fusion algorithms.

Part I maintains a top-down perspective of algorithm development that permits us to avoid the details that might well obscure a more bottom-up investigation. In addition, although we focus on tactical data fusion applications, the algorithm and database principles apply equally well to many forms of machine reasoning. In order that the arguments will be more easily followed by readers from a wide range of interests and backgrounds, unnecessary details will be avoided whenever possible.

As discussed in Chapter 2, algorithms that depend on *specific declarative* knowledge tend to be nonrobust. On the other hand, because they can potentially model the phenomenon under investigation, algorithms that employ *general declarative* knowledge tend to be more robust. Algorithms that rely on *rigid control* knowledge are generally simple but inflexible, while algorithms that use more *flexible control* knowledge tend to support more complex problems. Although *single level-of-abstraction control* can be entirely appropriate for relatively simple problems, *multiple level-of-abstraction control* tends to be more adequate for attacking highly complex problems or those that demand global solutions. While humans naturally attack complex tasks using higher order canonical form reasoning approaches (flexible, top-down, model-based reasoning), traditional data fusion algorithms have relied, in most cases, on lower order canonical form approaches.

Since they depend on *specific declarative* knowledge and *rigid, single level-of-abstraction* control, canonical forms I and II class algorithms represent relatively simple, rigid problem-solving approaches. Typical examples of canonical form I algorithms include:

1. General signal processing;
2. Pattern classification based on linear discriminant models;
3. Simple decision tree–based reasoning;
4. Optimization based on gradient descent.

Although conceptually simple to develop and implement, canonical forms I and II approaches often lead to highly combinatoric generation and/or evaluation operations. For finite-state problems that naturally generate relatively small state spaces, such approaches may be entirely adequate. However, since many tactical data fusion applications possess extremely large (or even nonenumerable) state spaces, canonical forms I and II approaches can lead to highly inefficient algorithm performance. Because they support *multiple level-of-abstraction control*, canonical forms III and IV tend to be more efficient than comparable canonical forms I and II approaches.

Canonical forms V and VI algorithms extend canonical forms I and II approaches by supporting *flexible* , *single level-of-abstraction* reasoning strategies (multiple hypothesis, multiple strategy, limited context-sensitive reasoning). With the addition of *multiple level-of-abstraction control*, canonical forms VII and VIII algorithms support both more sophisticated reasoning and more efficient problem solving than comparable canonical forms V and VI approaches.

Because they rely on *general declarative* knowledge, the eight higher order canonical form approaches support the development of more model-based reasoning approaches. Because they employ only *rigid, single level-of-abstraction control*, canonical forms IX and X support the use of only relatively simple models. With the addition of *multiple level-of-abstraction* control, canonical forms XI and XII class algorithms tend to support more complex applications, and they tend to be more efficient than comparable canonical forms IX and X approaches.

Because they offer more *flexible control* than canonical forms IX and X, canonical forms XIII- and XIV-based algorithms support flexible, single level-of-abstraction, model-based reasoning. With the addition of *multiple level-of-abstraction control*, canonical forms XV and XVI algorithms permit flexible and robust multiple level-of-abstraction problem solving.

The first half of this chapter presents examples of the eight generation-based canonical forms; the second half treats the eight hypothesis-based forms. The generation-based problem-solving forms are illustrated by a *target-tracking, target classification,* and *path-planning* application, while the eight hypothesis-based paradigms are illustrated by an *object detection* and *doctrinal-templating* application. Consideration of the full range of canonical form algorithms for a given application permits insight into the relationships between:

1. Algorithm effectiveness;
2. Algorithm complexity;
3. Computational efficiency.

Several of these same applications are reinvestigated in Part III (Chapter 13) following the consideration of important database principles presented in Part II.

3.1 GENERATION-BASED APPROACHES

3.1.1 Target Tracking

This section illustrates the application of each of the eight generation-based canonical form algorithm classes for tracking moving ground-based targets based on time and location information obtained from radar detection reports. Most conventional radar-tracking algorithms rely on statistical-based methods for track association and trajectory estimation.[1] In such algorithms, future target dynamics are normally assumed to be related to the object's past behavior and the object's performance parameters (maximum velocity, maximum acceleration). A typical canonical form I target tracker extends known tracks (*medium-term declarative* knowledge) based on smoothed estimates of the locations of new detections (*short-term declarative* knowledge). If the "distance" between the smoothed track location and a projected track update satisfies a particular metric test, the target location estimate is added to the appropriate track file. Because the recursive estimator (Kalman filter), represents *specific long-term declarative* knowledge, the tracking algorithm relies on *specific declarative* knowledge and *rigid, single level-of-abstraction control.*[2] Thus, the basic statistical tracker represents a canonical form I class algorithm.

---

1. For a summary of current data association algorithms see Table 6.2 in [2].
2. In a first-order Kalman filter, the track extension is based on the last position estimate of the target. In a second-order Kalman filter, it is based on both the last and next-to-last position estimates.

By adding *multiple level-of-abstraction control*, a canonical form III statistical-based tracking algorithm can be developed. For example, a pretrack filter could be added prior to the statistical estimator for the purpose of reducing the number of candidate tracks that must be extrapolated. The initial filter might restrict candidate tracks to only those existing track files that are within a certain distance of the new detection or that possess a certain range of translational velocities. By reducing the number of candidate tracks associations that are made, track assignment efficiency can be improved.

The *flexible, single level-of-abstraction control* associated with a canonical form V problem-solving approach supports the development of a simple multiple hypothesis tracking algorithm. For example, a bank of Kalman filters, each tuned to a particular combination of target class and operational parameters, can offer significant performance advantages over a single Kalman filter approach.[3] A number of researchers have used such techniques to accommodate several potential target accelerations, for instance. Over time the candidate association that best explains the observational data would be selected as the correct decision; competing hypotheses would be dropped.

With the addition of multiple level-of-abstraction control, a canonical form VII approach would support the development of an efficient recursive, multiple-resolution algorithm that recursively refines candidate tracks based on a set of competing, independently generated hypotheses.

While the declarative element of all four algorithms just outlined was *specific*, the declarative element associated with the next four algorithms to be outlined can be quite *general*. A canonical form IX approach, for instance, can support the use of simple target mobility strategies, such as minimum terrain gradient, minimum resistance, or road following. Due to its model-based character, algorithm performance can be enhanced by using relevant contextual knowledge (ground-based vehicles have various operational limitations, such as maximum speed under specific terrain, slope, soil type, and weather conditions), leading to much more effective algorithm performance.

By adding a *multiple level-of-abstraction control* element, a canonical form XI problem-solving approach can extend the previous approach by supporting the development of efficient, hierarchical-based algorithms. Figures 3.1 to 3.4 illustrate a top-down, recursive track assignment algorithm for associating a new detection with an existing track based on a minimum terrain-gradient target mobility model. Figure 3.1 shows the current position of an existing target track (location 1) overlaid on a topographic map. The algorithm objective is to determine whether a path exists between location (x_1, y_1, t_1) and location (x_2, y_2, t_2) that would permit a certain class vehicle to have reached location 2 within the time interval $(t_2 - t_1)$. Because a

3. In a single Kalman filter approach, the filter parameters represent a compromise across the anticipated target types and operational characteristics.

minimum terrain-gradient path has potentially low mobility resistance, a minimum terrain-gradient path is assumed to be the minimum-time path. In Figure 3.2, a coarse resolution minimum terrain-gradient path is generated between location 1 and location 2. In Figure 3.3, this path is refined by applying the same domain constraints but viewing the spatial database at twice the spatial resolution. This recursive refinement process generates progressively higher resolution paths that satisfy the problem constraints. The final figure in this series (Figure 3.4) depicts a high-resolution minimum terrain-gradient path between the last known location of a moving target (position 1) and the new detection at position 2. If the path meets certain consistency criteria (path travel distance, maximum vehicle speed), the detection at location 2 would be associated with the existing target track.

Since a canonical form XIII approach uses *general declarative knowledge* and *flexible, single level-of-abstraction control* knowledge, a target-tracking algorithm based on this canonical form can employ much more sophisticated models than the previous approaches. For instance, even though the sensor resolution and measurement accuracy may be inadequate to assure road association, vehicles can be inferred to be on a road if the target is:

1. Near a known road;
2. Moving in approximately the same direction as the road;

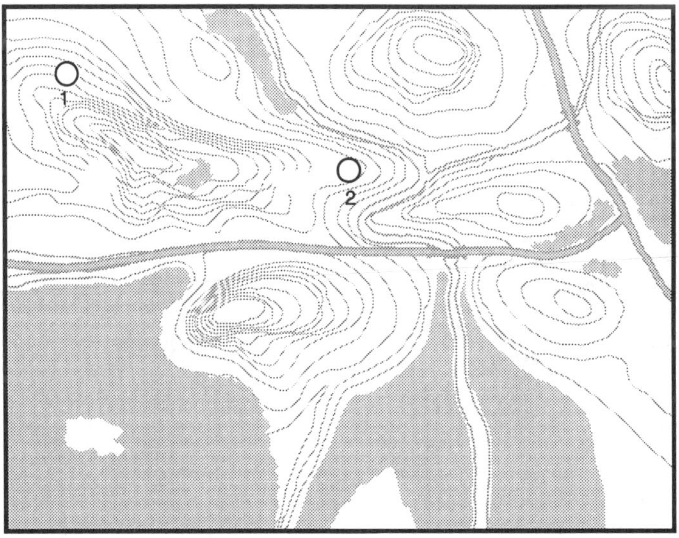

Figure 3.1 Database for multiple-resolution target-tracking algorithm.

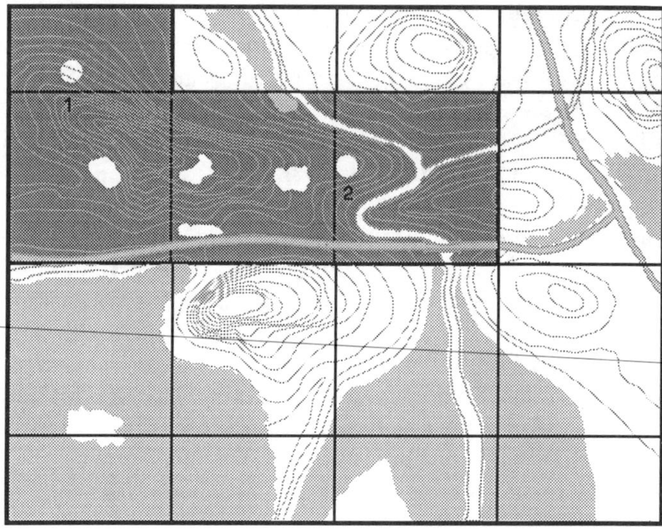

Figure 3.2 Coarse-resolution path between points 1 and 2 based on a minimum resistance path development.

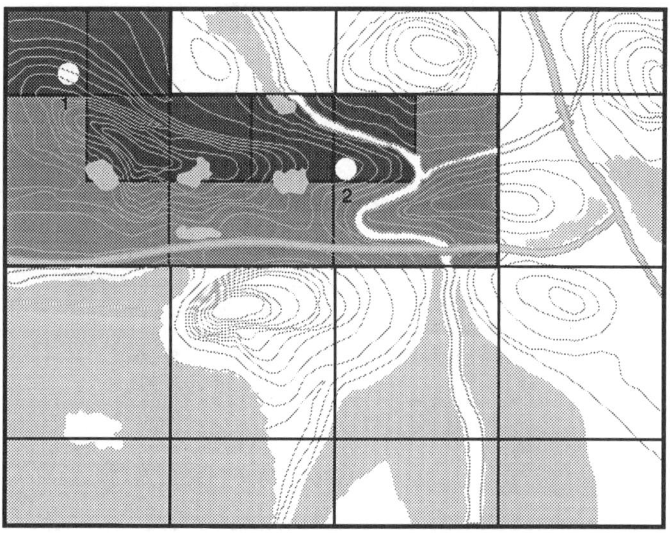

Figure 3.3 Refinement of the path developed in Figure 3.2.

Figure 3.4 Refinement of path developed in Figure 3.3.

3. Traveling at a speed commensurate with road travel.

The confidence of such inference can be enhanced by consideration of contextual information, such as the vehicle's relationship to other entities, the current weather conditions, and the specific vehicle class.

Figure 3.5 illustrates a multiple-hypothesis track-assignment problem that is similar to the canonical form XI problem just discussed. In this case, one of two closely spaced detections must be assigned to a single existing target track. T_1 represents a current target track, while P_1 and P_2 represent two observations that may be associated with the new position of track T_1. Because P_2 is geometrically closer to T_1 than P_1, a non-context-sensitive algorithm would likely select the nearest neighbor assignment. However, because of the very steep terrain between T_1 and P_2, when evaluated within context, it is infeasible for T_1 to have reached P_2 under the time and domain constraints. Thus, P_1 is clearly a more likely track update for T_1 than P_2.

Many fusion problems require the evaluation of alternative hypotheses under real-world domain constraints. Examples of such constraints include the following.

1. For long-distance movement, wheeled vehicles prefer road travel to off-road travel.
2. For high-speed movement, higher class roads are preferred to lower class roads.
3. For operations that depend on concealment, ground-based vehicles may prefer either lower class roads or limited off-road travel.

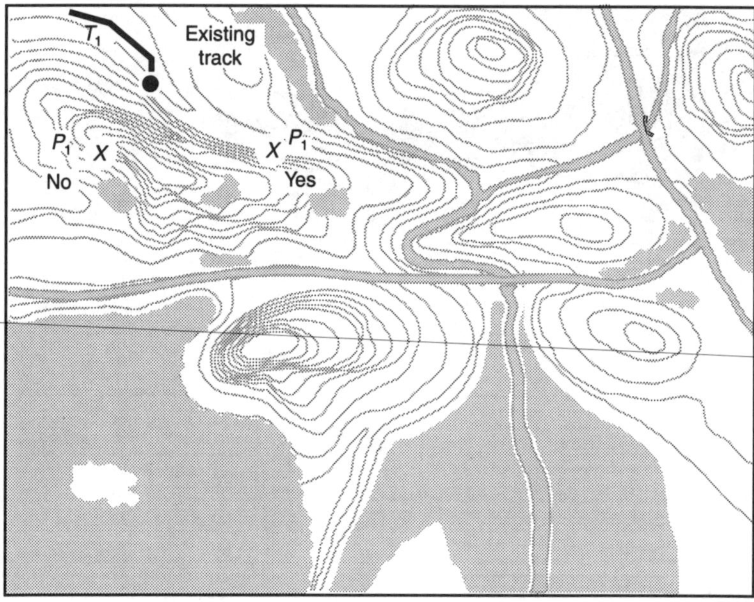

Figure 3.5 Context-sensitive, model-based evaluation of competing hypotheses (P_1 is more likely to be associated with T_1 than P_2).

4. A given class of wheeled vehicles can traverse certain terrain in dry weather, and only a subset of those classes in wet weather.
5. A low-speed target moving over terrain not negotiable by a wheeled vehicle can be inferred to be a tracked vehicle.
6. Wheeled vehicles are not generally located in waterways and marshy areas, on high mountain tops, or in regions that are not reasonably accessible to ground vehicles.

With the addition of *multiple level-of-abstraction control*, canonical form XV supports more sophisticated approaches to target tracking than any of the previous examples. For instance, target tracking could employ general multiple hypothesis reasoning based on multiple level-of-abstraction target-tracking models. Consider the target trajectory fragments depicted in Figure 3.6. In Figure 3.7, model-based reasoning has permitted the association of various segments of the overall trajectory with several local target behavior models: (1) "concealment," (2) "minimum terrain gradient following," (3) "road-seeking" objective, and (4) "road following." From a higher level perspective, *combinations* of these local hypotheses support the development of a more global interpretation of the target trajectory, such as a "river-crossing" strategy or a "road-following" strategy. These midlevel hypotheses, in turn, can be interpreted as elements of even higher level strategies, such as "reinforce a particular unit" or "reach a staging area." Thus, the target-tracking model effectively represents a hierarchical set of *goal-oriented* strategies. Apparent consistency between the track association process and "purposeful" target behavior tends to reinforce the credibility of the target association.

Figure 3.8 depicts the possible organization of a sophisticated model-based target tracker. Rather than a simplistic association of domain characteristics with trajectory segments, a hierarchically organized multiple hypothesis tracker would support the efficient development and maintenance of independent *composite hypotheses* based on a set of robust, context-sensitive models. At each level of abstraction, the hypotheses with the highest likelihood above some threshold would be carried forward. Since, as viewed here, target tracking is really a form of plan recognition, path planning and target tracking can be considered to be *dual* operations.[4] The characteristics of the various target-tracking algorithms just discussed are summarized in Table 3.1.

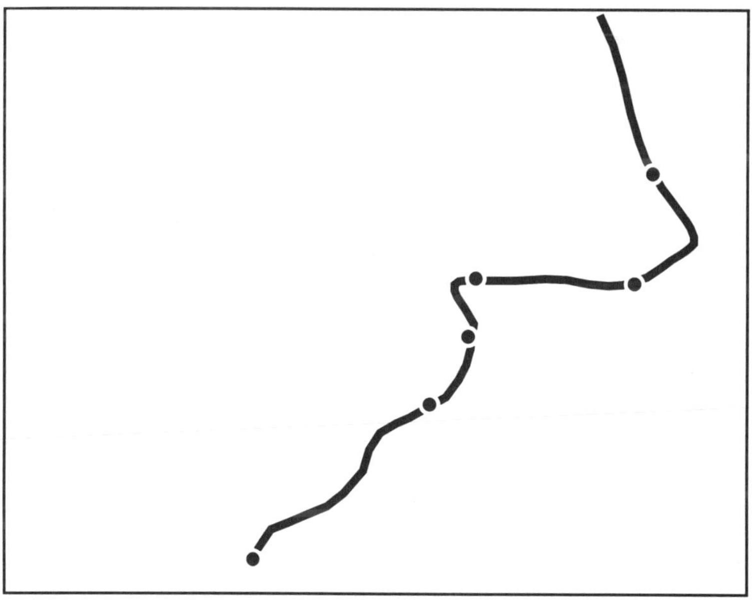

Figure 3.6 Target track example.

4. This important observation will be explored in more detail in Chapter 13.

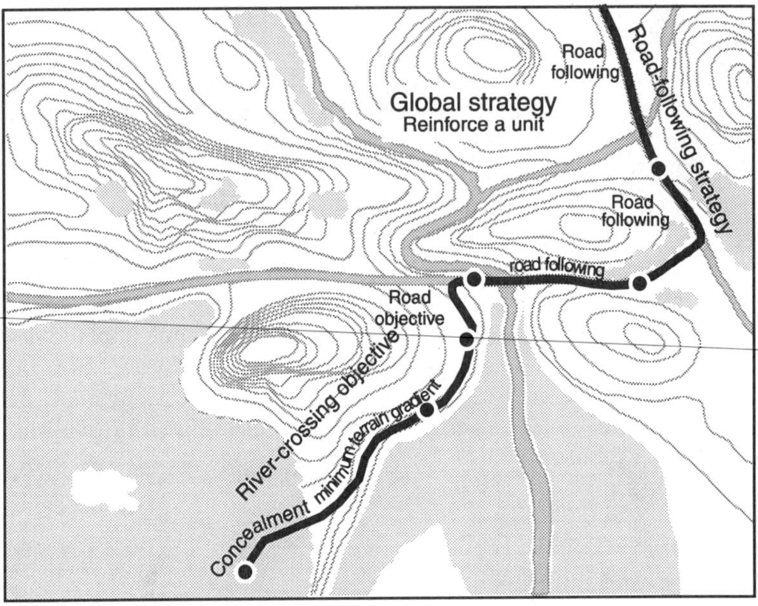

Figure 3.7 Interpretations of target detections within both local and global context.

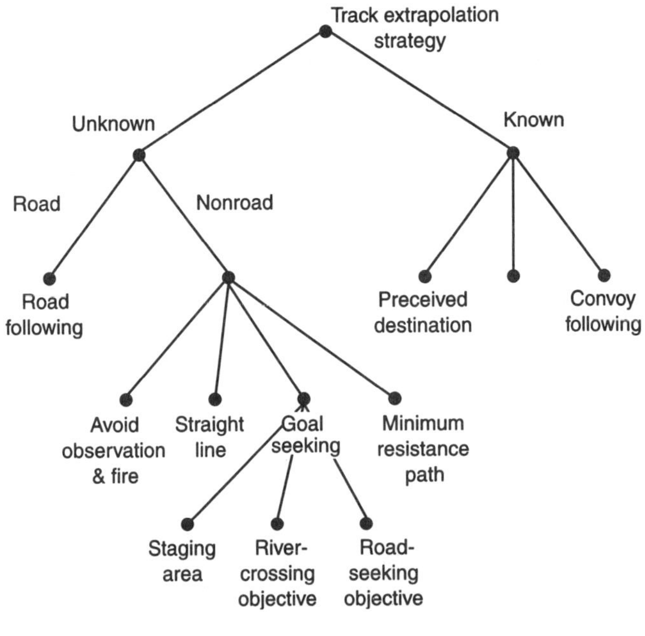

Figure 3.8 Multiple hypothesis-tracking strategies.

Table 3.1
Summary of the Target-Tracking Examples

Canonical Form	*Long-term Declarative Knowledge*	*Context Sensitivity*	*Control Element*	*Control*	*Level of Abstraction*
I	Statistical estimator	No	Rigid	Highly local	Single
III	Coarse pretrack filters, statistical estimators	No	Rigid	Highly local	Multiple-stage algorithm
V	Bank of statistical estimators tuned to various target classes	No	Flexible	Local	Single
VII	Coarse pretrack filters followed by bank of statistical estimators	No	Flexible	Local	Multiple-stage algorithm
IX	Target class-dependent mobility model	Yes	Rigid	Less local	Single
XI	Target class-dependent mobility model	Yes	Rigid	Less local	Multiple-stage algorithm
XIII	Multiple-dimensional, mobility-based constraint set	Yes	Flexible	More global	Single
XV	Multiple-dimensional, mobility-based constraint set	Yes	Flexible	Global	Multiple-resolution multiple hypothesis

3.1.2 Target Classification

Pattern recognition seeks to automate the recognition of patterns of features, symbols, objects, or events. *Classification* and *identification* represent higher level-of-abstraction forms of pattern recognition. The goal of *classification* is to associate data sets with *classes of patterns* (tanks, trucks, aircraft). The goal of *identification* is the *differentiation of intraclass data sets* (distinguishing among versions or instances of a certain class of tank or aircraft).

Pattern recognition methods will be classified as either *deterministic* or *nondeterministic.* Deterministic methods include *syntactic pattern recognition* approaches and *templating* approaches. In syntactic methods, patterns are viewed as sentences or other higher level logical constructs; the pattern recognizer is treated as a syntax analyzer that is based on a formal theory of languages. Model-based templating can be treated as a form of syntactic pattern recognition. Although model-based pattern recognition has received increasing interest in recent years, template-based approaches that rely on either fixed exemplar sets or rigid stereotypical patterns continue to be used.

Nondeterministic methods include both statistical and fuzzy set–based techniques. The following statistical-based methods are listed in order of decreasing a priori information requirements:

1. Bayesian decision theory;
2. Assumed density function form, plus training samples that permit estimation of density function parameters;
3. Estimation of the entire density function;
4. Linear or nonlinear discriminant functions;
5. Clustering and other forms of unsupervised learning.

While, at one extreme, Bayesian decision theory requires knowledge of all a priori and a posteriori probability density functions, at the other extreme, clustering requires no prior knowledge of the data.

A radar classification example will be used to illustrate the spectrum of canonical problem-solving approaches to pattern classification. In particular, the radar return from a ground-based Doppler MTI system is to be classified to distinguish among human, animal, and vehicular targets.[5] Suppose that a linear discriminant function is to be used to distinguish the three target types. A typical approach to developing such discrimination functions is first to identify and then to extract relevant features from the target backscatter. Next a weight matrix (or discrimination function) is developed that permits a mapping from the *m*-dimensional feature space to the *n*-dimensional decision space.

5. An MTI radar takes advantage of the relative motion between moving targets and the relatively stationary fixed background (ground, buildings, foliage) to enhance subclutter visibility [3].

Since canonical form I approaches depend on *specific declarative* knowledge and *rigid, single level-of-abstraction control*, only relatively simple classification algorithms are supported. In a typical canonical form I algorithm, the dot product between the *m*-dimensional feature vector (*short-term declarative knowledge*) and the ($m \times n$)-dimensional weight matrix (*specific long-term declarative knowledge*) is computed and the largest element of the output vector is selected as the most likely target class (*medium-term declarative knowledge*).

Because the paradigm is based on specific declarative knowledge and rigid, single level-of-abstraction control, the algorithm is relatively simple. As discussed in Chapter 1, even when the signal-to-clutter ratio is adequate to measure individual target features reliably, the overall information content of the radar backscatter may not be sufficient to support reliable target classification, regardless of the class of decision function or the extent of discriminator training.

Since a canonical form III approach supports *multiple level-of-abstraction control*, a canonical form III classification algorithm could employ efficient multiple-resolution feature extraction. The spectral regions associated with the target modulation could be isolated using traditional Fast Fourier Transform (FFT) techniques that could then be analyzed using super-resolution spectral techniques for resolving individual modulation components.

Although a canonical form V problem-solving approach represents a rigid pattern recognizer (the linear discriminant model represents *specific declarative knowledge*), the *flexible control* element of this canonical form permits limited context-sensitive performance. For example, if a target must be located within regions that support ground-target mobility (roads, relatively level terrain), target classifications that are inconsistent with such domain knowledge (tanks in lakes, wheeled vehicles in marshes) would be suppressed. In addition, simple multiple-hypothesis algorithms are supported as well. For example, rather than a simple *m*-out-of-*n* sequential classifier, a canonical form V classifier could develop multiple, independent assessments of the target backscatter. With the addition of *multiple level-of-abstraction control*, a canonical form VII classifier could employ multiple-resolution reasoning that might improve the computational efficiency relative to a comparable canonical form V algorithm.

A canonical form IX algorithm relies on *general declarative* knowledge and *rigid, single level-of-abstraction control*. A canonical form IX target classification algorithm was constructed by the author in the late 1970s [4]. The algorithm was based on the development of simple fuzzy set–based target models for humans, birds, animals, vehicles, and clutter that relied on a small number of physically meaningful, class-dependent spectral features derivable from the target signatures. As part of a modeling study, class-dependent fuzzy-set membership functions were developed for each of the features based on simple physical laws and/or empirical studies.

The traditional evidence combination rule used in fuzzy logic is the MINIMUM (MIN) operator. However, with this operator, a relatively unimportant feature with a low membership value can effectively "capture" the decision function.

Thus, rather than rely on the default fuzzy-set intersection operator, a target class-dependent linear discriminant model was used. The discrimination model computed the overall support for a target class as the sum of the likelihood that each feature value was generated by a particular target class weighted by the relative importance of that feature to the target class. Thus, for each radar detection, an independent assessment of the likelihood of the return being due to each target class was generated.

Figure 3.9 depicts the typical backscatter signature from a walking human, and Table 3.2 summarizes the selected spectral feature set. As can be observed from this table, all the target features have a highly intuitive, physical basis. The value of each feature is determined by properly associating the feature with a peak in the power spectral density function.[6] For example, the strongest target-related frequency component in Figure 3.9 is associated with the target's radial translation velocity. The strongest symmetrical sidelobe components, relative to the translational component, are associated with the target step rate.

Figure 3.10 summarizes the conditional fuzzy-set membership functions for the human target model that were developed during the modeling study. Once the value of the individual features are determined, the support for each feature conditioned on a human target is determined using Figure 3.10. The evidence combination rule then computes the linear combination of the product of the individual

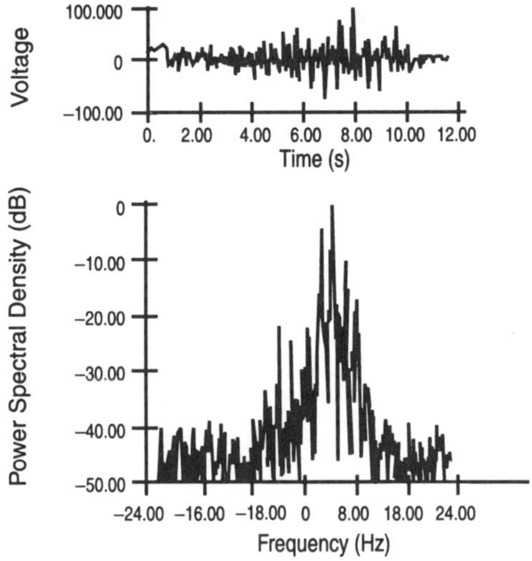

Figure 3.9 Doppler waveform and power spectral density function of the UHF radar return from a human walking radially in-bound (step rate = 100 steps/minute).

6. The values of relevant spectral features were extracted from the power spectral density function using a peak-finding algorithm.

Table 3.2
UHF Backscatter Features from a Human Target

Feature	*Assigned Variable*	*Feature Importance Factor*
Transitional velocity	V_T	0.1
Target natural frequency	F_n	0.5
Target stepsize	d	0.1
Target acceleration	a	0.05
Natural frequency rates of change	r	0.05
Number of natural frequency harmonics	n	0.2

evidence weights and the feature importance factor from Table 3.2. Figure 3.10 indicates the weights associated with each feature for the target signature shown in Figure 3.9. The linear weighted sum leads to the following confidence that the observed radar return was due to a human target:

$$C(human) = 0.1(1.0) + 0.5(0.9) + 0.1(1.0) + 0.05(0.9) + 0.05(0.8) + 0.2(0.95) = 0.925$$

Thus, the confidence that the return was due to a human is 92.5%.

Because the algorithm relied on physically meaningful features (target radial velocity, target step rate), only very limited classifier training was required. As a result, virtually all of the several hundred available target signatures were used to evaluate classifier performance. Table 3.3 summarizes the performance of each of the single-cell classifiers. The algorithm represents a canonical form IX approach because it employs general declarative knowledge (a true model of class-dependent target backscatter) and simple, rigid, single level-of-abstraction control.

By adding *multiple level-of-abstraction* control, multiple-resolution spectral techniques could be employed to evaluate relevant backscatter features beginning with the most general and proceeding to the most specific. Thus, the efficiency of a canonical form XI problem-solving approach would potentially be enhanced relative to the canonical form IX algorithm.

By replacing the rigid control of a canonical form IX approach with a more *flexible control* element, context-sensitive target models (canonical form XIII) that would adjust their performance to account for relevant domain context could be

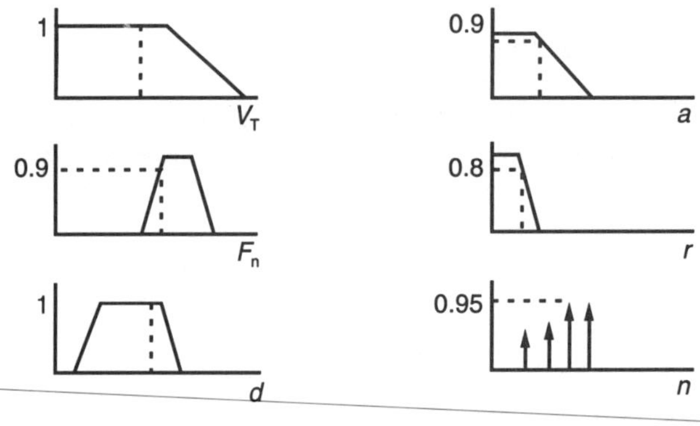

Figure 3.10 Conditional fuzzy-set membership functions for characteristic human radar backscatter features shown in Table 3.2.

Table 3.3
Single-Cell Classifier Performance

	Classification				
Actual Target	*Clutter*	*Vehicle*	*Human*	*Animal*	*Bird*
Clutter	1.0	0	0	0	0
Vehicle	0	0.95	0	0.5	0
Single man	0	0	0.92	0.08	0
Multiple men	0	0	0.78	0	0
Dog	0	0	0.04	0.88	0.11
Pheasant	0	0	0	0.11	0.89

utilized. With the introduction of *multiple level-of-abstraction control*, canonical form XV approaches would support powerful context-sensitive, model-based reasoning that could emulate the reasoning of expert analysts. Because canonical form XV approaches can support true global problem-solving control, such algorithms would tend to be both highly effective and highly efficient. The important characteristics of the various target classification algorithms are summarized in Table 3.4.

3.1.3 Path Planning

In this section, we consider the problem of generating optimal or near-optimal paths between arbitrarily selected locations in a given domain. Ground-based auto-

Table 3.4
Summary of the Target Classification Examples

Canonical Form	*Declarative Knowledge*	*Context Sensitivity*	*Control Element*	*Control*	*Level of Abstraction*
I	Linear discriminant function	No	Rigid	Local	Single
III	Linear discriminant function	No	Rigid	Local	Multiple-resolution feature extraction
V	Linear discriminant function	Limited	Flexible	Local	Single
VII	Linear discriminant function	Limited	Flexible	Local	Multiple-resolution reasoning
IX	Class-conditioned target backscatter model	Limited	Rigid	Less local	Single
XI	Class-conditioned target backscatter model	Limited	Rigid	Less local	Multiple-resolution analysis
XIII	Class-conditioned target backscatter and behavior model	Yes	Flexible	More global	Single resolution
XV	Class-conditioned target backscatter and behavior model	Yes	Flexible	Global	Multiple resolution; multiple level-of-abstraction reasoning

mated tactical route planners often employ a canonical form I algorithm that is based on an n-state ($n > 1$) map summarizing the mobility characteristics of a region of interest. These mobility maps are normally generated by computing the Boolean intersection among combinations of relevant natural (slope, soil type, vegetation) and cultural domain features (roads, buildings). The resulting mobility map typically consists of a large number of disjoint cells that represent regions of equal mobility resistance values.

In its most basic form, "optimal" path development often entails the exhaustive generation of all possible paths within a raster-based mobility map, followed by the selection of the "best" path based on a suitable objective function. The evaluation criteria, for example, might be the shortest overall Euclidean path length, the path that possesses the minimum total resistance, or the path that requires the minimum travel time.

Such a path development algorithm is based on (static) *specific declarative knowledge* (both the mobility map and the objective function) and *rigid, single level-of-abstraction control* (rigid recursion, exhaustive generation, and rank ordering of all candidate paths). Although conceptually simple to implement, path development based on exhaustive path generation and rank ordering represents a highly combinatoric form of global optimization. Thus, exhaustive generate-and-test-based path-planning algorithms tend to be highly inefficient. In addition, such algorithms produce path recommendations that depend on a static mobility map. As a result of this latter shortcoming, the algorithms are fundamentally insensitive to dynamic context (weather, current restricted travel zones, present enemy positions).

Because it supports *multiple level-of-abstraction control*, a canonical form III algorithm can employ more efficient, multiple-resolution path development strategies. Consider the top-down, multiple-pass algorithm shown in Figures 3.11 to 3.13. In the first pass (Figure 3.11), a path between the start and the goal state is determined by first generating all possible coarse-resolution paths and then selecting the shortest path. In each subsequent pass, the search-space size is confined to lie within the path developed during the previous pass but evaluated at the next higher resolution level. In this way, an initial rather coarse resolution path is recursively refined until the objective resolution has been achieved. In general, top-down, multiple-resolution approaches tend to be much more efficient than single-pass algorithms that operate at the objective resolution.

Although a canonical form V approach to path planning depends on the same *specific declarative* knowledge base (a mobility map), due to its *flexible control* element, it can support more sophisticated reasoning. For instance, rather than the exhaustive generation of all possible paths, simple context-sensitive reasoning and heuristic evaluation functions (for example, the A* search algorithm [5]) can be used. The mobility map can be augmented by dynamic domain constraints such as the current enemy surveillance envelopes and force locations. With the addition of *multiple level-of-abstraction control*, a canonical form VII algorithm could employ multiple-resolution path development strategies as well as multiple level-of-abstraction path

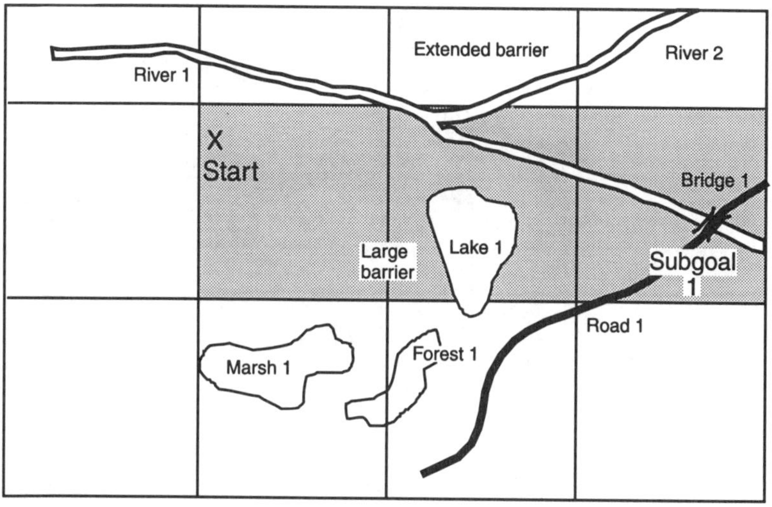

Figure 3.11 Coarse resolution path to Bridge 1.

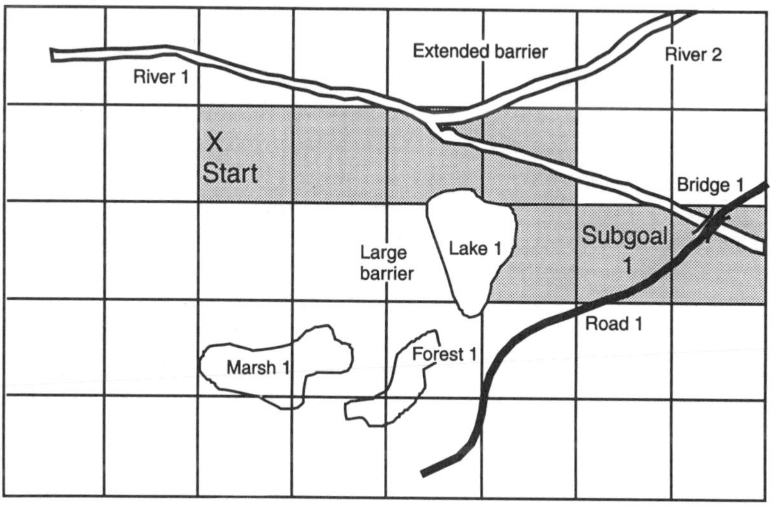

Figure 3.12 Refinement of path from Figure 3.11.

extension heuristics. The path extension heuristic at a coarse resolution, for example, might depend only on whether or not a given node contained a road, while at a higher resolution, it could depend on terrain and elevation information as well.

Whereas the four path development algorithms just outlined rely on *specific declarative* knowledge, the next four algorithms use *general declarative* knowledge that

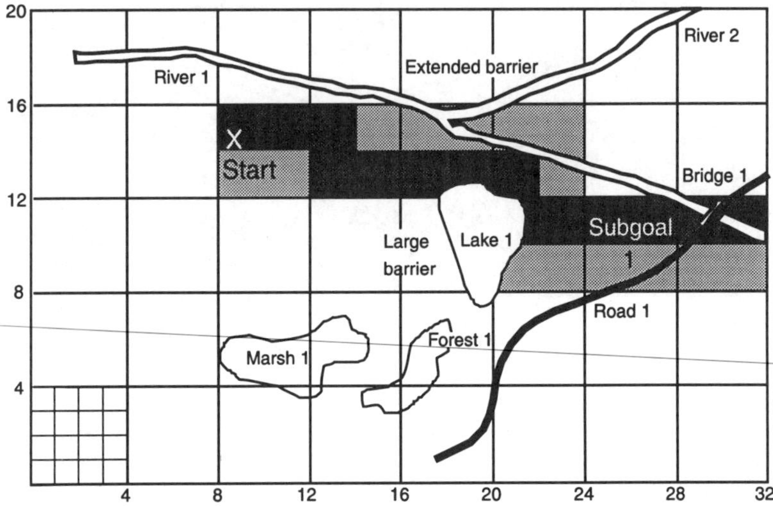

Figure 3.13 Refinement of path from Figure 3.12.

supports more model-based reasoning approaches. Because of the diversity of both geographic and cultural domain constraints that are relevant in many data fusion applications, model-based algorithms tend to be significantly more robust than non-model-based algorithms. For example, while a treeless desert with moderate elevation changes might represent a relatively simple route-planning challenge, route planning in hilly, heavily forested areas that contain numerous lakes, rivers, and restricted travel zones tends to be considerably more complex. With a canonical form IX approach, a path can be developed based on a vehicle mobility model that is dependent on a wide range of specific geographic and cultural features rather than on a static mobility map. Since the underlying knowledge base is potentially dynamic (restricted travel zones may change over time, rain changes the mobility characteristics of many soil classes), path planning is inherently temporally sensitive. In order to control their combinatorially explosive computational requirements, path development algorithms that rely on *rigid, single level-of-abstraction control* (both canonical forms I and IX approaches) often require the introduction of manually established "way points."

Since a Production System is a canonical form IX problem-solving form, a production rule-based path-planning algorithm could be developed that relies on the search of a mobility graph. An IF/THEN rule set would control the path development process. Unfortunately, a graph representation of two- or three-dimensional space might require a significant oversimplification of the problem domain. In addition, the search of the extremely large graph-based domain representations that might be required could lead to highly inefficient algorithms. With the addition of the *multiple level-of-abstraction control* perspective offered by a canonical form XI algorithm, a top-down, multiple-resolution algorithm could be used that would potentially reduce the problem state-space size from one of exponential complexity to one of logarithmic complexity.

A canonical form XIII algorithm offers a potentially *flexible*, model-based (*general declarative* knowledge), *single level-of-abstraction* approach to path planning. Since algorithms that employ general declarative knowledge and flexible control can support highly context-sensitive reasoning, a canonical form XIII path-planning algorithm could use class-dependent vehicle mobility models that depend directly on the underlying geographic, cultural, weather, and target database (*general declarative knowledge*). These models could employ both default constraints (wheeled vehicles are not found in lakes) and problem-specific constraints (favor highly concealed routes; pass through Region A; prefer road travel to off-road travel unless an alternate off-road path exists that is less than 1/3 the length and has a maximum slope of 15%).[7]

A canonical form XV problem-solving approach adds *multiple level-of-abstraction control* that can support the construction of more efficient path development algorithms. For instance, since the most serious ground mobility barriers tend to have a more significant and predictable impact on path development than less serious barriers, the domain impact on mobility can be represented in terms of a hierarchical set of mobility constraints. Path development can then be formulated in terms of top-down, multiple-pass constraint-satisfaction. For example, in the first pass, a set of gross resolution candidate paths can be developed that are based on a rather global evaluation criterion. On subsequent passes, all candidate paths can be recursively refined based on progressively higher resolution data and more local constraints. In this way, a canonical form XV algorithm could support true global control of the reasoning process.

To illustrate the potential advantages of a canonical form XV class approach, we outline a simple multiple-pass path development algorithm. In the first stage, low resolution spatial search from the "start" node to the "goal" node identifies potential vehicle class–dependent *extended barriers,* such as rivers, canyons, or ridge lines that must be breached in order to reach the goal state. If an extended barrier is detected, the closest barrier-crossing locations must be determined (bridges or fording locations for a river, passes for a mountain range). One or more of these crossing sites is (are) selected as potential path subgoals. Starting from each of these subgoals, in turn, the global planning process is repeated. The product of stage 1 is a set of low-resolution path fragments that satisfies the global constraint set.

Since roads are likely to have the next most significant impact on route selection, stage 2 develops road connectivity for all candidate paths that are near a road.

7. In general, default constraints are generic, context-insensitive facts or rules, while problem-specific constraints refine or over-ride the default constraints.

Upon completion of stage 2, the coarse path plan developed in stage 1 is refined with appropriate road-following segments.

In stage 3, overland routes between all remaining subgoal pairs not already connected by roads are refined. The top-down constraint set begins with the consideration of *large barriers* that exist between subgoal pairs. A *large barrier* is an obstacle that unlike an extended barrier *can* be circumnavigated if a path through the barrier cannot be discovered. Once a path around or through large barriers have been established for all subgoal pairs, all overland path segments are refined based on progressively smaller barriers until the objective path resolution is achieved. In addition to supporting the use of top-down constraint sets, the algorithm can maintain multiple alternative paths between subgoals that are based on multiple path-selection criteria.

Because the algorithm uses a top-down divide-and-conquer strategy that effectively reduces the problem size, hierarchical path planning:

1. Supports global path development strategies;
2. Is far more efficient than single level-of-abstraction algorithms;
3. Requires more straightforward heuristics than traditional plan-development algorithms.

Since each subgoal pair is known to be a potential member of the final solution set, recursive refinement of the paths between subgoals supports highly efficient path development. Once all alternative path segments from the start to the goal state are determined, complete paths can be generated from the alternative path fragments and ranked as to minimum travel time, minimum distance, and maximal concealment.

Path planning based on *specific declarative* knowledge and *rigid, single level-of-abstraction control* generally leads to local, nearest neighbor path-extension strategies and a highly combinatoric optimal path-evaluation process. Just as in maze-solving problems, path planning using a canonical form I approach tends to develop a large number of apparently promising paths that either terminate prematurely or wander excessively.

On the other hand, an algorithm based on *general declarative* knowledge and *flexible* and *multiple level-of-abstraction control* can support efficient, top-down, context-sensitive, model-based path development. The canonical form XV path-development algorithm just outlined effectively emulates the efficient problem-solving style of human problem-solvers. The highest level strategy locates and exploits potential "weaknesses" in the mobility barriers. The approach permits the development of sophisticated, multiple-strategy candidate paths based on complex, realistic constraint sets. All resulting candidate paths are recursively refined using progressively higher resolution constraints with the attendant advantages of tree-oriented search over sequential, nondirected search. While the lower order canonical form approaches to path development typically rely on highly myopic control, a canonical form XV approach can support true global control of the reasoning process. The outlined characteristics of the path-planning algorithms are summarized in Table 3.5.

Table 3.5
Summary of the Path-Planning Examples

Canonical Form	*Long-term Declarative Knowledge*	*Context Sensitivity*	*Control Element*	*Control*	*Level of Abstraction*
I	*n*-state mobility map	Limited	Rigid	Local	Single
III	*n*-state mobility map	Limited	Rigid	Local	Multiple-resolution reasoning
V	*n*-state mobility map; context-sensitive heuristics	Limited	Flexible	Local	Single
VII	*n*-state mobility map; context-sensitive heuristics	Limited	Flexible	Local	Multiple-resolution reasoning
IX	Vehicle-dependent mobility model	Yes	Rigid	Less local	Single
XI	Vehicle-dependent mobility model	Yes	Rigid	Less local	Multiple-resolution reasoning
XIII	Vehicle-dependent mobility model	Yes	Flexible	More global	Single
XV	Vehicle-dependent mobility model	Yes	Flexible	Global	Top-down multiple

3.2 HYPOTHESIS-BASED APPROACHES

Simple diagnostic reasoning (car repair, disease identification), a grading template for multiple-choice test scoring, object detection within imagery, image understanding, and numerous other forms of templating involve hypothesis-based reasoning. In hypothesis-based approaches, rather than *generating* output states from input states, a priori hypotheses or expectations are validated by identifying input or problem states that match, explain, or justify such expectations. For certain classes of problems, hypothesis-based algorithms can be more computationally efficient than their generation-based counterparts.

Canonical form II problem-solving paradigms support simple, rigid algorithms (*specific declarative* knowledge, *rigid, single level-of-abstraction control*). Thus, canonical form II approaches exhibit the same fundamental problem-solving assumptions and limitations as canonical form I approaches. Although the simplicity of canonical form II approaches tend to encourage their use, for complex fusion tasks, brittle and highly inefficient performance can be expected. With the addition of *multiple level-of-abstraction control*, canonical form IV algorithms support simple hierarchical templating and clustering. For example, rather than using a single, high-resolution template, the templating process could employ a set of progressively more refined templates that permit efficient hierarchical search-space pruning.

Although canonical form VI approaches are based on *specific long-term declarative* knowledge (fixed mobility maps, explicit exemplar patterns of vehicles of interest), their *flexible, single level-of-abstraction control* element supports the development of more context-sensitive algorithms than the two previous paradigms. With the addition of *multiple level-of-abstraction control*, canonical form VIII algorithms support efficient hypothesis-based reasoning, such as top-down windowing and multiple-resolution Boolean operations.

Because each of the next four hypothesis-based canonical forms employ *general long-term declarative* knowledge, these forms support more model-based problem-solving approaches. Canonical form X, for instance, supports rather *rigid, single level-of-abstraction* model-based reasoning. Backward-chained Expert Systems, model-based diagnostic reasoning, constraint-based search, and parametric templating are representative problem-solving forms. Although still employing rigid control, the addition of a *multiple level-of-abstraction control* element permits canonical form XII approaches to support simple top-down model-based templating.

Canonical form XIV algorithms employ *general declarative* knowledge and *flexible, single level-of-abstraction control*. Thus, canonical form XIV algorithms support general, context-sensitive, constraint-based reasoning. With the addition of *multiple level-of-abstraction control*, canonical form XVI represents the most powerful hypothesis-based problem-solving form. Just as its generation-based dual (canonical form XV), canonical form XVI class paradigms can support the development of algorithms that exhibit true global control of the reasoning process.

3.2.1 Object Detection In Imagery

Object detection using pulse-Doppler radar is accomplished by discriminating between targets and nontargets based on amplitude, spatial, spectral, and temporal characteristics of the one-dimensional backscatter signal. Similarly, the detection of objects in imagery can be accomplished by the analysis of amplitude, spatial, spectral, and temporal characteristics of two-dimensional images. While the detection of moving objects can rely on the relative motion between the objects of interest and the background, the discrimination between stationary objects and the stationary background tends to be more difficult.

Although target classification is generally an *optional* requirement in an MTI radar application, it is often virtually *mandatory* for detecting nonmoving targets in imagery. Many classes of sensors provide inadequate information to both detect and reliably classify stationary, nonemitting tactical targets. On the other hand, because of their higher bandwidth, imaging sensors (optical, FLIR, synthetic aperture radar) provide considerably higher dimensional information that can support both detection and recognition. However, since simple metrics based on amplitude and spatial characteristics of objects (which generally rely on image segmentation) can lead to both erroneous as well as missed detections, automatic target detection and recognition in complex, cluttered environments represent a highly challenging problem.

Regardless of the class of imaging sensor employed, early object detection algorithms typically employed a set of rigid two-dimensional templates that represented the set of objects of interest (*specific long-term declarative* knowledge) that were compared to (or convolved with) an input image. The object would be declared present if the correlation between this "moving-window" template and the underlying data set crossed a certain threshold. The control element of such algorithms is typically very *rigid*, and all analysis is performed at a *single level of abstraction*. Due to variations in the target range, view angle, relative target orientation, and other relevant parameters, a large number of object templates can be required. In effect, canonical form II algorithms perform two-dimensional matched filtering. For complex problems, moving-window templating tends to be highly inefficient.

Guided by a wide range of contextual clues and sophisticated reasoning models, trained imagery analysts can locate and identify even partially obscured objects of interest in aerial imagery. A consideration of the "Where's Waldo" books that are popular with children provide an interesting perspective on the object detection problem. Even trained analysts can have considerable difficulty locating the randomly positioned Waldo from among dozens or even hundreds of near look-a-likes. Because the pictures are purposely devoid of contextual clues that would help locate Waldo, humans are forced to abandon many of their most sophisticated spatial-reasoning techniques. In effect, humans are forced to rely on a canonical form II moving-window pattern-match process. Since canonical form II problem-solving approaches tend to be inherently inefficient (whether performed by a human or emulated by a machine), looking for Waldo can be a frustrating experience for adults. Fortunately, many problems of practical interest possess considerable exploitable context.

Ad hoc heuristics are often used to improve the efficiency of canonical form II algorithms or to attempt to generalize the rigid template match process (statistical degree of match, rubber mask templates). However, since such generalizations are insensitive to the internal character or dynamics of the particular problem, they offer little more than a "band-aid" on an inherently rigid problem-solving paradigm. As discussed in the last chapter, the choice of a more appropriate paradigm is generally preferable to attempting to "work around" the inherent deficiencies of a particular approach. Since a canonical form IV approach supports *rigid, multiple level-of-abstraction* control, the use of multiple-resolution search windows operating on multiple-resolution data sets can potentially improve problem-solving efficiency.

Because a canonical form VI approach supports *flexible*, context-sensitive control of the reasoning process, such algorithms can potentially be sensitive to a wide range of relevant domain constraints. For instance, in attempting to detect a wheeled vehicle in aerial imagery, all regions of the image that do not support wheeled vehicle mobility can be ignored. The remaining regions can be evaluated with respect to more detailed mobility constraints. For example, certain classes of wheeled vehicles may be expected to be found either on secondary roads or in regions with low mobility resistance that are adjacent to secondary roads.

With the addition of *multiple level-of-abstraction* control, a canonical form VIII class algorithm can support efficient top-down, multiple-resolution analysis. Thus, a canonical form VIII object-detection algorithm could evaluate regions of an image where a particular class vehicle might be located based on top-down, multiple-resolution evaluation of ground-target mobility or "just-above-tree-top" flight paths. With a top-down algorithm, each stage of analysis effectively prunes the search space prior to subsequent, higher resolution analysis. Since a canonical form VIII algorithm depends on *specific declarative* knowledge, the object template remains rather rigid. Anticipated variations in the observed objects (such as orientation and size) must be explicitly accounted for in the template set.

Because they employ *general long-term declarative* knowledge, the next four canonical form approaches support more model-based approaches to object recognition. Rather than exhaustive testing of explicit, rigid templates, the higher order canonical form approaches can utilize individual observables as well as all relevant relationships among the observables. For instance, a model-based algorithm can be sensitive to physically meaningful target parameters, such as the relative target dimensions (length-to-width ratio), predicted range, target center of gravity, target cross section, and the relationships among individual target scatterers on the object. Because they possess *rigid, single level-of-abstraction control*, canonical form X algorithms employ rather simple search mechanisms.

Because it supports multiple-resolution (*multiple level-of-abstraction control*) search-and-test, a canonical form XII algorithm can potentially enhance problem-solving efficiency. For instance, at a relatively low resolution, an object to be recognized may be adequately represented as a fixed-size rectangle; while at a higher resolution, the detailed vehicle subparts (body, tube, turret, and tracks of a tank) might be coarsely represented using a set of simple multiple-resolution target models that depend on target range and lighting conditions. Finally, at the objective resolution, the highest resolution sensor data would be evaluated against a high-resolution model of all candidate vehicles.

Canonical form XIV algorithms possess *flexible* but *single level-of-abstraction control* that can support more sophisticated approaches than all previous approaches. Object detection efficiency can be enhanced by first directing search to only those regions of the image where the vehicles of interest will most likely be found (on high ground, away from power lines) and not in parts of the image where such vehicles are either unlikely (although a region supports vehicle mobility, a unit in this region would be out of line-of-sight radio contact for an extended period) or not of current interest (depots, storage facilities). Once the data set is pruned, the remaining regions of interest can be evaluated against sophisticated models of all candidate target classes.

Imagery analysts routinely employ context-sensitive, model-based reasoning (*general declarative knowledge*) that exhibits both local as well as global problem-solving control (*flexible, top-down control knowledge*). Thus, sophisticated manual approaches to object detection represent canonical form XVI problem-solving approaches. The detailed model of a particular object might be sensitive to, for example, the target's physical size, shape, performance capabilities, the sensor look-down angle, surrounding terrain, and sensor frequency band. At a coarse resolution, the target model might be represented as a small area of high reflectance, while progressively more refined models would characterize the relationships among prominent target features.

Consider, for instance, a tank target detector that is based on a flexible, hierarchical model. First, regions that support tank mobility would be generated. Next, all candidate targets in regions of potential interest would be located by searching for locally bright image areas. The length-to-width ratio of the bright pixel regions would be tested. All candidate objects that pass both tests would then be reevaluated using more detailed models that include turret and gun-tube identification. Once all prominent features are discovered, the relative relationships among the features would be evaluated. In addition, the analysis can depend on the object's relationship to objects that have been previously discovered in the image. Thus, rather than simply employing progressively higher resolution rigid templates of an object, a canonical form XVI algorithm can employ a true multiple level-of-abstraction object model that is sensitive to appropriate semantic and spatial constraints.

Canonical form XVI supports recursive, hierarchical search-space pruning because each subsequent stage operates on a reduced search space size. Candidate targets are evaluated with respect to:

1. Progressively more detailed vehicle models;
2. Possible relationship with nearby vehicles;
3. Vehicle class mobility requirements;
4. Weather effects on image contrast ratio;
5. Doctrinal and operational requirements.

In this way, candidate targets are efficiently discovered using global search followed by object evaluation against multiple level-of-abstraction target models in full local context. Because the algorithm uses top-down model-based reasoning, once the most prominent features of all objects of interest have been located, a goal-directed search can attempt to locate and refine remaining identifying features.

The important characteristics of the various target detection algorithms just discussed are summarized in Table 3.6.

Table 3.6
Summary of the Object Detection in Imagery Examples

Canonical Form	*Long-term Declarative Knowledge*	*Context Sensitivity*	*Control Element*	*Control*	*Level of Abstraction*
II	Rigid template	No	Rigid	Local	Single
IV	Rigid template	No	Rigid	Local	Top-down multiple-resolution analysis
VI	Multiple-stage rigid templates	Limited	Flexible	Local	Single
VIII	Multiple-stage rigid templates	Limited	Flexible	Local	Top-down multiple-resolution analysis
X	Simple target model	Yes	Rigid	More global	Single
XII	Simple target model	Yes	Rigid	More global	Top-down multiple-resolution analysis
XIV	Sophisticated target model	Yes	Flexible	Global	Single
XVI	Sophisticated target model	Yes	Flexible	Global	Top-down multiple-resolution analysis

3.2.2 Doctrinal Templating

Doctrinal templating seeks to identify higher level-of-abstraction organizations of objects based on the observation and identification of their constituent parts. In the last section, a series of progressively more sophisticated approaches for detecting individual objects in imagery were outlined. Rather than the detection of individual objects, doctrinal templating can be thought of as a *composite object*-detection problem. The lower order canonical form approaches to doctrinal templating rely on rather rigid pattern-match techniques, while the higher order canonical form approaches typically employ more general constraint-based search.

Because canonical form II approaches employ *specific declarative* knowledge and *rigid, single level-of-abstraction control,* typical canonical form II doctrinal templating algorithms generate the exhaustive correlation between a set of rigid exemplar patterns and a single level-of-abstraction representation of the current situation database. Since a small change in the input data can produce a large change in the correlation with the exemplar set, rigid pattern-match paradigms tend to be inherently nonrobust. Figure 3.14 illustrates an observed pattern of entities and depicts a partial set of rigid templates that represent the acceptable variants of the unit to be recognized. Although the data set is very similar to template #3, no template precisely matches the data set.

To compensate for the rigid nature of the pattern-match process, a *similarity measure* is generally employed to permit the template that achieves the "closest match" to be selected as the most likely unit class. Since real-world problems often

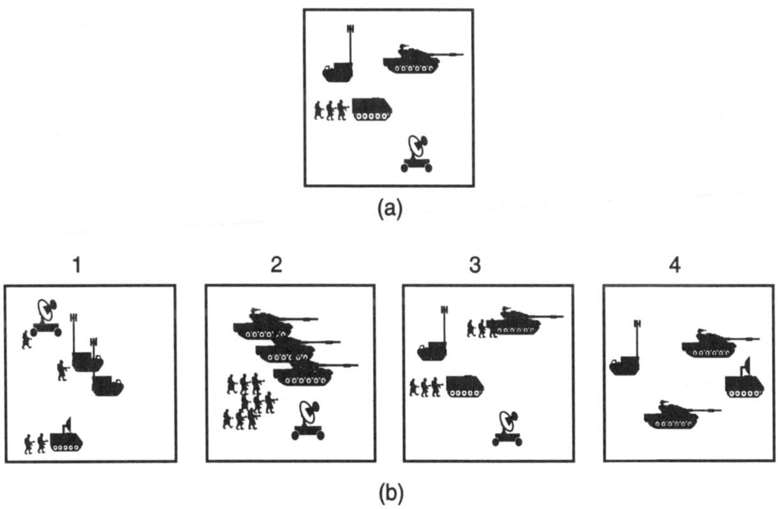

Figure 3.14 Observed pattern and explicit doctrinal template set.

possess a large number of *degrees of freedom*, template sets can potentially require a large number of exemplar patterns. Thus, in addition to being nonrobust, such techniques tend to be inherently inefficient.

Due to the simple, rigid nature of both the templates and the control element, rigid templates are insensitive to contextual knowledge, such as the weather, the surrounding terrain, nearby cultural features, nearby friendly, and enemy units, as well as important relationships among the individual elements of the unit. With a canonical form IV approach, multiple-resolution templates can be used that potentially improve the efficiency of the template instantiation process.

Although canonical form VI approaches rely on *specific*, rigid templates, the *flexible, single level-of-abstraction control* element associated with this problem-solving form supports limited context sensitivity. Figure 3.15 shows a set of data that closely matches one of the rigid exemplar patterns shown in Figure 3.14. However, when the template match is evaluated with respect to spatially local context, the unit is observed to be split by a river. Because doctrine may prohibit the separation of subordinate elements by such a maneuver barrier, an otherwise near-perfect template match would be invalidated. Since the lower order approaches do not support con-

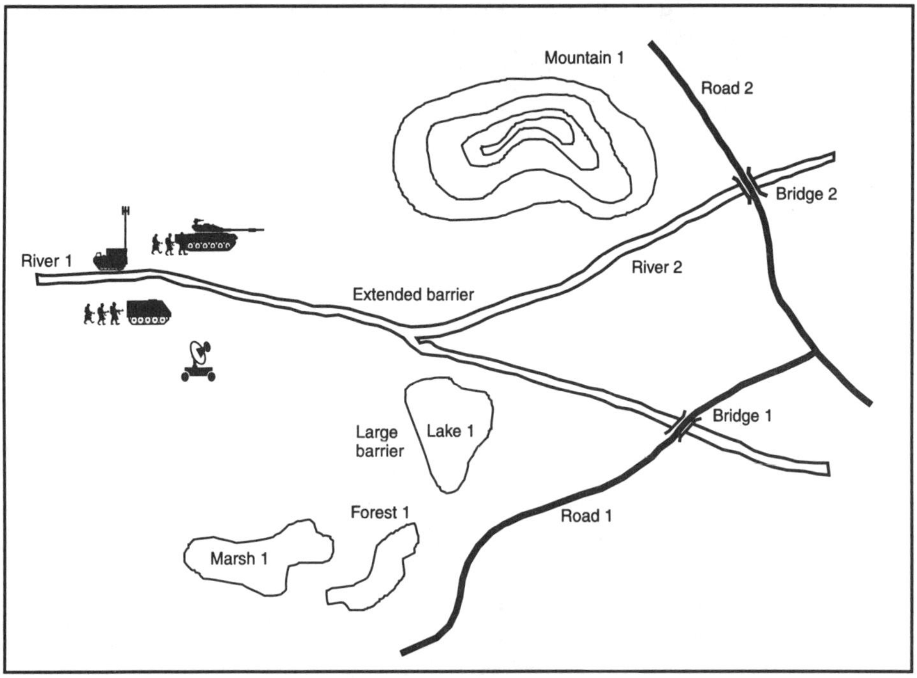

Figure 3.15 A rigid template paradigm could not detect that the unit has violated doctrine by being split by a maneuver barrier.

text-sensitive reasoning, a typical canonical form II approach could not provide such capability. By adding *multiple level-of-abstraction control*, a canonical form VIII algorithm would permit flexible, top-down reasoning.

Rather than relying on rigid templates, the next four canonical forms support much more powerful constraint-satisfaction-based templating. Although doctrinal templating based on rigid templates often relies on matching sensor-derived information against stereotypical spatial configurations of the units of interest, the higher order approaches permit composite objects to be recognized based on the functional, organizational, and logistical constraints that govern the make-up and relationships among the individual entities. For example, Figure 3.16 depicts a general constraint set for the recognition of a particular military unit that is in sharp contrast to the rigid template set depicted in Figure 3.14. If certain entities must always maintain line-of-sight communications, for instance, this condition can be evaluated as an integral part of the template instantiation process.

When the templating process demands only *simple, single level-of-abstraction* constraint-based reasoning, a canonical form X problem-solving approach would be appropriate. With the addition of a multiple level-of-abstraction control component, a canonical form XII approach could evaluate constraint satisfaction and violation using efficient top-down analysis. Because it supports *flexible, single level-of-abstraction*, model-based reasoning in full context, a canonical form XIV approach supports more flexible reasoning. With the addition of *multiple level-of-abstraction control*, canonical form XVI supports very general top-down problem solving.

	One per unit Located on high ground near secondary road Located 10 to 20 km from FEBA
	Up to two groups per unit Generally forward of command post
	One per unit Requires line-of-sight comms with all maneuver units All elements within 10 km Prefers locations with moderate concealment
	One or more groups per unit Rearward of tank groups No maneuver barrier with respect to APC units
	No maneuver barrier with respect to tank groups Associated with ground troops

Figure 3.16 Simple model-based doctrinal template.

A simple example of a canonical form XVI algorithm is outlined next to demonstrate the potential of top-down, constraint-based reasoning. The unit to be recognized consists of a number of wheeled vehicles, several tracked vehicles, and supporting equipment. The wheeled vehicles carry troops and supplies and communicate by radio; the tracked vehicles include tanks and construction equipment. The supporting equipment includes an air defense radar and its associated weapon system. While a traditional rigid template approach would attempt to characterize the physical organization of the unit (the relative spatial relationships of all unit sub-elements), a model-based template would attempt to evaluate the doctrinal, organizational, logistical characteristics among a set of targets in addition to the physical relationships among the constituent elements.

For example, suppose that although a number of vehicles have been detected, none have been classified. If the differentiation between units depends on whether most of the vehicles are "tracked" or most are "wheeled," an attempt must be made to classify the individual targets. If the vehicles in question had been observed for some time, an evaluation of their track histories might permit such inference. Suppose that the targets had been observed traveling through rather rough terrain and somewhat swampy regions; tracked vehicles would be suggested. Conversely, if the vehicles had been traveling at relatively high speeds along a hard-surfaced road, a wheeled-vehicle hypothesis would be more likely. Thus, nontraditional approaches to target classification (target behavior profile evaluation) can be a necessary element of the overall doctrinal templating process.

When doctrine demands that certain elements of the unit maintain direct line-of-sight (LOS) communication, this condition can be evaluated based on the positions of the two subelements and the knowledge of the underlying topography and vegetation. With a multiple-resolution spatial representation and the flexible control element of a canonical form XVI approach, the domain characteristics in the vicinity of the objects of interest can be efficiently evaluated using a top-down, multiple-resolution line-of-sight algorithm such as the one discussed in Section 8.3.

Thus, a model-based approach to doctrinal templating can utilize all the relevant characteristics of individual objects as well as all the known historical, logistical, and doctrinal relationships among the objects. In addition, the higher order canonical form approaches permit the use of more efficient search strategies than the lower order forms. For example, by locating the unit control node first, a constraint-based search for subordinate nodes can be employed with respect to this control node.

As a simple example of such a model-based strategy, suppose the control node of a particular unit is known to prefer establishing its command posts in lightly forested regions that are within 50m of a secondary road. Figure 3.17 shows the development of the refined object location of a candidate control node based on the successive intersection of two independently measured error ellipses, with all regions that are within 50m of secondary roads and "within" lightly forested regions. The resultant product provides potentially better object localization than is possible

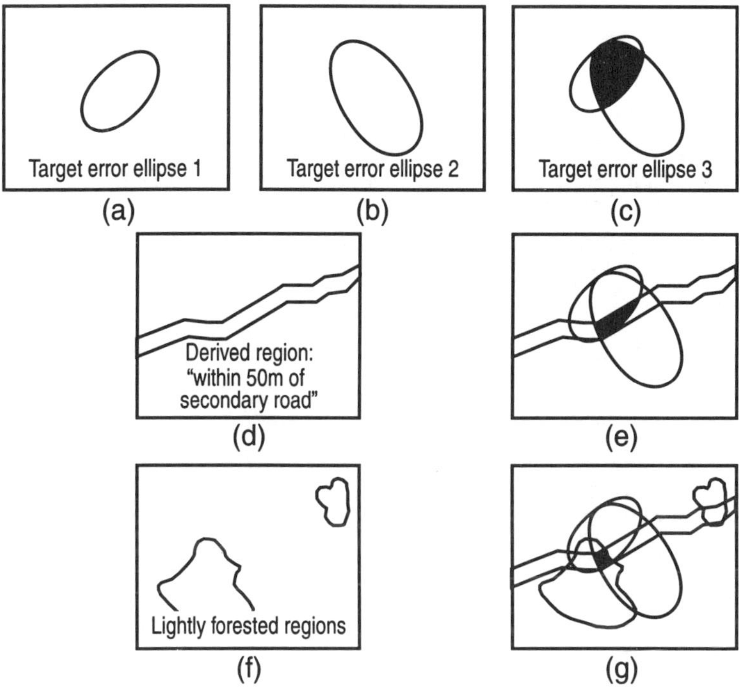

Figure 3.17 Initial target location (a) is refined by the intersection of an independently measured uncertainty function (b) as shown in (c); the resulting region is intersected with the region-grown road network (d) as shown in (e); the cumulative intersection with the lightly forested region (f) is shown in (g).

from the intersection of the sensor-derived error ellipses alone. Since the individual conditions can be applied as the conjunction of individual templates, the approach offers a natural method for combining evidence.

The characteristics of the various doctrinal templating algorithms discussed in this section are summarized in Table 3.7.

3.3 CONCLUDING REMARKS

This chapter has discussed a number of classic data fusion problems and outlined a wide range of algorithmic approaches for solving these problems. The next chapter maps conventional problem-solving paradigms into the problem-solving form taxonomy for the purpose of evaluating their potential support to higher order canonical form problem-solving approaches. Chapter 5 summarizes the general algorithm principles that have been developed throughout Part I.

Table 3.7
Summary of the Doctrinal Templating Examples

Canonical Form	*Long-term Declarative Knowledge*	*Context sensitivity*	*Control Element*	*Control*	*Level of Abstraction*
II	Rigid template	No	Rigid	Local	Single
IV	Rigid template	No	Rigid	Local	Multiple resolution
VI	Rigid template	Yes	Flexible	Less local	Single
VIII	Rigid template	Yes	Flexible	Less local	Multiple
X	Constraint-based model	Yes	Rigid	More global	Single
XII	Constraint-based model	Yes	Rigid	More global	Multiple
XIV	Constraint-based model	Yes	Flexible	Global	Single
XVI	Constraint-based model	Yes	Flexible	Global	Multiple

Although the higher order canonical form algorithms presented in this chapter appear to have considerable potential, the development of efficient, context-sensitive model-based algorithms is far from trivial. As will be demonstrated in Part II, many of the technical problems associated with the implementation of higher order canonical form algorithms can be partially overcome by the development of more powerful database management systems. Part II addresses a wide range of database issues that are driven by both algorithm development and computational issues as well as database search and manipulation requirements.

Armed with a high-level perspective of the shortcomings of conventional data fusion algorithms, the potential of higher order canonical form approaches, and a powerful database kernel designed to support both algorithm development and efficient algorithm performance, a number of the examples discussed in this chapter are revisited in Part III. The objective of Part III is to demonstrate that it is both possible and practical to develop data fusion algorithms that approach the performance of human analysts.

References

[1] Hall, D.L., *Mathematical Techniques in Multisensor Data Fusion*, Norwood, MA: Artech House, 1992.

[2] Waltz, E., and J. Llinas, *Multisensor Data Fusion*, Norwood, MA: Artech House, 1990.

[3] Skolnick, M.I. (ed.), *Radar Handbook*, New York: McGraw-Hill, 1970.

[4] Antony, R., "Target Classification for the Installation Security Radar System," U.S. Army Electronics Research and Development Command, Harry Diamond Laboratories Technical Report, HDL-TR-1976, 1981.

[5] Winston, P.H., *Artificial Intelligence*, 2nd edition, Reading MA: Addison-Wesley, 1984.

CHAPTER 4

Conventional Problem-Solving Paradigms

4.1 INTRODUCTION

Chapter 1 provided a general introduction to the tactical data fusion problem, discussed several data fusion metaphors, and emphasized the inherent difficulty of automating tactical situation awareness. Chapter 2 developed a biologically motivated model of the data fusion process and emphasized the importance of proper paradigm selection. Since no formal guidelines have existed for selecting promising problem-solving paradigms from among the large number of candidate approaches, algorithm developers have investigated a wide range of technical approaches in an attempt to meet performance requirements. Because of the inherent difficulties of developing a selection criteria based on *direct* comparisons among competing paradigms, an indirect comparison that is based on a taxonomical characterization of both problem-solving paradigms and problem-solving requirements was recommended. Chapter 3 presented a series of progressively more sophisticated algorithmic approaches to several common data fusion tasks. These examples demonstrated that the higher level canonical form approaches support both more sophisticated and more powerful reasoning than the lower order forms. Regardless of the level of effort expended to design, implement, and validate algorithms, performance shortfalls can be expected if the inherent capability of a selected problem-solving paradigm does not match the problem requirements.

Table 4.1 illustrates the primary decomposition dimensions of the canonical form taxonomy developed in Chapter 2 (*specific* versus *general* declarative, *rigid* versus *flexible* control, *single level of abstraction* versus *multiple levels of abstraction*). The number assigned to each canonical form is shown across the top of the table; listed down the left-hand column are general problem-solving characteristics. In effect,

Table 4.1 provides a mapping between problem requirements and canonical problem-solving forms. For instance, if a particular fusion task requires a dynamic, robust algorithm that is sensitive to local context, a canonical form IX or X class approach would be appropriate. On the other hand, if a multiple-resolution approach employing exhaustive generate-and-test is appropriate, a canonical form III would be selected.

Table 4.1
Summary Characteristics of the 16 Canonical Problem Solving Forms

	Algorithm Characteristics							
	I–II	*III–IV*	*V–VI*	*VII–VIII*	*IX–X*	*XI–XII*	*XIII–XIV*	*XV–XVI*
Problem Characteristics	*Specific Rigid Single*	*Specific Rigid Multiple*	*Specific Flexible Single*	*Specific Flexible Multiple*	*General Rigid Single*	*General Rigid Multiple*	*General Flexible Single*	*General Flexible Multiple*
Robustness	No	No	No	No	Yes	Yes	Yes	Yes
Context sensitivity								
• Declarative	Explicit	Explicit	Explicit	Explicit	Implicit	Implicit	Implicit	Implicit
• Control	Explicit	Explicit	Implicit	Implicit	Explicit	Explicit	Implicit	Implicit
Control	Rigid/ local	Rigid/ hier-archical	Flexible/ local	Flexible/ hier-archical	Rigid/ local	Rigid/ hier-archical	Flexible/ local	Flexible/ hier-archical
Dynamic behavior	No	No	Limited	Limited	Limited	Limited	Yes	Yes

The objective of this chapter is to populate the canonical form taxonomy b associating conventional technical paradigms with their most appropriate canonica form(s). Because the canonical problem-solving form taxonomy is based on a sim ple taxonomy of procedural knowledge, the association process is reasonabl straightforward. Once individual problem-solving paradigms are organized withi this taxonomy, Table 4.1 can be employed to establish a *one-to-several* mapping be tween problem requirements and candidate algorithm classes.

The closer two problem-solving paradigms appear within the taxonomy, th more similar are their capabilities. Conversely, the further apart the paradigms, th more dissimilar are their capabilities. Since two paradigms that share the same ca nonical form possess very similar characteristics, capabilities, and underlying limita tions, the relative advantages of selecting one paradigm over the other is generall less significant than the potential advantage of selecting a paradigm associated wit a higher order canonical form.

4.2 FUSION PROCESS MODEL EXTENSION

In order to facilitate paradigm assignment, the data fusion process model developed in Chapter 2 must be extended somewhat. As shown in Figure 4.1, the fusion process will be decomposed into three conceptually separate processes: (1) *composition of knowledge sources,* (2) *evidence accumulation,* and (3) *decision.* In many cases, problem-solving paradigms naturally divide into these three distinct, relatively independent stages, while in other cases, considerable dependency can exist between the stages. For instance, pattern recognition based on a pre-established discriminant function naturally possesses these three separate stages. The first stage is associated with feature extraction, the second with model evaluation, and the third with class assignment. In contrast, knowledge composition and evidence accumulation are not readily separable in a neural network paradigm.

All three stages of the extended fusion process model require the use of either *implicit* or *explicit* procedural knowledge. For a given algorithm or problem-solving class, the assigned canonical form class is determined by the highest order procedural knowledge demanded by *any* stage. For instance, in the model-based target classification example from Chapter 3, the target model (*evidence aggregation*) demands the use of higher order procedural knowledge (*general declarative*) than either the feature extraction or final decision stage. On the other hand, in traditional templating algorithms, all three stages might employ only *specific declarative* knowledge and *rigid, single level-of-abstraction control.*

The *knowledge-composition* stage of the fusion process is discussed in Section 4.2.1. In this first stage, the knowledge sources are brought together using various

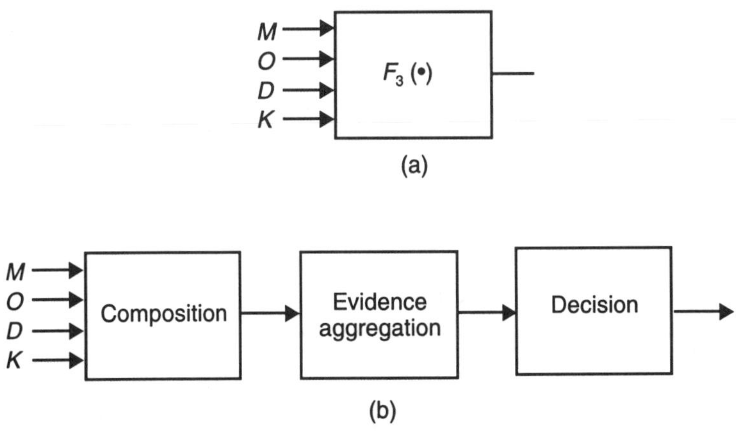

Figure 4.1 (a) Abstract data fusion model from Figure 2.6 and (b) fusion model extension.

numeric and symbolic techniques. In many lower order canonical form approaches, this stage is the prominent element of the algorithm; conversely, in higher order approaches, the other two stages typically take on increased importance. In a classic rigid templating paradigm, for example, knowledge composition involves the correlation between an input pattern and a set of exemplar patterns. In such approaches the *evidence aggregation* stage would be associated with metric evaluation and the decision stage might involve a simple thresholding operation.

Evidence aggregation is discussed in Section 4.2.2. While some algorithms possess an easily distinguished evidence aggregation element, other algorithms do not. For instance, in Bayesian statistical approaches, evidence accrual is often based on the product of several univariate probability density functions. In signal processing-based approaches, on the other hand, the evidence accrual stage might be indistinguishable from the knowledge composition stage. Because explicit evidence combination is an important element of many data fusion algorithms, a number of formal as well as ad hoc evidence accrual methods are reviewed.

The *decision* process associated with fusion algorithms is addressed in Section 4.2.3. Decision functions, such as "choose the filter with the maximum output" or "select the candidate solution with the smallest Hamming distance metric" represent relatively simple decision rules; semantic rule sets or semantic constraint sets support more sophisticated decision processes.

4.2.1 Stage 1: Composition of Declarative and Procedural Knowledge Sources

In Section 2.4, the general data fusion process was viewed abstractly as the composition among short-term declarative knowledge, medium-term declarative knowledge long-term declarative knowledge, and long-term procedural knowledge, that is,

$$F_3(O_i, D_{ij}, M_k) = (\text{short-term declarative knowledge}) \bullet (\text{medium-term declarative knowledge}) \bullet (\text{long-term declarative knlowledge}) \bullet (\text{long-term procedural knowledge}) \quad 4.1$$

Short-term and medium-term declarative knowledge always represent *specifi* knowledge, while long-term declarative knowledge can be either *specific* or *genera* The procedural knowledge used by an algorithm can be either *explicit* or *implicit.* Explicit procedural knowledge includes stated procedures, rules, or reasoning methodologies, while implicit procedural knowledge represents unstated (or default knowledge.

Eight classes of procedural knowledge were defined in Chapter 2, representing a taxonomy of both its declarative element (*specific* or *general*) and contro element (*rigid* or *flexible, single* or *multiple level of abstraction*). Thus, *explicit procedura knowledge* can range from *specific, rigid, single level of abstraction* to *general, flexible, multiple level of abstraction. Implicit procedural knowledge,* on the other hand, is always *specifi igid,* and *single level of abstraction.* Thus, implicit procedural knowledge is identified vith rigid composition operations, such as the cross-correlation of functions or the ntersection of sets.

For algorithms that use only implicit procedural knowledge, (4.1) reduces to

$$F_3\,(O_i, D_{ij}, M_k) = \text{(short-term declarative knowledge)} \bullet \text{(medium-term declarative knowledge)} \bullet \text{(long-term declarative knowledge)} \qquad 4.2$$

As will be seen, many classes of traditional data fusion algorithms rely almost xclusively on implicit forms of procedural knowledge.

For purely organizational purposes, the procedural element of an algorithm's omposition stage will be classified as either predominately *correlation-based, func- onal evaluation-based,* or *transformation-based,* as depicted in Figure 4.2. Since *correla- ion-based* operations compare two or more entities, such operations include the orrelation integral, set operations, and various similarity measures. The actual omposition function tends to be highly application-dependent. For instance, while ie correlation between two waveforms in the time or frequency domain might de- iand only a one-dimensional similarity metric, the correlation among three-dimen- onal objects may require a higher dimensional metric.

Functional evaluation includes those classes of operations where a function such ; an equation, a probability distribution, or a conditional fuzzy-set membership inction are evaluated. In addition, general search and constraint-based evaluation perations, such as *table look-up* and *decision-tree-based search,* are treated as functional valuation operations.

Transformations are those operations where data is directly manipulated or iapped into a new space. Transform operators include convolution, multiplication, rojection, and problem decomposition.

Individual fusion algorithms can involve more than a single dominant com- osition class. For instance, a feedback-based optimization algorithm might use ansformation-based composition for its internal processing model and a correla- on-based composition model for generation of the error signal. Although the para- gm-to-canonical problem-solving form assignment process does not rely on this assification scheme, the composition class categorization provides useful insights to the character of algorithms.

Figure 4.2 shows the classification of a number of traditional problem-solving aradigms with respect to the three classes of composition functions. Since the first ven *fusion classes* defined in Table 2.1 rely on default procedural knowledge, these gorithms tend to employ rather rigid composition operations, such as correlation, nvolution, and multiplication. The latter eight forms, on the other hand, can em- oy considerably more general forms of procedural knowledge.

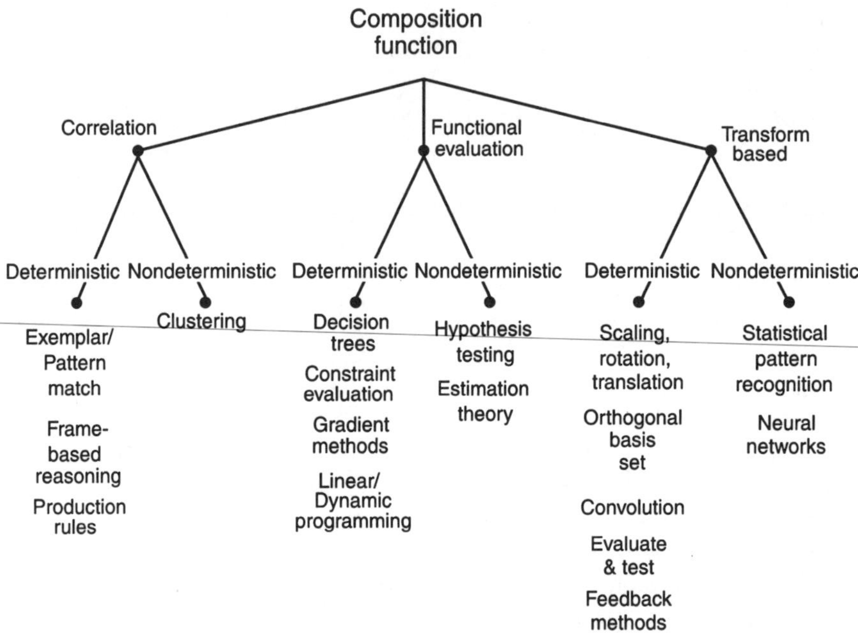

Figure 4.2 Classification of problem-solving paradigms based on the character of the compositio function.

4.2.2 Stage 2: Evidence Aggregation

A number of both mathematically formal methods (Bayesian, Dempster–Shafe fuzzy set) and somewhat ad hoc (Certainty Factors, Endorsements, voting, polling approaches can be used to combine uncertain data, rules, and decisions. In genera Bayesian, fuzzy logic, and Dempster–Shafer approaches have been the most widel used methods in data fusion applications.

Bayesian approaches to evidence accumulation are based on Bayes Theorem which permits a posteriori probability density functions to be computed from eithe the joint or conditional probability density function of the data, X, as shown i Equation 4.3.

$$p(H_j \mid X) = p(X, H_j)/p(X) = p(X \mid H_j)\, p(H_j)/p(X) \qquad 4.$$

where H_j represents a set of hypotheses. The probability density function $p(X)$ ca be computed as

$$p(X) = \sum p(X \mid H_j)\, p(H_j) \qquad 4.4$$

and $p(X \mid H_j)$ can be expanded as

$$p(X \mid H_j) = \sum p(X \mid f_i)\, p(f_i \mid H_j) \qquad 4.5$$

to effectively introduce free variables, f_i, permitting data associated with X to be combined with data that is associated with feature f_i.

The maximum a posteriori probability density function is the density function, $p(H_j \mid X)$, that is the largest over all j, for a given X. When the a priori probabilities, $p(H_j)$, are unknown, maximum likelihood estimates that do not rely on the prior probabilities are often used.

While Bayesian approaches assign probability to mutually exclusive and exhaustive singleton hypotheses, the Dempster–Shafer (D–S) method assigns probability mass to both individual hypotheses and to arbitrary Boolean combinations of hypotheses (in general, to the entire *power set*).[1] In particular, the D–S method establishes an interval-based uncertainty measure whose lower bound represents the direct *support* for hypothesis H_i and whose upper bound represents the measure of the evidence that fails to refute H_i (referred to as the *plausibility* of H_i). While a Bayesian approach does not directly handle ignorance, a D–S approach treats ignorance as uncommitted belief. Thus, the assignment of probability mass to the proposition that contains *all* hypotheses is an assertion of ignorance (in other words, the *lack* of knowledge about specific hypotheses). The D–S *rules of evidence combination* permit the development of a "fused" probability interval for each of the propositions.

In general, evidence (or the lack of it) can impact an arbitrary hypothesis, H_i, in one of four ways:

1. Support H_i;
2. Refute H_i;
3. *Not* support H_i (but not directly refute H_i);
4. *Not* refute H_i (but not directly support H_i).

Bayesian methods fundamentally treat only item 1 in the preceding list. Dempster–Shafer methods, on the other hand, consider both evidence that *supports* H_i (item 1) and the lack of evidence that refutes H_i (item 4). Dempster–Shafer handles direct evidence that *refutes* H_i (item 2) indirectly as the support to competing

1. The power set contains all mutually exclusive propositions as well as *all* possible disjunctions among those propositions. As a result, for n elemental propositions, the power set contains 2^{n-1} propositions.

hypotheses. In general, when both items 3 and 4 are true, the evidence is independent of H_i. When all hypotheses, H_i, are mutually exclusive and the set of hypotheses is exhaustive, Bayes and D-S methods produce identical results.

Fuzzy logic emulates the inexact nature of both human language and reasoning and has evolved as an alternative approach to probability-based evidence accrual. Fuzzy logic is based on fuzzy set theory originally introduced in 1965 by Lotfi Zadeh [1]. In 1983, Henri Prade [2] proposed using possibility theory as a basis for a logic of uncertainty. Fuzzy logic permits operations among fuzzy sets analogous to set operations on "crisp" Boolean sets. As such, fuzzy logic offers a link between symbolic processing and numeric computations [3].

In fuzzy logic, conditional fuzzy weighting functions (fuzzy-set membership functions conditioned on one or more problem attributes) are manipulated by fuzzy operators within a well-defined certainty calculus. Consider, for example, the "fuzzy models" of the two linguistic variables in the following production rule: A *high-flying* aircraft is *near* a friendly command post.

Figures 4.3 and 4.4 show fuzzy-set membership functions for both variables. The first figure defines the relationship between the fuzzy modifier *high-flying* and *altitude* conditioned on the type of aircraft. The second figure defines the relationship between the fuzzy quantifier *near* and the air-to-surface *range* of each aircraft's missile system.

A number of general comparisons can be made between probability theory and fuzzy set theory. First, probability theory supports reasoning related to uncertainty about the world, while fuzzy set theory supports reasoning based on vagueness and imprecision. Whereas probabilistic reasoning rests on the concept of *set measures* (probability density functions over a set), possibilistic reasoning based on fuzzy sets rests on the concept of *similarity measures* (distance metrics).

Although considerable controversy remains as to the relationship between fuzzy set theory and probability theory, interest in fuzzy-based reasoning is increasing due to both the intuitive nature of fuzzy logic and the success of numerous commercial products that employ fuzzy-based control.[2]

The *Certainty Factor* approach [5] uses an ad hoc assignment of hypothesis certainty along with an ad hoc calculus for combining *supporting* (measures of belief) and *conflicting* (measures of disbelief) information. The composite certainty factor expresses a relative degree of belief. Rules, data, and hypotheses are all treated as uncertain.

Endorsements [6] are fundamentally a data structure for summarizing reasons for believing or disbelieving data, rules, tasks, and conclusions. A fundamental

---

2. Jim Bezdek, a leading researcher in fuzzy systems, estimates that over 500 papers have been written on the topic of fuzziness and probability. A special issue of the IEEE Transactions on Fuzzy Systems [4] featured a debate-like format to reconsider this very matter. This special issue provides a good source of pointers to the literature for readers interested in a more detailed treatment of these matters.

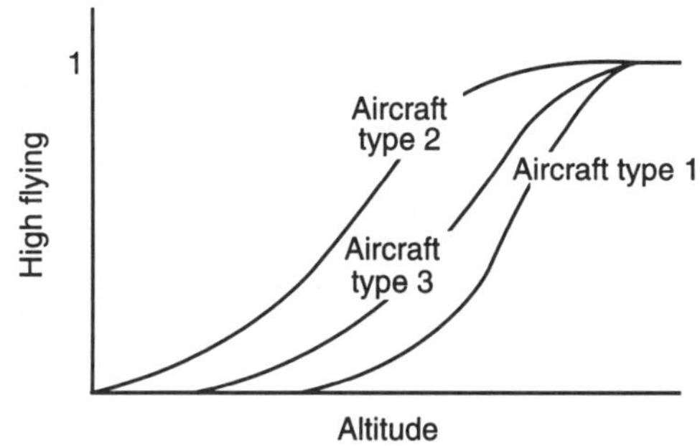

Figure 4.3 Relationship between "high flying" and altitude for three aircraft types.

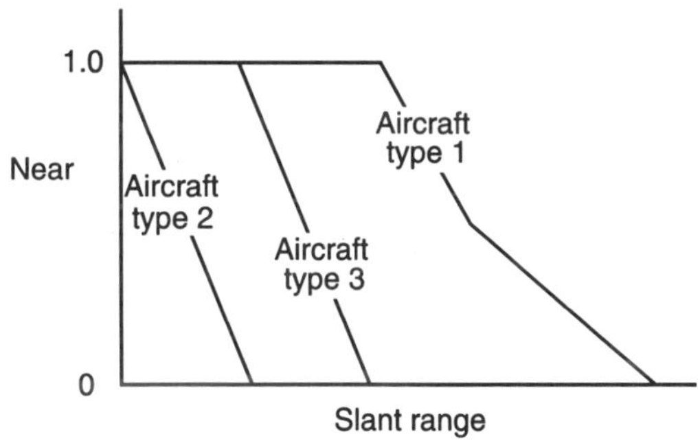

Figure 4.4 Relationship between "near" and slant range for each missile carrying aircraft.

premise of the theory of endorsements is that it is necessary to explicitly represent and maintain different kinds of uncertainty and to select the most appropriate for a given resolution task. Endorsements on data, for example, include its source, type, and accuracy; endorsements on conclusions denote whether they corroborate, conflict, or are redundant.

In *voting and polling* approaches, the candidate decision receiving the majority of independent votes is selected as the process outcome.

In general, each of the above evidence-combination approaches have both strengths and shortcomings. The following weaknesses are paraphrased from [7].

- *Probability theory* (1) does not directly deal with ambiguity, inconsistency, incompleteness, and irrelevance; and (2) represents only uncertainty associated with atomic elements.
- *Dempster–Shafer theory* requires highly combinatoric evidence assignment and combination (although relatively efficient approximate techniques exist).

- *Possibility theory* offers somewhat weaker uncertainty distributions than other evidence accumulation techniques.
- *Certainty Factor methodology* makes modular, closed-world assumptions, such that
  (1) either all rules must be predictive or all must be diagnostic (the two forms cannot be mixed), and
  (2) all dependent evidence must be brought together in the same rule.
- *Endorsement theory* lacks a method for propagating, combining, and ranking individual endorsements.

Evidence accumulation remains a dynamic and somewhat volatile research area. In spite of extensive theoretical and empirical studies, considerable disagreement remains as to the relative merits of the various approaches. Although the more ad hoc methods are often criticized because they lack axiomatic underpinnings, the assumptions required by the more formal methods (independence of evidence, determination of a priori probabilities, assignment of consistent absolute, or relative weights to the entire power set of hypotheses) are problematic, as well. In addition, regardless of the certainty calculus or methodology employed, all heuristic decision processes require the assignment of subjective conditional weights.

Data fusion is sometimes viewed as nearly *synonymous* with *evidence accumulation*. However, since data fusion seeks to *synergistically* combine disparate sensor and non-sensor-derived information, the aforementioned evidence accumulation methods must be viewed as merely a supporting technology.

4.2.3 Stage 3: Decision

Many fusion algorithms possess an *explicit*, easily identified decision element. Both target track association and target classification algorithms, for example, possess such a distinct decision element. On the other hand, some algorithms employ an *implicit* or embedded decision function. Higher order canonical form algorithms that generate the "best" ground-based route to a given town or that identifies preferred locations for deploying a particular class of weapon systems possess embedded decision functions.

In the case of explicit or nonembedded decision processes, the algorithm developer is free to choose the type (numeric, symbolic, or hybrid), the complexity and the sophistication of the decision function. For instance, target class assignment based on linear discriminant functions or neural network classifiers typically apply a simple threshold test to the output of the evidence accrual stage. Alternatively, a much more sophisticated decision rule could be used that depends on the available evidence for each possible target class, an estimate of the current level of hostilities, the time of day, and the apparent objective of the target.

The three-stage fusion process paradigm is summarized in Figure 4.5.

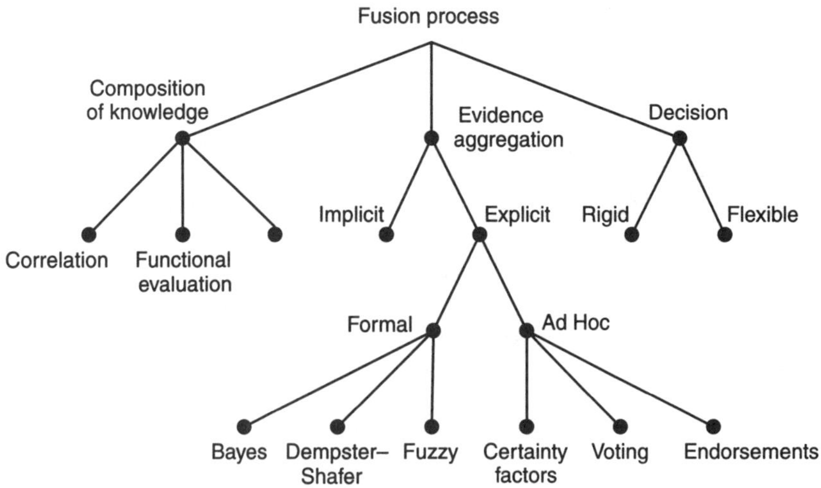

Figure 4.5 Decomposition of the three-stage fusion process model.

4.3 PARADIGM REVIEW

This section reviews a wide range of conventional problem-solving paradigms from the perspective of the three-stage fusion process model. Due to the large number of paradigms covered, the treatment of the individual approaches is necessarily brief. However, since all the problem-solving techniques have been extensively discussed in the literature, the interested reader will find no shortage of reference material. As with most chapters, a categorically organized bibliography is offered at the end of the book as a starting point for further study.

Because the essence of data "fusion" is the "bringing together of similar and dissimilar sources of information," our discussion will focus on the *knowledge-composition* stage of the fusion process model. Knowledge composition, for the purpose of fusing information, is metaphorically similar to the bringing together of deuterium atoms to produce nuclear fusion; in both data fusion and nuclear fusion, the first stage is the critical element of the process. In the case of data fusion, the latter two stages merely interpret the outcome of the composition process.

For some of the paradigms described below, only the compositional stage of a fusion algorithm will be apparent. In such cases, explicit evidence accumulation and decision functions are required to "complete" the algorithm. In other cases, the paradigm clearly includes both composition and evidence accrual stages. In some of these cases, these stages will be quite distinct, while in others, the two stages are inextricably intertwined. As we will see, the nature of the composition element tends to have a significant bearing on the canonical form assignment process. At the highest level of abstraction, composition operations are treated as either *deterministic* or *nondeterministic.*

4.3.1 Deterministic Techniques

Deterministic methods, as used here, loosely identify those composition operations that do *not* assume an underlying random or fuzzy phenomenon. The techniques to be discussed include a broad range data analysis and decision-making paradigms. Some of the composition operations discussed involve rather distinct mathematical operations, while others are less separable from the other two stages of the process. Some classes of algorithms employ two or more distinct composition forms.

Transform-Based Approaches

Transform-based paradigms include those problem-solving approaches where data and information (sensor-derived and non-sensor-derived) are mapped from one space to another space.

Scaling, Rotation and Translation. A wide range of relatively simple data transformations, such as scaling, rotation, and translation, are commonly used in data fusion algorithms. For instance, radar return must be scaled with respect to the proper range law. Images can be enhanced by rescaling brightness based on the pixel-intensity histogram of the original image. Transformations of the output from spatially-distributed sensors may be required to bring the data into full spatial registration. Because time delays exist in both data acquisition and processing systems, sensor data must be temporally registered.

Projections. Projections are specialized data transformations that modify the perspective or dimensionality of a data set. Since it maps (or projects) multiple dimension data into a single linear dimension, indexing in a relational database management system is a form of projection. Coordinate transformations map data sets from one coordinate system to another. An important class of projections map data onto an orthogonal basis set. The Fourier transform of time-domain information, for instance, creates a frequency domain representation of a signal by mapping the signal onto a set of complex exponentials. Conversely, the inverse Fourier transform maps a frequency domain representation into the time domain.

Convolution. The convolution integral is associated with a broad range of transform-based problem-solving paradigms, including signal and image processing. The convolution of $x(t)$ and $m(t)$ is defined as

$$\int x(\tau)\, m(t-\tau)\, d\tau = x(t) * m(t) \qquad 4.6$$

where $*$ is the convolution operator. The output of a linear system $y(t)$, for instance, is the convolution between the input $x(t)$ and the filter impulse response function $h(t)$, that is,

$$y(t) = x(t) * h(t) \qquad 4.7$$

The *convolution theorem* states that the Fourier transform of the convolution of two functions can be computed as the product of the Fourier transforms of the individual functions. Thus, the output of a linear filter can be computed as the inverse Fourier transform, $\mathcal{F}^{-1}$, of the product of the Fourier transform of the input signal and the Fourier transform of the filter impulse response function, that is,

$$y(t) = \mathcal{F}^{-1}\left[\mathcal{F}\{x(t)\} \cdot \mathcal{F}\{h(t)\}\right] \qquad 4.8$$

where $\mathcal{F}\{\cdot\}$ represents the Fourier transform of its argument. While $h(t)$ represents the filter response to an impulsive input in the time domain, $H(f) = \mathcal{F}\{h(t)\}$ represents a window in the frequency domain through which the signal $X(f) = \mathcal{F}\{x(t)\}$ is "viewed."

The linear filter that achieves the maximum signal-to-noise ratio for the detection of a known signal in additive Gaussian noise is referred to as the *matched filter*. Although the matched filter is presented in Section 4.2.1.3 as a correlation-based paradigm, because of the intimate relationship between convolution and correlation [8], the approach could equally-well be characterized as a transformation-based composition operation.

Image Compression. Numerous image compression coding schemes have been developed. In general, good image compression requires good image decomposition. Since images, speech waveforms, and other classes of data generally possess highly localized characteristics, a good decomposition scheme should be capable of dealing with local correlations. Because the Fourier basis set consists of an infinite set of sinusoids (or more generally, complex exponentials), the decomposition is best suited for data that is *temporally or spatially dispersed,* but *frequency localized.* Thus, the transform is not particularly well suited for decomposing highly localized spatial correlations. The wavelet decomposition [9], on the other hand, is well suited to data that is *time or spatially localized* and *frequency dispersed.* As a result, there has been considerable recent interest in wavelet-based compression of speech, video, and imagery.

A wavelet basis set consists of a set of waveforms that are generated from a single wavelet through a series of translation and dilation operations. In the case of imagery, the image to be compressed is broken into a *low-resolution image* and the *difference image* that is formed by taking the difference between the original image and the low-resolution image. All spatially localized correlations that exist in the image (or speech waveform) are, for the most part, preserved with the low-resolution image. Because the difference image contains only the high-resolution "corrections," the difference image possesses a zero-mean histogram that can be efficiently coded with relatively few bits.

A multiresolution wavelet decomposition was introduced by Mallat in the late 1980s [10]. Since the low-resolution image may still contain pixel-to-pixel correlations, this image can be broken into an even lower resolution image, plus its difference image. By repeating this recursive decomposition, a highly efficient multiple-resolution decomposition results.

Vector Quantization. Vector quantization represents the natural generalization of scalar quantization to multiple dimensions.[3] Although vector quantization has traditionally been concerned with data compression [11], because a vector can be used to describe a general pattern, vector quantization can be treated as a form of pattern recognition. In recent years, extensive speech recognition work has been based on this particular technology [12].

Functional Evaluation-Based Approaches

When there exists a straightforward functional relationship between an algorithm's input and output, composition can be based on direct functional evaluation. Both search-based and constraint satisfaction–based paradigms can be treated as forms of functional evaluation.

Search-based. Search-based problem-solving paradigms can be expressed functionally as

$$Y_j = D(X_j) \qquad 4.$$

where the output of the jth stage of a search process is a function D of the search variable X_j. Decision-tree analysis employs a very simple form of search. As the name implies, decision trees are an exhaustive, hierarchical representation of a problem solution-space. For relatively well-defined tasks such as equipment diagnosis (water pump, hydraulics system, internal combustion engine) and programmed learning

3. One-dimensional analog-to-digital conversion is an example of scalar quantization.

(when a student makes a logical error solving a problem, remedial exercises are offered; otherwise, a slightly more difficult example is supplied), decision tree–based algorithms offer a simple and natural problem-solving strategy. An example of a decision tree for diagnosing the cause of engine overheating is shown in Figure 4.6.

Many artificial intelligence–based approaches depend heavily on search-based methods. For instance, a semantic network is an entity-relationship model of knowledge where the nodes represent concepts and the arcs represent semantic relationships between nodes, such as "is-a" (that is, an example of) and "kind-of" (that is, a class of). Semantic relationships between nonadjacent nodes are deduced via search of the network and analysis of the relevant attribute-inheritance properties.

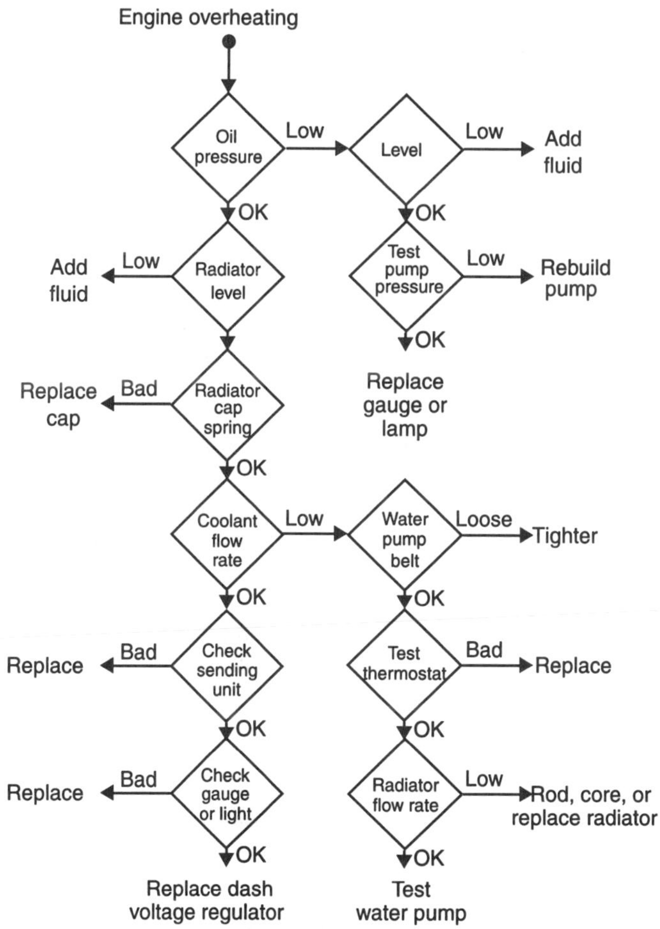

igure 4.6 Example of decision tree for diagnosing engine-overheating problem.

Finite state-space problems can often be solved using search-based methods. Because exhaustive search tends to be impractical for many large-scale problems, search-based methods often employ heuristic evaluation functions in an attempt to reduce the size of the required search space.

Constraint-Based. In general, constraint satisfaction–based approaches require that a data set or an hypothesis be evaluated with respect to a set of constraints. A wide range of problems lend themselves to constraint-satisfaction based reasoning, including:

1. Stress analysis in a bridge design;
2. The selection of a new family car;
3. The search for all enemy units that are within 1 km of small villages *and* within a lightly forested region.

Correlation-Based Approaches

A common form of nonstatistical-based pattern recognition involves the correlation between a data set X_i and a finite set of explicit exemplars, patterns, or templates $\{T_j\}$, that is,

$$Y_i = f(X_i \cdot T_j) \tag{4.10}$$

where f is an arbitrary function and • is the correlation operator. The degree of correlation Y_i is often expressed in terms of a generalized "distance" metric. The smaller the "distance" between the data and an exemplar, the higher the correlation. Typical distance metrics include Euclidean, Manhattan, Hamming, mean-squared error, and nearest neighbor.

Exemplar or Pattern Match. Signal correlation, matched filtering, and character, speech, and image recognition are examples of problem-solving approaches that recognize or classify information based on the correlation between data and an a priori developed set of patterns. For example, in a common approach to optical character recognition, a two-dimensional binary image of an unknown character is compared to binary images of all upper- and lower-case letters, numbers, and special symbols. The unknown symbol is assigned to the character whose exemplar pattern best matches the input pattern.

For a known signal $s_j(t)$ embedded in additive, colored Gaussian noise $n(t)$ the optimal detector $y_j(t)$ is the *matched filter* $h_j(t)$ [13]

$$y_j(t) = x(t) * h_j(t) \tag{4.1}$$

where $x(t) = s_j(t) + n(t)$. The ideal matched filter can be shown to be equal to the time-reversed signal normalized by the power spectral density function of the noise process, that is,

$$h_j(t) = (1/\alpha)\, s_j(-t) \tag{4.12}$$

Thus,

$$\begin{aligned} y_j(t) &= (1/\alpha)\, x(t) * s_j(-t) \\ &= (1/\alpha)[s_j(t) + n(t)] * s_j(-t) \end{aligned} \tag{4.13}$$

Because of the relationship between convolution and correlation, (4.13) can be rewritten as

$$\begin{aligned} y_j(t) &= (1/\alpha)\, [s_j(t) + n(t)] \cdot s_j(t) \\ &= (1/\alpha)\, [s_j(t) \cdot s_j(t) + n(t) \cdot s_j(t)\,] \end{aligned} \tag{4.14}$$

where • represents the correlation operator.

If the signal and noise are uncorrelated ($n(t) \cdot s_j(t) = 0$), the output of the matched filter reduces to the single nonzero correlation product

$$y_j(t) = (1/\alpha)\, s_j(t) \cdot s_j(b) \tag{4.15}$$

Therefore, the matched filter output is effectively the autocorrelation of the desired (*matched*) signal.

Frame-Based Reasoning. Frames [14] are stereotypical representations of knowledge that support expectation-based reasoning. A frame-based representation provides a partial template of an object or concept and represents the natural generalization of a property-list in LISP, a record in PASCAL, and a relation in a database. In practice, a frame often takes a form similar to the one depicted in Figure 4.7, which consists of predefined *slots* that maintain numeric or symbolic values of attributes as well as associated context-sensitive procedural code that triggers under specified conditions (during frame instantiation, given certain slot values, under external activation). Slots are typically initialized with default values. The organizational, functional, and relational aspects of frames must be explicitly maintained.

Entity name
Classification (unknown, class, subclass, id)

Relations (subordinates, superordinates, laterals)

Location/time
Present ________
Historical _______

Measurable features

Derived features (road association, staging area association)

Function/intention

Historical

Figure 4.7 Frame-based representation of an entity.

Figure 4.8 illustrates the fusion of ELINT-derived information with the overall situation representation using the frame paradigm. In this example, relevant parameters maintained in an ELINT-specific frame are compared to the key parameters in an All-source frame. If the ELINT information either supports, contradicts, or can update attributes of the all-source frame, the appropriate inferences would be initiated.

Object-oriented reasoning (OOR). OOR represents the logical generalization of frame-based reasoning where the rather rigid frame representation becomes a general object representation. *Scripts* [15] represent the natural extension of single

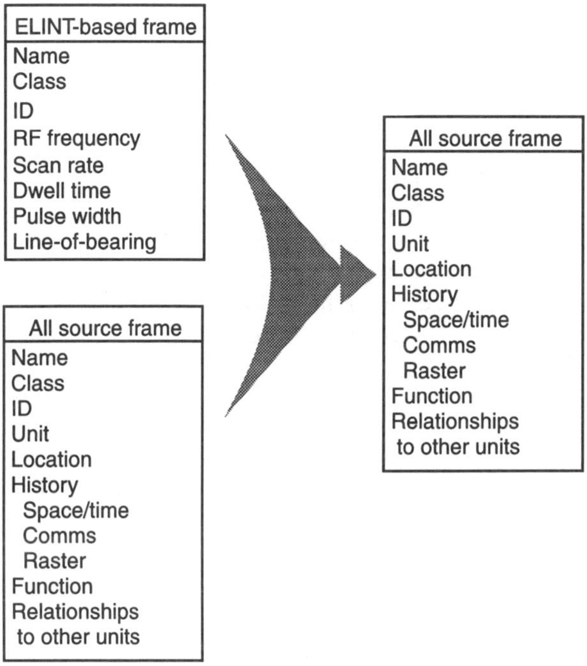

Figure 4.8 Conceptual view of the fusion of the information contained in an ELINT-based frame and an all-source-based frame.

frames to a sequence of logically connected frames. Because of the sequential character of the individual representations, scripts are often used to support temporal reasoning. *Case-based reasoning* [16] is another closely related concept where a situation is compared to a set of historical or exemplar *cases* for the purpose of identifying the closest match (the most similar case).

Production Rules. Expert Systems use modular IF/THEN rules solicited from human experts to automate complex, narrow-domain tasks. In its most basic form, the production rule paradigm performs the recursive evaluation of the entire rule-set against both the current input (short-term knowledge) and the current problem state (medium-term knowledge). Following changes to working memory or new input to the system, the rule-evaluation process begins anew. Execution terminates when there is no new input or when no further changes occur to the current problem state following one complete pass of the rule-set.

A typical rule set consists of a large number of expressions of the form:

$$IF(conditions) \quad THEN\,(action) \tag{4.16}$$

The left-hand side of a rule describes the conditions under which the right-hand side decision or action should occur. The conditions are typically expressed as a set of Boolean relations (for example, $x_1 \, \varepsilon A$ AND $x_2 \, \varepsilon B$). Production rules can represent both formal, high-confidence knowledge as well as heuristic, lower confidence knowledge. In the latter case, a confidence level is generally associated with each rule conclusion. Individual conditions within a single rule generally carry equal weight.

Production rules represent a symbolic-oriented variant of the If/Then conditionals employed by more conventional procedural-oriented approaches. The two forms of If/Then rules can be distinguished by the fact that an Expert System rule condition set represents a template through which the evolving problem state is viewed, while traditional If/Then rules test conditions to control the flow of processing. In terms of the eight classes of procedural knowledge defined in Chapter 2, a production rule represents highly modular *general long-term declarative* knowledge and *rigid, single level-of-abstraction control*, while the conventional If/Then rule represents *specific long-term declarative* knowledge and *rigid, single level-of-abstraction control.*

Production rules can be employed in either a data-driven or a goal-driven fashion. In a data-driven application, the left-hand side of all rule conditions represent templates that are tested against both the system input data and the current problem state (often referred to as "working memory"). If a rule condition set is satisfied, the action described by the right-hand side is taken. In general, rule firing changes the current problem state. In a goal-driven application of IF/THEN rules, the rule right-hand side actions represent templates that are tested against both the input states and the existing problem states. When a match occurs, the left-hand side conditions can be tentatively inferred. If more than a single rule is instantiated during each pass through the rule set, a resolution process is required to determine which rules, or in what order, the rules should be fired. When conflicting rules are instantiated, the resolution process can be nontrivial.

Set Operations. *Sets* are arbitrary collections of objects; *subsets* are partial collections of such objects. Subsets of the set of all real numbers, for instance, include the natural numbers, the integers, the rational numbers, and the irrational numbers. Important operations on sets include *union, intersection, difference,* and *complement.* Whereas *Boolean sets* possess a hard boundary between inclusion and noninclusion, *fuzzy sets* describe a degree of membership within a set. As will become apparent in the following chapters, set operations play an important role in many data fusion applications.

Optimization

Optimization-based methods attempt to achieve the "best" solution from among a set of admissible solutions. Although listed as a separate category, optimization-based approaches generally employ one or more of the aforementioned three categories of composition functions.

Gradient-Based Methods. The maxima and minima of continuously differentiable functions can be determined by setting the derivative (or gradient) of the function to zero. However, few realistic fusion applications lead to simple closed-form functions that permit direct computation of all maxima and minima. Because search along the maximal gradient of a function tends to move toward the function's critical values, the general notion of gradient-based problem-solving—that of finding extrema using the gradient of functions and surfaces—remains valid. Approaches that seek a maximum of a function or surface are typically referred to as *hill-climbing* strategies; those that seek a minimum are referred to as *steepest descent* strategies.

Because gradient-based techniques are fundamentally local optimization strategies, achieving global optimization using such algorithms requires significant extensions. One such extension is to introduce a control mechanism that forces the algorithm to sample the entire space before settling into a single region of the function or surface. *Simulated annealing* is a gradient-based algorithm that uses a stochastic control strategy analogous to the annealing process used to free glass and metals from internal stresses. The material to be annealed is first heated and then cooled at a controlled rate. If the material is cooled either too slowly or too rapidly, the desired crystalline state will not be achieved. Because there exists a probabilistic relationship between the energy level of an individual molecule and the material's temperature, the annealing process can be effectively controlled by regulating the cooling rate.

By using a control structure that is analogous to that of controlled cooling, simulated annealing first applies gradient search at a coarse resolution level (representing a high-temperature state) that effectively forces the algorithm to widely sample the entire surface. Once the algorithm reaches the vicinity of a global maximum, however, the stepsize is reduced (the temperature must be lowered) to ensure that the next point evaluated remains in the vicinity of the global maximum. The objective of the simulated annealing process is to achieve a tradeoff between excessively slow convergence (temperature reduced too slowly) and too rapid a convergence (temperature reduced too rapidly), both which may lead to nonglobal solutions. In the former case, the algorithm may find, but later abandon, the region containing the global optimum before convergence occurs; in the latter case, the algorithm is more likely to settle into a local maximum without adequately exploring the entire space.

Generate-and-Test. For finite-state problems, a common approach to global optimization (sometimes referred to as combinatoric optimization) is the three-stage evaluate-and-test paradigm:

1. Generate all possible solutions;
2. Rank-order the solution set against a selected evaluation metric;
3. Choose the "best" solution with respect to the evaluation metric.

The paradigm divides naturally into the three separate stages of *composition of knowledge sources*, *evidence accumulation*, and *decision*. While the "generate" stage (stage 1) of the algorithm might entail a transformation-based or functional evaluation-based composition process, the "test" stage (stage 2) could involve a correlation-based composition process.

The Euclidean Traveling Salesman Problem (TSP) is a classic combinatoric optimization problem. The objective of the TSP is to find the order in which a set of cities should be visited so as to minimize the travel distance. In the Euclidean TSP, the path between all cities is assumed to be a straight line. For n cities, the optimal tour is the globally shortest among all $(n/2)!$ possible tours. Only exhaustive tour generation and evaluation ensures that the globally optimal solution is found. The development of an optimal path between a start and goal state in a raster-based mobility map is another common example of an evaluate-and-test paradigm (recall the example in Chapter 3). The optimization criteria might be the minimum distance or minimum travel time for the set of all admissible paths. Because global optimization based on an exhaustive evaluate-and-test paradigm is highly combinatoric, approximate solutions are often required for problems that inherently possess a large state space.

Linear and Dynamic Programming. *Linear programming* [17] and *dynamic programming* [18] are iterative computational techniques that address the problem of finding an optimal solution to a set of equations conditioned on a set of constraints. In linear programming, the optimal solution to a set of linear equations is determined using the simplex algorithm (which is effectively a screening process for eliminating solutions that are not promising candidate solutions) or one of its variants. In dynamic programming, a problem is typically divided into stages that are handled separately and the results combined.

Feedback Methods. In feedback-based control, an "error signal" is generated that is the difference between the current system output and the desired output. This error signal is fed back to the system, which then adjusts its processing to attempt to reduce the magnitude of error signal and thus drive the system to the desired state.

This same feedback strategy is employed in many forms of "learning." For instance, in *backward error propagation* training of neural networks, least-squared error decision boundaries are generated based on repeatedly applying the training set while supervising the decision process. Biological organisms make extensive use of feedback control (medium-term knowledge) as well as feedback learning (long-term knowledge) [19]. For problems that exhibit a dynamic character, feedback-based methods represent a natural extension of static optimization techniques. Since the development and maintenance of a tactical situation awareness inherently involves dynamic problem solving, feedback-based reasoning can be an important element of data fusion algorithms.

Figure 4.9 shows the classification of a number of deterministic problem-solving approaches.

4.3.2 Nondeterministic Techniques

Nondeterministic approaches have been distinguished from deterministic approaches by the fact that these problem-solving approaches are generally associated with an underlying probabilistic, stochastic, or fuzzy process. However, since none of the underlying data fusion principles rely on this assignment process, the distinction between deterministic and nondeterministic is merely a convenient organizational characterization. Because neural networks are often treated as a nondeterministic approach, they are included in this section, although, once trained, they could equally well be treated as a transformation-based deterministic paradigm.

Nondeterministic problem-solving approaches will be classified as either parametric or nonparametric. Parametric approaches include those problem-solving approaches where data is analyzed or decisions are made based on either known or estimated conditional probability density or conditional fuzzy-set membership functions. In a nonparametric approach, rather than modeling the underlying process, a decision function or decision surface is directly generated.

Parametric

Bayes Theorem. When a decision process is appropriately represented in probabilistic terms and all relevant probability distributions are known, Bayes theorem permits direct evaluation of the theoretically optimal decision, that is,

$$p(H_j \mid X) = p(X \mid H_j)\, p(H_j)/p(X) \qquad 4.17$$

where $p(H_j|X)$ is the a posteriori probability of hypothesis j conditioned on X (in other words, the probability of hypothesis H_j after the measurement X is made), $p(X|H_j)$ is the conditional distribution of the parameter space given hypothesis H_j, $p(H_j)$ is the a priori probability of hypothesis H_j, and $p(X)$ is the density function associated with the measurement space.

Four principle shortcomings of Bayesian inference methods [20] are:

1. Prior probabilities are typically unknown and may be difficult to estimate;
2. The existence of multiple conditional hypotheses tend to complicate the analysis;
3. No facility exists to represent general uncertainty (see Dempster–Shafer discussion in Section 4.2);

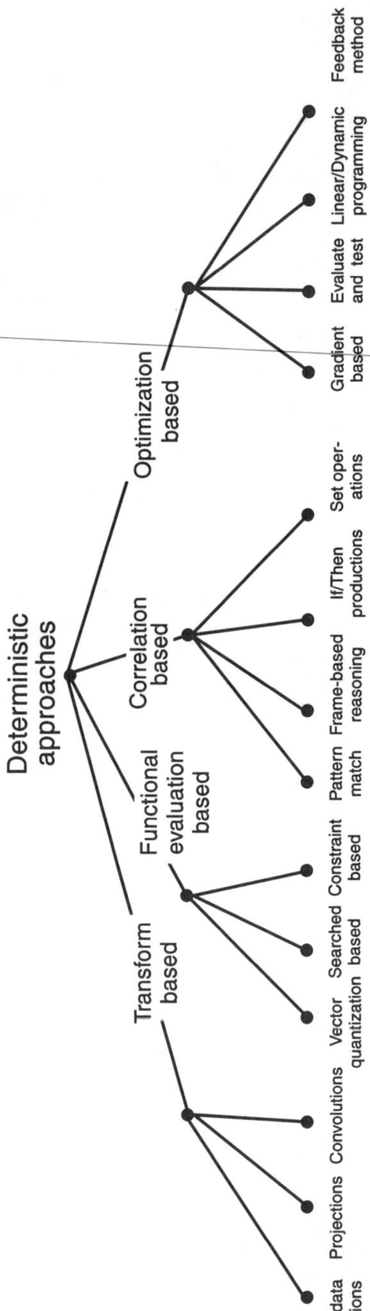

Figure 4.9 Classification of deterministic problem-solving approaches.

4. Whether appropriate or not, competing hypotheses are generally assumed to be mutually exclusive.

Since neither $p(H_j)$ nor $p(X)$ are typically known, the maximum likelihood estimator is often employed because it depends only on $p(X|H_j)$. If only the forms of the density functions are known, the parameters of the density functions (for example, first- and second-order statistics) must be estimated.[4] If even the form of the class conditional probability density functions are not known, the density functions themselves must be estimated.

Hypothesis Testing. Classical statistical hypothesis testing requires that all relevant conditional probability density functions be known. In the case of a binary decision (either hypothesis H_0 or hypothesis H_1 is true), the likelihood ratio

$$p(X \mid H_1) \;/\; p(X \mid H_0) \qquad 4.18$$

forms the basis for optimal decision making. If the ratio is greater than 1, hypothesis H_1 is selected; otherwise hypothesis H_0 is selected. By applying Bayes Theorem, the likelihood ratio can be reformulated in terms of a posteriori density functions.

Statistical Estimation. Most statistical estimation techniques depend critically on the density functions of the underlying statistical processes. For example, given two random vectors, the minimum mean-squared error (MMSE) estimate of Y in terms of X is the conditional expectation $E\{Y|X\}$. The MMSE estimate of a parameter is the mean of the a posteriori density function $(E\{p(H_i|X)\})$. The *maximum* a posteriori estimate of a parameter is the maximum value of $p(H_i|X)$ over all X.

In the classical radar-tracking application, the estimation of the current location of a target based on a noisy measurement is required. The first-order *Kalman filter* [21] is an optimal linear least squares estimator for a Markov sequence s_n, in terms of past samples of the process $x_n = s_n + q_n$, where q_n is additive noise.[5] [6] The Kalman filter is a recursive (infinite impulse response) linear filter. The estimate of the current location of the target $\hat{s}_n$, is dependent on the noisy radar report x_n and the estimate of the previous track update $\hat{s}_{n-1}$, that is,

4. The maximum likelihood estimator achieves the minimum probability of error when the prior probabilities are equally likely.

5. In order to provide optimal performance, the physical phenomenon must be properly defined by linear mathematical models, initial conditions must be accurately known, and the noise process must be white Gaussian.

6. In a first order Markov sequence, the estimate of s_n is dependent on only the most recent value of the sequence x_{n-1} (and thus is independent of all previous values).

$$\hat{s}_n = \alpha_n x_n - \beta_n \hat{s}_{n-1} \qquad 4.19$$

In general, the parameters α_n and β_n depend on the dynamic target behavior and the dynamic noise process. The so-called $\alpha - \beta$ tracker is the nonadaptive, fixed coefficient ($\alpha_n = \alpha$, $\beta_n = \beta$) filter solution for position-only input states.

Higher order filters are often used to achieve better tracking performance for maneuvering targets. For instance, a second-order Kalman filter provides an optimal estimator for a second-order Markov sequence (a sequence that depends on the previous two estimates). Because it is based on a single previous location estimate, a first-order (or linear) estimator assumes constant target velocity, while a second-order (or quadratic) estimator, which depends on the previous two position estimates, accommodates targets that exhibit constant acceleration.

Fuzzy Rule–Based Systems. Fuzzy rules are similar to their crisp counterparts except that the constraints are fuzzy rather than crisp conditionals. Because fuzzy rules are typically acquired from human analysts, fuzzy rules tend to be highly intuitive. For instance,

IF (the target speed is *fast*) AND (the elevation of the target is *low*) AND (the terrain is *hilly*) THEN (the aircraft is attempting to avoid radar detection)

expresses potentially complex and abstract relationships in a very simple and straightforward fashion. Clearly, *fast, low,* and *hilly* are "vague" modifiers that can be represented by fuzzy-set membership functions. If *hilly* terrain features, for example, are not explicitly maintained in the database, the terrain knowledge base must be abstracted to support this assessment.

In conventional rule-based systems, a single pass (or search) through the rule set typically results in only a few rules being instantiated. By contrast, in a fuzzy rule-based system, numerous rules typically fire due to the inherent overlap among the fuzzy conditionals. While condition-set overlap can be undesirable in conventional rule-based applications, such overlap is considered a strength of fuzzy rule-based approaches. The final product from a fuzzy rule set evaluation is generally "defuzzified," producing a crisp result.

Fuzzy rule-based controllers have been incorporated in a wide range of household appliances, in the shifting logic for automobile transmissions, and even in the control of a high-speed train in Sendai, Japan. In this latter application, the controller maintains constant train speed and provides very smooth accelerations and decelerations [22]. Employing only two dozen rules, the system has been shown to outperform conventional numerical control approaches and is considered to perform as well as skilled human operators.

Nonparametric

Nonparametric methods are natural for handling phenomenon that have a nondeterministic character but for which the underlying phenomenon are not readily modeled using statistical or fuzzy-based reasoning. These methods directly "train" a discrimination or decision function using "labeled" input samples. Methods that are both nonparametric and nondeterministic possess some of the same problem-solving forms as the deterministic problem-solving approaches discussed in Section 4.2.1. Since in nonparametric approaches the underlying decision-making process is not actually modeled, the identification of a problem-solving form as either deterministic or nondeterministic is somewhat arbitrary. Since the taxonomical association process presented in the next section does not rely on the categorization used in this section, we have used the classification that appears to be most consistent with convention.

Discrimination Function Methods. A typical nonparametric pattern-classification paradigm involves the three-stage process:

1. Feature extraction;
2. Feature weighting;
3. Decision.

For a linear classifier, the discrimination function D_j can be represented as

$$D_j(x) = \sum w_{i,j} x_i \qquad 4.20$$

where x_i are the values of the input features derived from the data and $w_{i,j}$ is a transition matrix. Without loss of generality, nonlinear decision surfaces can be represented as $D_j(X) = w_{i,j}\, g(x_j)$, where $g(\bullet)$ is an arbitrary function.

Neural Networks. Although a wide range of problems have been attacked using artificial neural systems, the two dominant uses to date have been the classification (supervised learning) and clustering (unsupervised learning) of network inputs. Consider a basic three-layer, fully forward-connected artificial neural system. The input states are designated x_i and the outputs of the second and third layers are designated f_j and g_k, respectively, as shown in Figure 4.10. Assuming that network connections carry multiplicative weights and that all nodes are linear summers, the output of the second layer (or *Kohonen slab*) is

$$f_j = \sum h_{i,j} x_i \qquad 4.21$$

and the output of the third layer (or *Grossberg slab*) is

$$g_k = \sum w_{j,k} f_j \qquad 4.22$$

By substituting for f_j and summing over j

$$\begin{aligned} g_k &= \sum w_{j,k} \sum h_{i,j}\, x_i \\ &= \sum x_i \sum w_{j,k}\, h_{i,j} \\ &= \sum x_i\, z_{i,k} \end{aligned} \qquad 4.23$$

Thus, the network output represents the dot product between the input and a weight matrix. Equation (4.23) demonstrates that for a strictly linear, fully forward-connected network, any multiple-layer neural network is equivalent to the output of a two-stage network with a properly redefined weight matrix. Note that (4.23) is identical to the form of the linear discrimination model (4.20) and the discrete form of the linear filter model (4.7).

In general, neural networks employ a nonlinear weighting function (the sigmoid function, a hard-limiting function) following the linear summation in at least one stage of the network. Such nonlinearities have both a strong biological motivation as well as practical advantages, permitting the network to represent arbitrarily complex decision surfaces. The actual impact of the nonlinearity on network per-

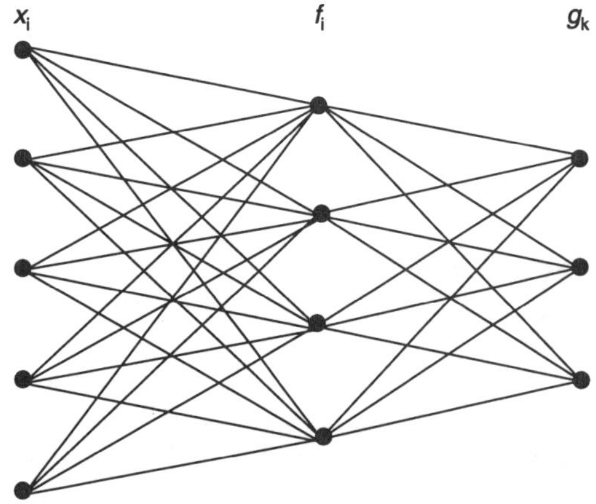

Figure 4.10 Forward-connected three-layer neural network.

formance, however, is often a second-order effect. For example, in a Hamming network, if the representative set of training exemplars differ by a large Hamming distance, the network will tend to train easily and perform quite well. On the other hand, if the exemplars differ by only a small Hamming distance, network training can be very slow and network performance might be poor regardless of the characteristics of the nonlinearity employed. In addition, for highly correlated exemplars, network stability can be a concern. With embedded nonlinearities, a neural network is analogous to a nonlinear filter.

The selection of the network *size* represents a rather critical design parameter. The number of nodes must be large enough to form a decision boundary that is as complex as is required by the problem but not so large that the network weights can't be reliably estimated from the available training set. An additional consideration is that smaller networks tend to "generalize" better than larger networks [22]. Because the network complexity tends to increase exponentially with the network size, the selected network dimensionality should ideally be matched to the problem dimensionality.[7]

In a typical classification network, weights are repeatedly adjusted during the training process using *backward error propagation* (often referred to as *back propagation*), which creates the optimal least-squared error decision function for the training set. For many problems, a large number of training samples may be required to achieve adequate performance. In addition, the training samples must typically be passed through the network many times before acceptable performance is achieved. Despite their limitations, a large number of successful neural net applications have been developed in recent years [24].

Clustering. Whereas, in supervised learning, a constraint or cost function is minimized across a set of labeled data sets, in unsupervised learning, unlabeled input is associated based on the "distance" (according to some metric) between the data and all currently existing clusters. If this distance is above a threshold, a new cluster is formed with the new data sample as its first member. Clustering algorithms have been used in many disciplines to support feature extraction, dimensionality analysis, and pattern classifier design. As noted, neural networks can perform data clustering.

4.3.3 Hybrid Approaches

While the identification of approaches as purely deterministic or purely nondeterministic is admittedly somewhat arbitrary, such distinctions can be conceptually useful. This section attempts to recognize that there exist some problem-solving

---

7. As used here, dimensionality refers to the number of independent (or at least orthogonal) features present in the data that supports decision making. For inherently low dimensionality problems, a relatively low-dimensionality network implementation may provide the best performance. The use of higher order network than is appropriate might introduce artifacts into the decision process.

approaches that clearly possess both deterministic and nondeterministic components. *Genetic algorithms* [25], for instance, generally possess a deterministic generation-and-test element as well as a nondeterministic element for altering the "genetic code" prior to the next recursion. Other hybrid approaches merge two very distinct paradigms. Fuzzy-neural hybrids, for example, seek to complement each other's weaknesses [26].

Bayesian belief networks [27] fit within our definition of a hybrid technique. Belief networks use a directed, acyclic graph to represent probabilistic dependencies among a set of variables. Graph nodes represent random variables, and the arcs between nodes represent the associated conditional probability density functions. Thus, Bayesian networks combine a static decision network (deterministic problem-solving paradigm) with a statistical inference mechanism (nondeterministic problem-solving paradigm). Since a Bayesian network graphically depicts all explicit relationships among random variables, such representations facilitate the development of statistical models for rather complex problems.

Given one or more pieces of evidence, the network provides a computational architecture for propagating belief so as to generate the impact of evidence on the a posteriori probabilities associated with the various nodes (for example, the probability of a class 1 target, given one or more pieces of evidence). Although this technique tends to lead to highly combinatoric evaluation operations [28] for even modest-sized networks, the recent development of a number of highly efficient approximate belief propagation algorithms, for example, [29] has spurred renewed interest in Bayesian networks. Because a Bayesian network fundamentally implements Bayes rule, it possesses the limitations of more conventional Bayesian approaches, as previously discussed.

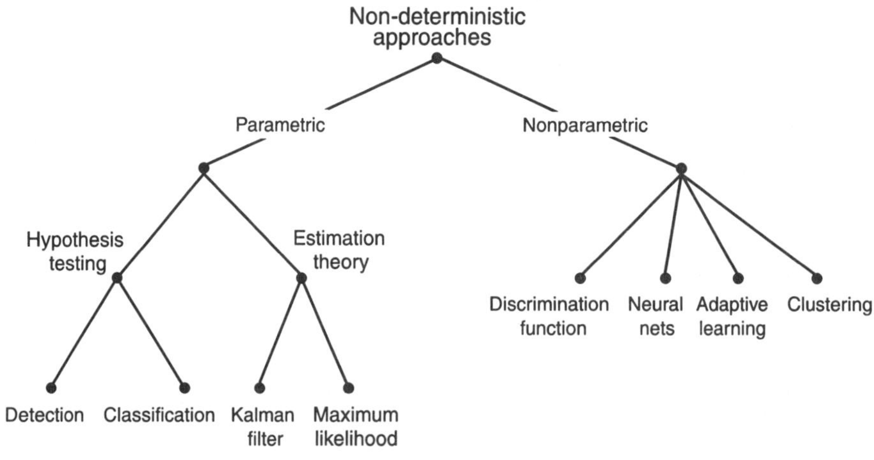

Figure 4.11 Classification of nondeterministic problem-solving approaches.

Figure 4.11 shows the classification of a number of nondeterministic problem-solving methods.

4.4 PARADIGM ASSIGNMENT

Table 4.2 summarizes the key characteristics of the problem-solving paradigms discussed in the last section. The first column identifies whether the procedural knowledge employed is *implicit* or *explicit*. The next three columns associate a *composition class*, an *evidence accumulation* technique, and a *decision process* with each paradigm. Finally, the last two columns identify the canonical form most closely associated with typical applications as well as the potentially applicable canonical forms.

Table 4.2
Summary of the Characteristics of Conventional Problem-Solving Paradigms

	Procedural Knowledge					*Canonical Form*	
Paradigm	*Default*	*Explicit*	*Composition Class*	*Evidence Accumulation*	*Decision*	*Typical*	*Applicable*
Simple data transformations		•	Transformation	(Implicit)	—	I	I, V
Projections		•	Transformation	(Implicit)	—	I	I
Convolution	•		Transformation	(Implicit)	—	I	I
Vector quantization		•	Transformation	—	Threshold	I	I
Generate & test		•	Transformation	Comparison	Selection	I	All odd-numbered forms
Exemplar or rigid pattern match		•	Correlation	Compute distance metric	Choose minimum	I	I, III, V, VII
Model-based pattern match		•	Correlation	Model evaluation	Threshold	IX	All odd-numbered forms
Frame-based reasoning		•	Correlation	Rule based	Activate appropriate rules	I, II	All forms

Table 4.2 (continued)

Paradigm	*Procedural Knowledge*		Composition Class	Evidence Accumulation	Decision	*Canonical Form*	
	Default	*Explicit*				*Typical*	*Applicable*
IF/THEN rule-based systems		•	Correlation	Boolean AND	Rule activation and inference	IX, X	IX–XII
Fuzzy rule–based systems		•	Correlation	Fuzzy AND	Rule activation and inference	IX, X	IX–XII
Set operations	•		Correlation	—	—	I	I
Feedback methods		•	Correlation transform	Compute error signal	Feedback control	I	All odd-numbered forms
Constraint satisfaction		•	Correlation	Boolean	Threshold	II	All even-numbered forms
Clustering		•	Correlation	Compare distance metrics	Inside or outside cluster	II	All even-numbered forms
Decision tree	•		Functional evaluation	(Implicit)	(Default)	II, IV	I–V
Search-based reasoning		•	Functional evaluation	Evaluate evidence	Threshold	II, VI	All even-numbered forms
Linear/dynamic programming	•		Functional evaluation	Find minimum	—	II	II
Gradient-based methods		•	Functional evaluation	Select max or min	—	II	II
Bayesian inference		•	Functional evaluation	Product of probabilities	Threshold	I	I
Hypothesis testing		•	Functional evaluation	Likelihood ratio test	Threshold	I	I

Paradigm	*Procedural Knowledge*		*Composition Class*	*Evidence Accumulation*	*Decision*	*Canonical Form*	
	Default	*Explicit*				*Typical*	*Applicable*
Statistical estimation		•	Transform	(Implicit)	—	I	I
Discrimination function methods		•	Transform	E.g., linear aggregation	Selection	I	I, III, V, VII
Neural networks	•		Transform	(Implicit)	Selection	I	I

Since algorithms utilize both factual knowledge (*long-term declarative*) and procedural knowledge (*long-term declarative knowledge plus an associated control element*), in ome cases, distinguishing purely declarative knowledge from procedural knowledge can be problematic. In most cases, the declarative element associated with procedural knowledge in some way defines the fundamental *functional-evaluation, ransformation,* or *correlation* process involved, while strictly declarative knowledge has nore of a supporting role (Tables-of-Equipment, natural and cultural feature databases, organizational information).

To clarify the assignment process used to construct Table 4.2, several of the ntries are discussed in detail. Depending on the nature of the procedural knowldge employed, a *generate-and-test paradigm* can be associated with any of the oddumbered (generation-based) canonical forms. However, since it computes an xhaustive set of candidate solutions that are rank-ordered with respect to a simple valuation metric, the typical generate-and-test algorithm is a canonical form I approach. On the other hand, the canonical form XV generate-and-test-based pathlanning example in Chapter 3 employed a flexible, hierarchical model that led to he generation of a small set of highly likely candidate solutions. The generate-andest paradigm relies on *transformation-based* knowledge composition.

In the typical forward-inference *production rule paradigm, general declarative* nowledge (left-hand side conditions) is tested against *specific declarative* knowledge working memory) using a *rigid, single level-of-abstraction control* structure. Consequently, the paradigm represents a canonical form IX approach. Similarly, in hyothesis-based applications, the right-hand side rule conclusions are tested against orking memory. If the conditions are satisfied, the left-hand side conditions can be cautiously) concluded to be true. Thus, the paradigm represents a canonical form

X approach. Since the rule conditions are matched with the current problem state, the principle composition class is *correlation.*

Neural networks effectively form a (nonlinear) dot product between an input vector and a weight matrix. Thus, the procedural knowledge is *implicit.* Since the weight matrix is *specific declarative* knowledge, the control knowledge is *rigid* and *single level of abstraction* , and the algorithm is *generation-based,* as argued previously, neural networks are a canonical form I problem-solving form. The basic compositional form is *transformation.*

An *exemplar-based pattern-matching* algorithm typically correlates input data with an exhaustive set of rigid patterns. The output of the process is an indication of the pattern (or template) that best matches the input. The procedural knowledge is considered to be implicit if the algorithm relies on a straightforward mathematical correlation operation; however, when a nonconventional "correlation" operation is required, the procedural knowledge would be considered to be explicit. In either case, the procedural knowledge would be *specific, rigid,* and *single level of abstraction.* Thus, a canonical form I approach is indicated. The primary composition class is *correlation.*

A *decision tree* employs tree-structured problem-solving knowledge (*specific declarative*) that when coupled with a simple search mechanism (implicit control) produces an algorithm. Thus, the tree representation associated with the paradigm is the declarative element of procedural knowledge. The primary composition class is *functional evaluation.* Because of its tree-based character, the algorithm inherently supports a multiple level-of-abstraction reasoning. Since the paradigm is used principally as in hypothesis-based algorithm, decision tree–based algorithms typically represent a canonical form IV problem-solving approach (deduce a disease from a set of symptoms). Since a decision tree can be employed in a generation-based manner (generate a list of common symptoms for different diseases), a decision-tree analysis can be associated with canonical forms I and III as well.

In a *constraint-based reasoning* paradigm, a measurement, system state, or hypothesis is evaluated with respect to a set of constraints. The algorithm output, for instance, might be those elements of the medium-term knowledge base that satisfy the constraint set. In this case, the primary composition class is *functional evaluation.* The evaluation process can range from very simple tests with respect to specific, rigid, single level-of-abstraction constraint sets (intersect a rigid constraint set with the database) to flexible, multiple level-of-abstraction evaluation of general constraint sets (search-based evaluation of complex context-sensitive semantic constraints). Although a constraint evaluation–based paradigm supports the full range of hypothesis-based canonical problem-solving forms, typical automated approaches represent canonical form II class algorithms.

As discussed at the beginning of Chapter 2, taxonomical representations permit the similarities and the dissimilarities among member objects to be readily recognized. Since the various paradigms have been assigned to a true taxonomy of canonical problem-solving form classes, Table 4.2 effectively summarizes both th imilarities and the dissimilarities of the problem-solving techniques listed. Problem-olving paradigms that share a common canonical form can be expected to have imilar performance in terms of robustness, context sensitivity, and efficiency. Al-hough algorithms that are based on the lower-order forms of procedural knowl-dge tend to offer simpler implementations, the higher order forms tend to support nore powerful reasoning approaches.

Algorithms that employ explicit procedural knowledge can potentially use the ull range of canonical form approaches. However, as observed in these tables, many raditional problem-solving paradigms rely almost exclusively on *specific, rigid, single evel-of-abstraction* procedural knowledge (canonical form I and II algorithms). Be-ause the lowest order canonical form approaches are inherently nonrobust, non-ontext-sensitive, as well as potentially inefficient, serious performance shortfalls an occur when applied to complex problems. As mentioned previously, this obser-ation provides at least a partial explanation for the difficulty of automating many nanual fusion functions.

Other research efforts have attempted to compare competing problem-solving pproaches [30]. In one such study, statistical classifiers, fuzzy-set classifiers, and eural networks were compared using an abstract algebra formulation [31]. Table .3 summarizes a number of the conclusions from this study that are seen to be in ubstantial agreement with the results shown in Table 4.2.

Table 4.3
Comparison of Three Classification Techniques (*after:* [31])

riteria	*Statistical Classifier*	*Fuzzy Classifier*	*Artificial Neural Classifier*
opology	Single layered	Single layered	Multiple layered
ecision functions	Probability or mass function	Possibility distribution functions	Composition of linear functions
onnection weight $W_{i,j}$	$W_{i,j} = 1$ $i = 1, \ldots, I; j = 1, \ldots, J$	$W_{i,j} = 1$ $i = 1, \ldots, I; j = 1, \ldots, J$	Real numbers
ecision rule	Select maximum	Select maximum	Select maximum
raining/learning	Estimation of conditional probability density functions a) Parametric approaches b) Nonparametric approaches	Estimation of conditional possibility distributions	Modification of connection weights

Table 4.3 (continued)

Criteria	*Statistical Classifier*	*Fuzzy Classifier*	*Artificial Neural Classifier*
Pros	Minimum probability of error or cost	Estimation of membership functions is less restrictive and easier than probabilistic approaches	Mathematical structure less restricted than statistical or fuzzy, and no restrictions on number of layers or value of weights
Cons	Difficult to estimate high-dimensional probability density functions	Performance evaluation is weak	Most steepest descent training approaches require monotonic nondecreasing, differentiable functions; linear decision functions at all nodes; no performance criteria; stability unpredictable

4.5 SUMMARY AND CONCLUSIONS

A wide range of technical approaches have been used to develop data fusion algc rithms. Since problem-solving paradigms are often selected using highly subjectiv criteria, one of the primary objectives of Part I is to offer guidelines for this selectio process. This chapter briefly reviewed a wide range of traditional problem-solvin approaches that have been used to build data fusion applications. By identifying th dominant class of procedural knowledge used by a technical paradigm, individu problem-solving paradigms were associated with the taxonomy of algorithmic forn developed in Chapter 2. The association of competing technical approaches pe mits an objective comparison of the similarities and differences of problem-solvin approaches.

Although human analysts routinely attack complex problems using highly i tuitive, multiple level-of-abstraction, model-based reasoning that naturally incorp rates relevant domain knowledge, many automated problem-solving approach rely on rather simple, context-insensitive algorithms. While lower order canonic form algorithms may be entirely adequate for certain relatively simple application more complex problems can require more general and flexible algorithms.

Machine recognition of alphanumeric characters provides a simple "case-i point." Based on the current success of optical character recognition software, igid pattern-match paradigm appears to be an appropriate choice for handling ma-hine-generated characters. On the other hand, attempts to recognize handwritten haracters based on the cross-correlation of the input data with all possible variants f the character set has not been nearly as successful. Because handwritten charac-ers exhibit enormous variability that is related to an individual's writing style, letter ize, line thickness, and character length-to-width ratio, attempting to capture this vari-bility with a rigid template set tends to be both inadequate and highly inefficient.

Because most letters and numerals possess important invariant properties that ould be used in the recognition process, model-based approaches can potentially vercome many of the shortcomings of explicit templating.[8] For instance, the letter H" can be modeled as "two nearly vertical parallel lines interconnected in the ap-roximate center by a perpendicular line."

Rather than simply switching from one canonical form I to another, if the task undamentally *requires* a higher order canonical form approach, an algorithm devel-per must seek a more appropriate problem-solving paradigm. For example, if a oncerted effort to construct a statistical classification algorithm for a given problem as failed to achieve desired performance, switching to a neural network implemen-ation that represents the same canonical problem-solving form is unlikely to chieve significantly better performance.

A procurement contract for a large data-processing system typically specifies a et of system-level requirements that must be met by both the hardware and the soft-are. All proposals received are evaluated against these requirements. Because un-atisfactory algorithm performance can be expected when the requirements of an pplication exceed the capability of the selected problem-solving paradigm, the al-orithm development process requires this same discipline.

References

[1] Zadeh, L.A., *Fuzzy Sets and Systems*, Amsterdam: North-Holland Press, 1978.

[2] Prade, H., "Databases with Fuzzy Information and Approximate Reasoning in Expert Systems," *Proc. IFAC Int. Symp. on Artificial Intelligence*, Leningrad, 1983, pp. 113–120.

[3] Pao, Y.-H., *Adaptive Pattern Recognition and Neural Networks*, New York: Addison-Wesley, 1989.

[4] *IEEE Transactions on Fuzzy Systems*, Vol. 2, No. 1, February 1994.

[5] Buchanan, B.G., and E.H. Shortliffe, *Rule-Based Expert Systems: The MYCIN Experiments of the Stanford Heuristic Programming Project*, Reading, MA: Addison-Wesley, 1984.

[6] Cohen, P., *Heuristic Reasoning About Uncertainty: An Artificial Intelligence Approach*, London: Pitman, 1985.

[7] Krause, P., and D. Clark, *Representing Uncertain Knowledge: An Artificial Intelligence Approach*, Boston, MA: Kluwer Academic, 1993.

[8] Lathi, B., *Signals, Systems and Communications*, New York: Wiley, 1965.

Although conceptually straightforward, to date, model-based approaches to hand-written character recognition have not been fully successfully.

[9] Koornwinder, T.H. (ed.), *Wavelets: An Elementary Treatment of Theory and Applications*, River Edge NJ: World Scientific, 1993.

[10] Mallat, S., "Multiresolution Approximations and Wavelet Orthonormal Bases of L2(R)," *Trans Amer. Math. Soc.*, Vol. 315, 1989, pp. 69–87.

[11] Gersho, A., and R.M. Gray, *Vector Quantization and Signal Compression*, Boston, MA: Kluwer Academic Press, 1992.

[12] Shikano, K., K. Lee, and D.R. Reddy, "Speaker Adaptation Through Vector Quantization," *IEE Int. Conf. on Acoustics, Speech and Signal Processing*, April 1986.

[13] Papoulis, A., *Probability, Random Variables, and Stochastic Processes*, New York: McGraw-Hill, 1965.

[14] Minsky, M., "A Framework for Representing Knowledge," *The Psychology of Computer Vision*, P. Winston (ed.), Hillsdale, NJ: McGraw-Hill, 1975.

[15] Schank, R.C., and R.P. Ableson, *Scripts, Plans, Goals, and Understanding*, Hillside, NJ: Lawrenc Erlbaum Associates, 1977.

[16] Kolodner, J., *Case-Based Reasoning*, San Mateo, CA: Morgan Kaufmann, 1993.

[17] Dantzig, G.B., *Linear Programming and Extensions*, Princeton, NJ: Princeton University Press, 1963.

[18] Bellman, R., and S. Dreyfus, *Applied Dynamic Programming*, Princeton, NJ: Princeton Universi Press, 1962.

[19] Albus, J.S., *Brains, Behavior and Robotics*, Peterborough, NH: BYTE Books, Subsidiary of McGraw Hill, 1981.

[20] Hall, D.L., *Mathematical Techniques in Multisensor Data Fusion*, Norwood, MA: Artech House, 1992.

[21] Brammer, K., and G. Stiffling, *Kalman-Bucy Filters*, Norwood, MA: Artech House, 1989.

[22] Yasunobu, S., and S. Miyamoto, "Automatic Train Operation by Predictive Fuzzy Control," *Industrial Applications of Fuzzy Control*, M. Sugeno (ed.), Amsterdam: North-Holland, 1985, pp. 1–18.

[23] Rummelhart, D.E., B. Widrow, and M.A. Lehr, "The Basic Ideas in Neural Networks," *Comm ACM*, vol 37, No. 3, March 1994, pp. 87–92.

[24] Widrow, B., D. Rummelhart, and M. Lehr, "Neural Networks; Applications in Industry, Busine and Science," *Comm. ACM*, Vol. 37, No. 3, March 1994, pp. 93–112.

[25] Goldberg, D.E., "Genetic and Evolutionary Algorithms Come to Age," *Comm. ACM*, Vol. 37, No. March 1994, pp. 113–119.

[26] *IEEE Transactions on Neural Networks*, Vol. 3, No. 5 (Special issue of Fuzzy Logic and Neural Networks), September 1992.

[27] Pearl, J., *Probabilistic Reasoning in Intelligent Systems: Networks of Plausible Inference*, San Mateo, C Morgan-Kaufmann,1988.

[28] Cooper, G.F., "Probabilistic Inference Using Belief Networks in NP-hard," KSL-87-27, Stanfo University, 1987.

[29] Shachter, R.D., and M.A. Peot, "Simulation Approaches to General Probabilistic Inference in B lief Networks," *Proc. of the Workshop on Uncertainty in Artificial Intelligence*, Windsor, Ontario, 1989.

[30] Post, S., and A.P. Sage, "An Overview of Automated Reasoning," IEEE Trans. Systems, Man an Cybernetics, Vol. 22, Jan./Feb. 1990, pp. 202–224.

[31] Jordan, J., and H. Choe, "A Comparative Analysis of Statistical, Fuzzy and Artificial Neural Patte Recognition Techniques," *SPIE Conf. on Signal Processing, Sensor Fusion, and Target Recognition*, C lando, FL, April 20–24, 1992.

Chapter 5

Algorithm Issues Summary

'he last four chapters of this book have focused on issues associated with the devel-pment of algorithms that can support real-world data fusion applications. In Chap-er 1, a biological situation awareness metaphor was presented to offer insights into ne shortcomings of current approaches to automated situation development. In hapter 2, an abstract, biologically motivated fusion process model was introduced nat treated data fusion abstractly as the composition among short-term, medium-erm, and long-term memory. Short-term and medium-term memory maintain dec-rative knowledge, while long-term memory holds both declarative and procedural nowledge. As shown in Table 2.1, the composition among three classes of declara-ve knowledge and one class of procedural knowledge led to the identification of 15 istinct data fusion classes.

Procedural knowledge can be represented as long-term declarative knowledge nd associated control knowledge. As discussed in Chapter 2, procedural knowledge nat relies on a *specific declarative* element leads to non-model-based algorithms. Al-nough they may be relatively simple to develop and use, such algorithms are often oth too brittle and too inefficient for many complex fusion tasks. Procedural nowledge that possesses a more *general declarative* element, on the other hand, can ipport more robust, model-based reasoning. Procedural knowledge that possesses rictly *rigid control* tends to be simple to implement as well as adequate for many outine tasks that do no require sophisticated reasoning. However, procedural nowledge that possesses *flexible control* can support more powerful reasoning ap-roaches. Additionally, control knowledge can be either *single level of abstraction* or *ultiple level of abstraction.* With two classes of long-term declarative knowledge and ur classes of control knowledge, there exist eight canonical forms of procedural nowledge. When these eight canonical forms (or classes) of procedural knowledge e combined with two basic problem-solving paradigms (*generation-based* and *hy-*

pothesis-based), 16 canonical problem-solving (or algorithm) forms result, as definec in Table 2.4.

The examples presented in Chapter 3 demonstrated that the 16 canonica problem-solving forms represent a wide spectrum of possible approaches fron rather simple to highly sophisticated. Chapter 4 extended the fusion process mode and associated conventional problem-solving paradigms with the canonical forn taxonomy. Once individual problem-solving approaches are organized within thi taxonomy, the general characteristics of the various approaches can be readily com pared and contrasted. This chapter discusses the general characteristics and limita tions of each of the canonical problem-solving forms.

In manual data fusion systems, sensor-derived information is routinely com bined and interpreted using extensive non-sensor-derived derived knowledge in th form of maps, tables, and images. To achieve performance comparable to manua data fusion systems, data fusion algorithms must likewise be sensitive to these sam classes of long-term domain knowledge.

Although the biological fusion metaphor provided a number of important ir sights, significant differences exist between biological and artificial situation aware ness systems. In particular, in biological systems both dynamic informatio (*short-term knowledge*) and relatively static domain context (*long-term knowledge*) are es tablished by sensory input [1], whereas, in artificial situation awareness system (whether manual or automated), non-sensor-derived long-term declarative inform tion must be relied upon to establish much of the domain context.

5.1 SUMMARY OF THE 16 CANONICAL PROBLEM-SOLVING FORMS

5.1.1 Canonical Forms I and II

Canonical forms I and II represent simple *generation-based transformation* and *hypoth sis-based templating* processes, respectively. Both of these canonical forms employ sp cific declarative knowledge and simple, rigid, single level-of-abstraction contro Algorithms represented by these forms generally:

1. Support very limited temporal reasoning;
2. Are sensitive to only explicit inputs;
3. Treat all problems at a single level of abstraction.

Conventional signal processing, correlation-based analysis, rigid patter match, and artificial neural systems are typical examples of these canonical form Because canonical forms I and II are relatively straightforward to implement, exar ples of these forms abound.

Early speech recognition systems employed relatively simple canonical form class algorithms. The waveform of individual spoken words was typically correlat with a set of prestored exemplars of all words in the recognition system's vocabulary. Because the exemplars were obtained during a training phase from the same person used to test its performance, these early systems tended to be highly *speaker-dependent.* The exemplar that achieved the highest correlation above some threshold would be selected as the most likely candidate word. The algorithm used *specific long-term declarative* knowledge (specific exemplars) and *rigid, single level-of-abstraction control* knowledge (exhaustive correlation and rank ordering). Although easy to implement and adequate in certain idealized environments (speaker-dependent, high signal-to-noise ratio signals, nonconnected-word speech applications), the associated exhaustive generation-and-test operations makes such an approach too inefficient for large vocabulary systems and too brittle for noisy, speaker-independent, or connected-speech applications.

Artificial neural systems are one of the newer emerging information analysis technologies. Although such systems are motivated by their biological counterpart, the current capabilities of *undifferentiated* artificial neural systems (ANS) fall short of the performance of even simple biological organisms.[1] In contrast to a human who is capable of complex, context-sensitive, multiple level-of-abstraction reasoning based on robust world models, an ANS effectively filters or classifies a set of input states (see Section 4.3.2). While humans can learn as they perform tasks, the ANS weight matrix is typically frozen (except for certain forms of clustering) during the state-transition process.

Thus, regardless of the type of training, the nature of the nonlinearity, or the details of the connection network, a pretrained ANS transfer function represents static, *specific long-term declarative* knowledge, and the associated control element is clearly static, *rigid,* and *single level of abstraction.* In addition, since a neural network transforms a set of input to generate its output (*generation-based*), artificial neural systems possess all the characteristics of a canonical form I problem-solving form.

Typical of canonical form I approaches, neural network performance tends to be brittle for problems of general complexity (since they are non-model-based) and non-context-sensitive (because they generally employ either a context-free or highly context-specific weight matrix). Widely claimed properties of neural networks such as robustness and ability to generalize tend to be highly problem-specific. In general, the ability to build networks that can generalize from examples has proven more difficult than originally expected [2].

While the performance of a neural network solution for a particular application might degrade gracefully with decreasing input signal-to-noise ratio, this phenomenon can often be attributed to a relatively large Hamming distance between exemplar patterns rather than an inherent property of the network. Unless the input data is invariant under rotation, translation, scaling, and deformation, the ANS

---

1. As used here, undifferentiated neural networks refer to those neural systems that have a single set of input and output neurons with no internal logical organization imposed by the developer.

weight matrix may stabilize during training to a "weighted" average across the parameter space. As a result, the net may possess limited discrimination potential. Although some success has been reported with building networks that are invariant to rotation, translation, and scaling, achieving such "invariance" can be computationally expensive.

Most neural networks are trained using some variant of *backward error propagation* learning that generates least-squared error decision boundaries for the training set. Network training using *back propagation* can require hundreds (or even thousands) of iterations through the training set in order to achieve acceptable performance. Asymptotic convergence of the network weights during training is generally not guaranteed and, in certain cases, weight nonconvergence can actually be proven.

The ability of an autonomous system to learn a desired behavior in response to input and then to generalize from such examples is often considered to be an indication of true "learning." From this perspective, *generalization* and *learning* are really duals. Although true "learning" has been the goal of artificial intelligence for several decades, in general, that goal has not yet been achieved. While general learning using feed-forward networks may be "theoretically achievable... for all practical matters, (it is) not feasible" [3].

Artificial neural systems are often applied to complex, ill-defined problems for which other approaches have been unsuccessful. Since the problems are often poorly understood, a *complete* set of realistic training data is generally not available. Without adequate training, acceptable performance is unlikely for complex problems. Thus, the lack of an adequate training set can represent an inherent problem that further limits the applicability of artificial neural systems to real-world fusion tasks.

Just as in the case of many other problem-solving disciplines, strong similarities exist between neural systems and more conventional technologies. "In one sense neural networks are little more than nonlinear regression and allied optimization methods... (in) most real-world problems there is other knowledge that can be used to guide modeling and so enable much more convincing generalization, in ways that can never be done by 'black-box' learning systems" [4]. Whereas there often exists no well-defined relationship between problem constraints and ANS node values, in other network-based processing models (for example, signal filtering), there are generally well-understood relationships between network parameters and physical constraints (the corner frequency of a low-pass filter is related to the product of a resistor value and a capacitor value).

Although most problem-solving approaches associated with canonical form I tend to be inefficient, an ANS lends itself to inherently high-concurrency hardware realizations. Performance issues remain, however, because before committing a network to hardware and periodically during evolutionary enhancements, extensive retraining and testing may be required. Although clearly possessing its own niche,"it is highly unlikely that neural networks will replace either knowledge base or database processing" [5]. Thus, as opposed to a universal solution to complex data fusion problems, undifferentiated artificial neural systems appear to be most appropriate for providing potentially real-time solutions to appropriate canonical form I class fusion tasks.

Despite its limitations, neural networks remain one of the most popular modern information processing technologies. Their popularity has been enhanced by the existence of numerous commercial tools that facilitate rapid prototyping. Neural systems have been successfully employed in the banking industry, the investment community, and elsewhere. Some neural network applications have used a network of relatively simple neural systems that each perform a well-defined function [6]. Thus, rather than a single "undifferentiated" network, such neural systems are organized into a network of networks.

5.1.2 Canonical Forms III and IV

Canonical forms III and IV algorithms depend on *specific declarative* knowledge and *rigid, multiple level-of-abstraction control* knowledge. Although such algorithms possess most of the limitations of canonical forms I and II approaches, by virtue of their multiple level-of-abstraction control, canonical forms III and IV algorithms tend to be considerably more efficient. Simple recursive multiple-resolution, scale-space, and relaxation-based algorithms are typical examples of these forms.

5.1.3 Canonical Forms V and VI

As with the previous four problem-solving forms, canonical forms V and VI algorithms rely on *specific declarative* knowledge. However, rather than *rigid control*, these algorithms possess *flexible, single level-of-abstraction control* that can support *multiple-hypothesis* approaches, dynamic reasoning, and limited context sensitivity. In the path-planning example from Chapter 3, for instance, the static mobility map was augmented with temporally sensitive contextual information, such as the current location of all enemy units.

5.1.4 Canonical Forms VII and VIII

Canonical forms VII and VIII approaches employ *specific declarative* and *flexible, multiple level-of-abstraction control* knowledge. Although still non-model-based, these forms support flexible, hierarchical-based reasoning. In the path-planning example presented in Chapter 3, for instance, a canonical form VII algorithm supported efficient top-down, multiple-pass path development that is dynamically sensitive to the present location of nearby enemy forces as well as highly localized weather conditions.

5.1.5 Canonical Forms IX and X

Because they are based on *specific declarative* knowledge, the eight lower order canonical form approaches generally provide adequate support to numeric-based reasoning. On the other hand, since human reasoning tends to be largely semantic and symbolic, the lower order form approaches tend to be inadequate for mimicking human-level reasoning. Since the eight higher level canonical forms use more general forms of procedural knowledge, they provide significantly better support for semantic- and symbolic-based reasoning.

Canonical forms IX and X, which use *general declarative* knowledge and *rigid, single level-of-abstraction control*, represent simple model-based transformations and model-based constraint set evaluation approaches, respectively. Because these two forms rely on general procedural knowledge, significantly more powerful and robust reasoning is possible than with the lower order forms. In addition, the use of general declarative knowledge supports both more dynamic and more context-sensitive reasoning than do approaches that depend on specific declarative knowledge. However, because these canonical forms rely on rigid, single level-of-abstraction control, such algorithms tend to be relatively inefficient. In the path-planning example in Chapter 3, for instance, a detailed parametric mobility model that *explicitly* depends on the terrain, weather, and doctrine can lead to highly combinatoric computational requirements.

In the last two decades, a number of symbolically oriented reasoning approaches employing general declarative knowledge have been developed in an attempt to overcome the limitations of the more numeric-based paradigms. Logic programming, for example, reasons directly from a fundamental set of principles or axioms. In spite of its theoretical appeal, strict logic-based approaches have generally proven inadequate for many complex, real-world problems.

Expert Systems, on the other hand, have been successfully used to support a large number of data fusion applications. Expert Systems attempt to emulate a human expert's reasoning process in a restricted problem domain. The rule set generally consists of both formal knowledge as well as heuristics and "rules-of-thumb" that have been gleaned from experienced human experts. Although Expert Systems can accommodate very general rule condition and action sets, the control structure is normally rather rigid (sequential condition set evaluation, followed by a simple resolution of which instantiated rule should fire). In fact, the separation of procedural knowledge into modular IF/THEN rule sets (*general declarative knowledge*) that are evaluated using a rigid, single level-of-abstraction control structure (*rigid control knowledge*) *is* the hallmark of the pure production rule paradigm. By demanding rule modularity and a uniform control structure, Expert Systems are relegated to the two lowest order model-based problem-solving approaches.

5.1.6 Canonical Forms XI and XII

Canonical forms XI and XII problem-solving paradigms use procedural knowledge that employs a *general declarative* element and a *rigid* but *multiple level-of-abstraction control* element. With multiple level-of-abstraction control, these forms support both top-down and bottom-up model-based reasoning. Production rule paradigms that employ a hierarchical rule set can emulate canonical form XI or XII approaches.

5.1.7 Canonical Forms XIII and XIV

Canonical forms XIII and XIV employ procedural knowledge that possesses a *general declarative* element and *flexible, single level-of-abstraction control*. As a result, these canonical forms represent the class of flexible, single level-of-abstraction, model-based algorithms. In the path-planning example from Chapter 3, for instance, a canonical form XIII algorithm supports dynamic, multiple-hypothesis, model-based path development using highly context-sensitive domain constraints.

5.1.8 Canonical Forms XV and XVI

Because canonical forms XV and XVI support procedural knowledge that is *general, flexible*, and *multiple level of abstraction*, these forms represent the highest order generation-based and hypothesis-based problem-solving approaches, respectively. As suggested by the examples in Chapter 3, the implementation of sophisticated canonical forms XV and XVI class algorithms can be conceptually straightforward. While the canonical form XV path-planning algorithm outlined in Chapter 3 utilized a very natural, yet powerful and efficient, recursive problem-solving approach, few automated path-planning algorithms have employed such global strategies.

The HEARSAY speech understanding system [7] was an early attempt to develop a canonical form XV algorithm. As mentioned in Chapter 1, this system treated speech recognition as both inherently context-sensitive and multiple level of abstraction. HEARSAY employed a hierarchy of models that were appropriate at a particular level of abstraction (signal processing to perform formant tracking, spectral analysis for phoneme extraction, symbolic reasoning to extract meaning). Higher level processes, with their broader perspective and higher level knowledge effectively controlled the lower level processes. Although a small number of errors committed by the lower processing levels could be recognized and corrected by higher level processes, a large numbers of errors at lower levels would rapidly overwhelm the capability of the higher level processes. Thus, the speech-understanding process was treated in a holistic manner where each level of the processing hierarchy was a critical component of the overall system.

The canonical form XVI doctrinal templating example in Chapter 3 demonstrated a very natural and efficient goal-oriented reasoning approach that emulates the analytical approach of an image analyst. First, attention was directed to the most promising regions of the image. Next, individual units were located based on:

1. Highly efficient top-down/multiple-resolution search;

2. Reasoning that was sensitive to a wide range of spatially oriented context knowledge;
3. Robust models of the units to be recognized.

The unit models characterized both the individual entity attributes as well as the relative relationships among entities within a unit, between units, and with respect to relevant natural and cultural domain features.

5.2 CHARACTERISTICS OF THE HIGHER ORDER CANONICAL FORMS

The following four key algorithm development issues have surfaced throughout the last four chapters:

1. Robustness;
2. Context sensitivity;
3. Extensibility and associated database maintainability;
4. Efficiency.

Each of these issues are now discussed in detail.

5.2.1 Robustness

Robustness measures the fragility of a problem-solving approach to changes in the input space. Since many data fusion problems exhibit complex and possibly dynamic dependencies among relevant features, achieving robust performance virtually demands the use of model-based reasoning. Because the eight lower order canonical forms I to VIII approaches rely on specific procedural knowledge, they cannot model complex interactions and dependencies. As a result, these approaches represent inherently nonrobust problem-solving forms. On the other hand, because the eight higher order canonical forms employ general procedural knowledge, they support the development of robust, model-based algorithms.

The robustness of an algorithm depends on both the quality and efficacy of the models employed. The nature of "an adequate" model, in turn, depends on the complexity of the process being modeled. A problem that intrinsically exhibits few critical degrees of freedom would be expected to require a simpler model than one that possesses many highly correlated features.

Consider the handwritten character recognition problem mentioned at the end of Chapter 4. Although handwritten characters possess a large number of degrees of freedom (line thickness, character orientation, style, location, size, color darkness, contrast ratio), a simple model was seen to capture the salient attributes of the character "H" (two parallel line segments connected at their approximate centers by a third line segment). Thus, although the handwritten character intrinsically possesses many degrees of freedom, most of these attributes are not highly relevant for distinguishing the letter "H" from other handwritten characters. On the other hand, with a non-model-based approach, each character must be compared with a complete set of exemplar patterns of all possible characters. Thus, a non-model-based approach must accommodate all combinations of both the *relevant* and the *nonrelevant* attributes. Consequently, model-based reasoning can potentially simplify algorithm development and substantially improve algorithm performance.

5.2.2 Context Sensitivity

In the broadest sense, *context* refers to both the static domain constraints (*natural and cultural features, physical laws*) and dynamic domain constraints (*current location of all air defense batteries*) that are relevant to the problem-solving process. Dynamic short-term and medium-term knowledge are generally context-sensitive, while a priori long-term reasoning knowledge may or may not be sensitive to context. *Context-sensitive* long-term knowledge (both declarative and procedural) is *conditional* knowledge that must be specialized by static or dynamic domain knowledge (mobility map, current dynamic Order of Battle). *Context-insensitive* knowledge is generic, absolute, relatively immutable knowledge that is effectively domain-independent (terrain can obscure radar coverage; wide rivers are obstacles to ground-based vehicles); such knowledge is fundamentally unaffected by the underlying context. *Context-specific* knowledge is long-term knowledge that has been specialized for a given, fixed context. *Context-free* knowledge simply ignores any effect due to the underlying context.

Thus, *context sensitivity* is a measure of a problem's dependency on domain knowledge and problem constraints. Canonical forms I to IV are most appropriate for tasks that require either *context-insensitive* or *context-specific* knowledge. Because they possess a flexible control, the control element of canonical forms V to VIII can be *implicitly* sensitive to context. Since *general declarative* knowledge can be dependent on non-sensor-derived domain knowledge (a mobility map, the weather, the current ambient light level, the distance to the nearest river), the declarative element of canonical forms IX to XII can be *implicitly* context-sensitive. Finally, since canonical forms XIII to XVI support both context-sensitive declarative and context-sensitive control knowledge, these forms support *fully context-sensitive* problem solving.

5.2.3 Extensibility and Maintainability

Extensibility and *maintainability* are two closely related concepts. *Extensibility* measures the "degree of difficulty" of extending knowledge to accommodate domain changes or to support related applications. *Maintainability* measures the "cost" of storing and updating knowledge. Because canonical forms I to VIII use procedural knowledge that rely on a specific declarative knowledge, significant modifications to the algorithm can be required for even relatively minor domain changes. On the other hand, since they employ *general procedural knowledge,* canonical forms IX to XVI are inherently much more extensible.

The various canonical forms possess varying degrees of domain sensitivity. Canonical forms I to IV approaches, for example, tend to exhibit brittle performance as the problem context changes. *Context-specific* knowledge may be of little value (a mobility map that is based on dry conditions cannot be used to support analysis during a period of flooding) when the problem context changes. On the other hand, *context-insensitive* knowledge (road, bridge, terrain-elevation databases) is unaffected by context changes. Since context-insensitive knowledge is still valid, while context-sensitive knowledge may need to be redeveloped, *database maintainability* is enhanced by separating the context-sensitive and context-insensitive knowledge bases.

Because they are intrinsically non-model-based approaches, the lower order canonical form paradigms typically rely on *context-free* and *context-specific* knowledge. The use of context-free knowledge can lead to ineffective algorithms that are non-extensible; supporting context-sensitive reasoning using context-specific knowledge can lead to significant database maintainability problems. Thus, the use of the lower order canonical forms can lead to the development of algorithms that exhibit both *extensibility* and *maintainability* problems. Since the eight highest order canonical form paradigms can use *context-general* procedural knowledge, such algorithms tend to be much more extensible than the lower order algorithms. In addition, such knowledge tends to impose limited database maintenance requirements. Thus, *algorithm extensibility* can be enhanced by the use of model-based algorithms that employ context-general and context-insensitive knowledge and *knowledge base maintainability* can be enhanced by the logical separation of context-sensitive and context-insensitive knowledge.

5.2.4 Efficiency

Efficiency measures the relative performance of algorithms with respect to computational and/or search requirements. The canonical form I path-planning algorithm discussed in Chapter 3, for example, represents a highly combinatoric, and therefore, highly inefficient problem-solving approach. Although more complex, the alternative canonical form XV algorithm that was outlined would be expected to be significantly more computationally efficient. While exceptions exist, the following generalizations are offered.

1. *Model-based reasoning* tends to be more efficient than *non-model-based reasoning.*
2. *Multiple level-of-abstraction reasoning* tends to be more efficient than *single level-of-abstraction reasoning.*

Since efficiency is a the major focus of Part II, both of the above observations will be explored in considerable detail.

The general characteristics of the 16 canonical forms are summarized in Figure 5.1.

5.3 A FEW OBSERVATIONS

This section offers a few observations that set the stage for our exploration of the database issues in Part II.

5.3.1 Observation 1

As suggested by the preceding chapters, attempts to automate many complex, real-world fusion tasks have not been entirely successful. One explanation for this lack of success is the obvious disparity between manual and algorithmic approaches to data fusion. Humans, for example:

1. Effectively update or modify existing beliefs to accommodate new information as it becomes available (which supports *dynamic reasoning*);

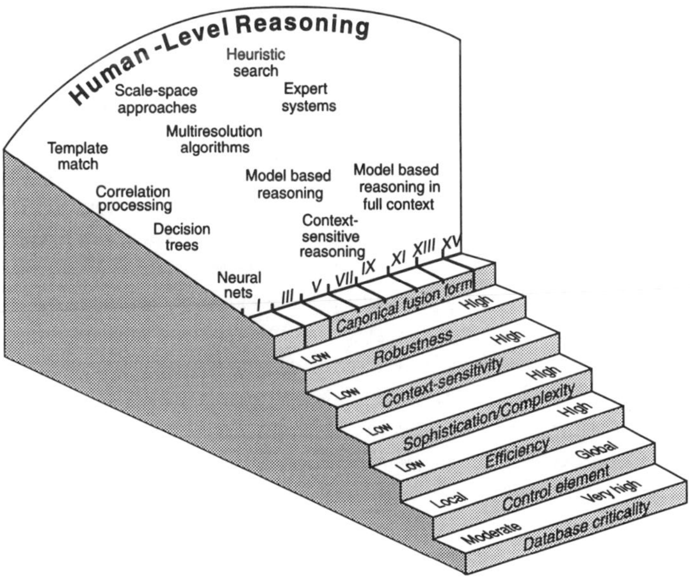

Figure 5.1 General characteristics of the 16 canonical fusion forms and associated problem-solving paradigms.

2. Are adept at model-based reasoning (which supports *robustness and extensibility*);
3. Naturally employ domain knowledge to augment formally supplied information (which supports *context sensitivity*);
4. Intuitively differentiate context-sensitive from context-insensitive knowledge (which supports *maintainability*);
5. Control the reasoning process in a highly focused manner (which enhances *efficiency*).

Thus, manual approaches to data fusion tend to be inherently dynamic, robust, context-sensitive, and efficient. On the other hand, as we have seen, many of the paradigms that have been used to develop data fusion algorithms are inherently static, nonrobust, non-context-sensitive, and inefficient. Thus, the automation of many fusion tasks clearly demands the use of more powerful problem-solving paradigms.

5.3.2 Observation 2

While a global phenomenon must be interpreted within a relatively global context, local phenomenon can require interpretation within both a local as well as a global context. Because target track assignment is a relatively local phenomenon, the track assignment process may benefit from the use of both a local and global perspective. With the conventional canonical form I target tracker, track assignment is based primarily on the most recent target behavior. As compared to the strong constraints afforded by a "road-constrained target extrapolation strategy," for example, ground-based target tracking that relies strictly on the vehicle's historical trajectory and the vehicle's performance capabilities provide rather weak constraints on future target motion. Thus, for ground-based targets whose behavior can be strongly constrained by relatively deterministic domain characteristics, the traditional statistical-based tracking algorithm generates highly underconstrained solutions.

Although a local viewpoint can adequately explain the local behavior of a target's trajectory (constant velocity travel along the road), a more global viewpoint is required to explain or interpret the behavior of a target over time (that is, the entire target track). Figure 3.4 demonstrated such a multiple level-of-abstraction interpretation of a target trajectory. The development and maintenance of a multiple level-of-abstraction perspective is often an unstated, but underlying, requirement of the tactical data fusion process.

5.3.3 Observation 3

Production systems have historically performed better against static, well-behaved, finite-state diagnostic-like problems than against highly context-sensitive problems that possess time-varying behavior. Since production systems rely on *rigid, single level of-abstraction control* that is often not sensitive to domain context, such shortcomings are not surprising. Yet, Expert Systems have been routinely used in dynamic, highly context-sensitive problem domains, often with disappointing results. In addition to such limitations, the modular nature of traditional IF/THEN rule sets opens the paradigm to the possibility of circular reasoning [8] and to problems due to correlated evidence sources [9]. As with any problem-solving paradigm, both its strengths and limitations must be fully understood by the algorithm developer.

When an appropriately-constrained task is successfully automated using an Expert System, it has been frequently observed that the now well-understood problem might be more efficiently implemented using more conventional procedural methods (for example, as a decision tree). In effect, either:

1. An alternative, canonical form IX or X problem-solving approach is indicated; or
2. A lower order, non-model-based approach is identified that can adequately handle the task.

When an Expert System paradigm appears inadequate for a given problem, artificial neural systems have often been nominated as an alternative, or even a preferred, approach. Since neural nets require no programming, they appear to be appropriate for ill-defined or poorly understood problems. While Expert Systems tend to have real-time performance problems, artificial neural systems offer the promise of very high speed hardware implementations. In addition, the adaptive nature of the neural net-learning process seems to match real-world, dynamically evolving problem-solving requirements. However, as already discussed, current artificial neural systems operate more like a statistical or fuzzy pattern recognizer than as a sophisticated reasoning system capable of generalization, reasoning by analogy, and abstract inference. While an Expert System represents a lower order *model-based reasoning* approach, a neural network represents a lower order *non-model-based* approach.

5.3.4 Observation 4

Indications and warnings, as well as many other forms of expectation-based analysis, have traditionally relied on relatively rigid doctrinal and tactical knowledge. Modulated by the intent of an adversary, the strongest constraints predictably tend to have the most significant impact on behavior. However, since future data fusion applications will be required to support limited tactical operations against adversaries that employ flexible, ill-defined, and highly creative tactics and doctrine, recent world events have effectively exacerbated the difficulty and credibility of any analysis that is based on rigid tactical and doctrinal information. As a consequence, future tactical domain analysis (whether manual or automated) must necessarily rely on somewhat weaker expectations and constraints. Although the lack of strong, reliable a priori knowledge handicaps all forms of expectation-based reasoning, the addition of relevant logical, physical, and logistical domain constraints can at least partially offset the loss of highly institutionalized tactics and doctrine. As a result, future data fusion automation must rely more heavily on the higher order canonical form approaches that can accommodate nontraditional information sources.

5.4 SUMMARY AND CONCLUSIONS

Developing robust, context-sensitive, and efficient fusion algorithms demands a clear understanding of the problem requirements at the outset of the development effort. Although selecting a higher order canonical problem-solving form than is actually required may represent "overkill," the use of a lower order form than is required will almost certainly fail to achieve acceptable results.

For highly constrained, relatively static problems, conventional pattern-match or templating approaches might be entirely appropriate (matched filtering, signal sorting, statistical parameter estimation). However, complex, real-world data fusion problems often require true model-based reasoning in full context (*general declarative* knowledge and *flexible control* knowledge). In addition, tactical data fusion systems demand real-time performance capability that necessitates highly efficient reasoning methods (top-down, multiple level-of-abstraction reasoning, mixed spatial and semantic reasoning). Thus, many real-world fusion tasks have requirements that are fully supported only by the highest order canonical form approaches. Simply committing more resources (dollars, people, time) to improve the performance of a data fusion algorithm or fusion system cannot be expected to overcome fundamental deficiencies in the selected problem-solving approach.

The implementation of more powerful fusion algorithms requires both a better understanding of problem requirements and algorithm capabilities as well as the development of critical tools and infrastructure. An important element of that infrastructure is the development of advanced database management systems. Chapters 6 to 10 address a wide range of issues associated with the development of a database system to support next-generation data fusion algorithms. Part III reviews the salient algorithm and database principles, discusses three important classes of generic algorithm support (spatial, temporal, and hierarchical reasoning), and provides comprehensive examples of several higher order canonical form fusion algorithms.

References

[1] Clark, J.J., and A.L. Yuille, *Data Fusion for Sensory Information Processing Systems*, Boston, MA: Kluwer Academic Publishers, 1990.

[2] Rummelhart, D.E., B. Widrow, and M.A. Lehr, "The Basic Ideas in Neural Networks," *Comm. ACM*, Vol. 37, No. 3, March 1994, pp. 87–92.

[3] Geman, S., E. Bienenstock, and R. Doursat, "Neural Networks and the Bias/Variance Dilemma," *Neural Computation,* Vol. 4, 1992, pp. 1–58.

[4] Ripley, B., "Statistical Aspects of Neural Networks," Invited Lectures for Seminaire Eurpoeen de Statistique, Sandbjerg, Denmark, April 25–30, 1992, London: Chapman and Hall.

[5] *DARPA Neural Network Study,* AFCEA International Press, Fairfax, VA, November 1988.

[6] Pao, Y.-H., *Adaptive Pattern Recognition and Neural Networks*, New York: Addison-Wesley, 1989.

[7] Erman et al., "Hearsay-II Speech Understanding System: Integrating Knowledge to Resolve Uncertainty," *Blackboard Systems,* R. Engelmore and T. Morgan (eds.), Reading, MA: Addison-Wesley, 1988, pp. 25–86.

[8] Pearl, J., *Probabilistic Reasoning in Intelligent Systems: Networks of Plausible Inference*, San Mateo, CA: Morgan-Kaufmann, 1988.

[9] Krause, P., and D. Clark, *Representing Uncertain Knowledge: An Artificial Intelligence Approach,* Boston, MA: Kluwer Academic, 1993.

Part II

Database Issues

Part I discussed a wide range of issues associated with the development of data fusion algorithms. Chapter 1 presented a high-level view of the objectives and the functional requirements of tactical data fusion. Chapter 2 introduced a biologically motivated fusion process model that led to the identification of 15 classes of fusion problems and 16 canonical problem-solving forms. Chapter 3 presented examples of several common data fusion tasks to illustrate the qualitative differences between the various canonical problem-solving approaches. Chapter 4 associated a wide range of traditional technical problem-solving paradigms with one or more canonical problem-solving forms. Finally, Chapter 5 summarized the key characteristics of each of the 16 canonical problem-solving forms with respect to algorithm performance criteria.

The effective automation of complex data fusion tasks was seen to be enhanced by algorithms that:

1. Employ adequate general problem knowledge;
2. Are sensitive to relevant domain conditions and constraints;
3. Employ a global control element.

Because the lower order canonical problem-solving forms do not adequately support these requirements, algorithms based on these classes of reasoning tend to be nonrobust, noncontext-sensitive, and relatively inefficient. The higher order canonical form algorithms, on the other hand, were seen to be potentially robust, context-sensitive, *and* efficient. However, achieving such capability generally requires the use of more complex algorithms, that, in turn, can place heavy demands on the supporting database management system.

Because tactical sensor systems have been designed primarily for the detection of "target-oriented" information, they generally do not supply the contextual knowledge that can be required by the higher order canonical form algorithms. As a result, domain knowledge, such as surface material, terrain, cloud ceiling height, and cultural features, required to support context-sensitive reasoning must be provided by other means. In most cases, such domain context knowledge is collected in advance of its use, analyzed, and stored in a database to be accessed as needed by the fusion algorithms. Since such knowledge bases can be extremely large, the supporting database system can be a critical analysis bottleneck in sophisticated data fusion applications.

Although a comprehensive treatment of database management systems is beyond the scope of this book, Part II focuses on those database issues that are critical to data fusion algorithm development.[1] As will be seen, not only do conventional database management systems represent a potential *performance bottleneck* for complex fusion applications, they can represent a significant barrier to the *development* of more sophisticated algorithms as well.

Chapter 6 discusses fundamental limitations of conventional database management systems with respect to data fusion application requirements. The relational database model, for instance, is shown to provide inefficient search and association among data sets that are implicitly related in both space and time.[2] Although commercially available Geographic Information Systems (GIS) systems permit general search and manipulation of spatially organized data, such systems generally do not support:

1. Both efficient *and* high-resolution Boolean set operations among *all* combinations of point, line, and region features;
2. Two-dimensional *spatially anchored search*;[3]
3. Optimal search for complex mixed spatial and nonspatial queries;
4. The efficient maintenance of highly dynamic data sets.

Chapter 7 discusses representation issues and evaluation criteria associated with the design of a general-purpose database management system kernel intended

---

1. The actual construction of a database management system entails numerous technical issues that are not discussed here, such as locking, access control, tertiary storage, incremental indexing, and multilevel security. However, numerous books, journals, and conferences are dedicated to the myriad of issues associated with the detailed design issues. The bibliography at the end of the book provides a starting point for such investigations.

2. Although three-dimensional spatially organized data can be required by data fusion applications, we restrict our discussion to two-dimensional representations for two reasons. First, such a restriction significantly simplifies the treatment; in most cases, the extension of the concepts and principles to three dimensions is reasonably straightforward. Second, two-dimensional spatial information fully satisfies the requirements of numerous applications.

3. *Spatially anchored search* is *true two-dimensional* search relative to a point, line, or region feature. In essence, true two-dimensional search mimics the virtues of human-level analysis of conventional analog two-dimensional spatially organized representations. A more formal description is offered in Chapter 7.

to facilitate complex data fusion applications. Since the amount of spatially organized information that can be required to support sophisticated applications can be orders-of-magnitude greater than all other classes of knowledge combined, spatial data considerations dominate the database design. A top-down database design methodology that achieves an effective compromise among both algorithm development and performance considerations is presented.

Chapter 8 outlines the key elements of a database kernel that embraces the design principles offered in Chapter 7. The database model effectively integrates a conventional semantic object representation with an object representation of two-dimensional space. A hybrid spatial representation that permits hierarchical spatial indexing of vector-represented point, line, and region features is introduced. In addition to supporting efficient true two-dimensional spatial search, the spatial-indexing scheme permits efficient top-down, low-resolution spatial reasoning. The database kernel is shown to overcome many of the shortcomings of both relational database management systems as well as current GIS systems.

Chapter 9 discusses several extensions to the proposed database kernel, including query optimization and high-level database interfaces. Chapter 10 summarizes the key technical issues and recommendations for satisfying the database requirements of real-world data fusion algorithms.

CHAPTER 6

Database Requirements

6.1 INTRODUCTION

Traditional file-oriented approaches to data management require a user to interact with the data at a very low level of abstraction, dealing directly with physical representation, file management, interrupt handling, and data sorting. Current database management systems (DBMS), on the other hand, permit interaction with data at a logical or conceptual level. In addition, such systems provide a wide range of application support designed to enhance user productivity.

Historically, data fusion algorithms have made limited use of domain constraints. Although more context-sensitive algorithms have been investigated in recent years, most of these applications have been proof-of-principle technology demonstrations. The size of the domain knowledge base typically associated with such technology demonstrations has tended to be relatively small. For example, a particular algorithm might be sensitive to a single feature, such as the underlying terrain elevation data (for example, the Digital Terrain Elevation Data (DTED) provided by the Defense Mapping Agency). If the area of interest is relatively small, the spatial database might be easily managed by conventional means. The objective application, on the other hand, might demand many more feature classes and a much larger area of interest. Thus, while database management system performance might not surface as an issue in a technology demonstration setting (reduced data set, simplified scenarios), serious performance problems can occur when highly context-sensitive algorithms are applied against full-scale operational databases.

The following example illustrates the magnitude of the potential problem. Suppose that a database stores road networks using a *vector-based representation* (as endpoint pair tuples of a piecewise-linear approximation of the roads). A target-tracking algorithm that depends on the distance between a detected target and the closest road might require exhaustive search and analysis of a large portion of the road database. Since the search-space size tends to increase exponentially with the size of the road database, the data-set size need not get very large before serious performance problems develop. Because a real-time tracking algorithm must operate at the track-update rate, with extremely large data sets, database search time requirements may exceed the track-update rate. Similarly, with region features represented by their *bounding polygon* (the list of endpoint pair tuples associated with a piecewise-linear approximation of the region boundary), the intersection of a target uncertainty ellipse with lineal and region-based cultural and geographic features can lead to highly combinatoric search-and-test operations as well. Even highly parallel algorithm implementations may not overcome performance bottlenecks associated with database search and manipulation operations.

The importance of controlling the size of the problem search, state, and solution spaces can be illustrated by consideration of the Euclidean *traveling salesman problem* (TSP). A considerable effort has been directed at this particular problem, in part, because a wide range of optimization problems map into this formalism. As mentioned in Part I, the objective of the traveling salesman problem is to develop the minimum distance path through a network of n nodes, such that every node is visited once. From the first city, there are $(n-1)$ potential cities that could be on the minimum distance tour; from each second city in a path, there are $(n-2)$ third cities, and so forth until the tour culminates at the final city. The set of all paths forms a tree with a single root node and a branching factor at level j of $(n-j)$, for $j \geq 1$. Because no local evaluation criterion can guarantee global optimality, the identification of the globally optimal path requires the generation and testing of all possible paths.

The apparent simplicity of the TSP problem statement and the straightforward nature of the optimization algorithm, however, belies its potentially staggering computational requirements. For a given starting city, on the order of $(n-1)!$ paths (one for each leaf node in the tree) must be evaluated. For 21 cities, the optimal solution requires the evaluation of all 10^{18} (that is, 20!) possible routes. If each path length is computed and ranked-ordered in 10 microseconds, 1 million years would be required to discover the optimal path on a serial machine.[1] Since the number of legal move sequences is on the order of 10^{40} in checkers and on the order of 10^{120} in chess, optimal solutions to even familiar board games via exhaustive generation-and-testing is for all practical purposes infeasible. The generation of global solutions based on exhaustive evaluation and testing is generally referred to as *combinatorial optimization.*

Although faster processors and parallel implementations can improve performance, highly combinatoric problems tend to overwhelm any hardware implementation.[2] The use of suboptimal heuristic approaches is an obvious alternative.

1. The exclamation mark for factorial notation is really quite appropriate!

2. "Few algorithms with exponential running time are likely to be appropriate for practical use, though

Numerous approximation algorithms have been developed that achieve near-optimal solutions [2] or attempt to develop optimal solutions using a reduced search space [3]. Consider the class of heuristic path selection criteria that require consideration of only relatively nearby cities. For instance, suppose the selection of the "next city" to visit is based on the k nearest cities not yet visited. The reformulation of the optimization process is $MIN(\Delta_{jk} \sum d_{ij}))$, where Δ_{jk} represents the set of k nearest cities to city j.

In this case, from the starting city, there are k possible second cities in the minimum distance tour; from each of these cities, there are k possible third cities; and so forth. Near the leaf nodes of the solution-set tree, the number of remaining cities is less than k. Thus, the set of all possible paths generated by the heuristic algorithm generates a k-dimensional tree to depth $(n - k)$, with a monotonically decreasing branching factor below this level. The resulting search tree contains $k^{n-k}(k-1)!$ leaf nodes. If $(n - 1)$ is substituted for k in this expression, the total number of paths is equivalent to the classic optimization requiring evaluation of $(n - 1)!$ paths.

The computational complexity of the optimal tour algorithm relative to the complexity of this simple approximate tour strategy is:

$$\frac{(n-1)!}{k^{n-k}(k-1)!} \tag{6.1}$$

Using the factorial approximation $m! = e^{-m} m^m (2\pi m)^{1/2}$, (6.1) can be shown to be approximately proportional to $e^{-(n+k)} (n/k)^n$. For $n > k$, the quotient approaches $(e^{-1}\, n/k)^n$, which grows exponentially with increasing values of n. Thus, for large n, the approximate solution is dramatically more efficient than the optimal solution. For example, for $k = 3$, the heuristic algorithm generates on the order of only 10^9 $(3^{18}2!)$ possible paths (approximately 1 hour execution time for a 10 microsecond path evaluation time) or, for $k = 2$, on the order of only 10^6 (2^{19}) paths (approximately 10 second execution time). In addition, a wide range of similar path development heuristics can be employed, such as "all cities within a fixed distance" or "all cities within a fixed distance of the closest city."

Thus, for algorithms that inherently possess exponential running time (on the order of 2^n where n is related to problem size), either approximate solutions or problem reformulation may be required to achieve acceptable performance.[3] While many of the lower order canonical form problem-solving approaches presented in Chapter 3 generated highly combinatoric solutions, the corresponding higher or-

---

such algorithms arise naturally as 'brute-force' solutions to problems" [1].

For instance, n could be the degree of a polynomial, the size of a file to be searched, or the number of nodes in a graph.

der canonical form approaches generally offered not only more robust and context-sensitive performance but considerably lower order computational requirements as well.

While the computational requirements of the higher order canonical form algorithms might be considerably lower than comparable lower order approaches, the context-instantiation process can lead to significant database performance problems. Thus, the algorithm recommendations offered in Part I tend to trade extensive *explicit* computational requirements for extensive *implicit* search and evaluation requirements. The next several chapters demonstrate that search and computational requirements associated with the context-instantiation process can be minimized by careful design of the supporting database system.

6.2 DATABASE MODELS

Since knowledge can be either *declarative* or *procedural*, databases must potentially support the storage, maintenance, and query of both declarative and procedural knowledge. Declarative knowledge is passive factual or relational knowledge that can be stored using a wide range of representations, such as tables, semantic networks, rigid templates, decision trees, organizational diagrams, entity-relationship graphs, property lists, frames, and scripts. Procedural knowledge can be treated as declarative knowledge that includes an explicit control element specifying when and how the declarative knowledge is to be employed. Procedural knowledge can be represented as pattern/action rules, parametric algorithms, and mathematical transformations. Since the size of the procedural knowledge base required to support context-sensitive fusion algorithms is typically dwarfed by the size of the supporting declarative knowledge base, our discussion focuses on database issues related to the storage, search, manipulation, and maintenance of declarative knowledge.

6.2.1 Traditional Models

The *entity-relationship* (E-R) *model* [4] is perhaps the most fundamental declarative data representation structure. In a graphical interpretation, nodes represent *entities* and arcs represent *relationships* between nodes. Figure 6.1 shows an example of a general entity-relationship model.

Network and *hierarchical representations* are specialized entity-relationship models. In a network model, nodes can have an arbitrary number of parents and offspring. *Semantic networks* [5] are a special form of the network model where the *nodes* represent physical entities or concepts and the *arcs* represent semantic functions that relate them. In a hierarchical data model, data relationships are depicted as a tree structure where each node has only a single parent but can have an arbitrary number of offspring. A binary tree, a family tree, and a general taxonomy are common examples of tree-struc

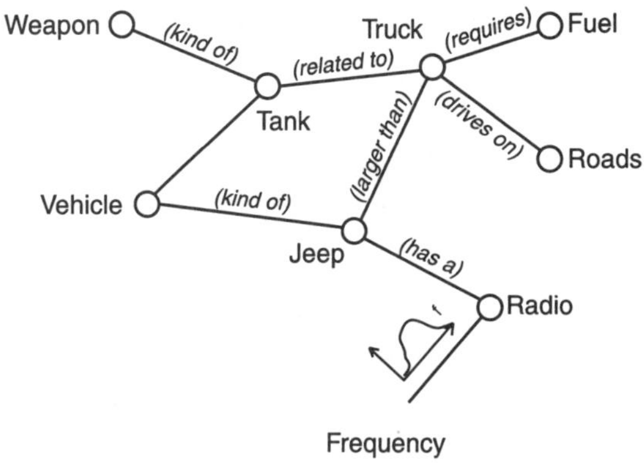

igure 6.1 Simple example of an entity-relationship data model demonstrating both semantic and numeric relationships between nodes.

ured representations. Hierarchical data models naturally preserve parent-child re-ationships between nodes and support efficient property inheritance. Figure 6.2 hows an example of network and hierarchical data organizations.

.2.2 Relational Data Model

he relational data model was originally proposed in 1970 [6]. The basic data struc-ure in the relational model is a *table* (or a *relation* in database jargon) containing ows and columns. Each row stores one occurrence of an entity, while each column aintains a separate *attribute* (or *field*) of the entity. Database management systems hat use the relational data model are referred to as relational database manage-ent systems (RDBMS). Most current commercial database management systems re based on this model.

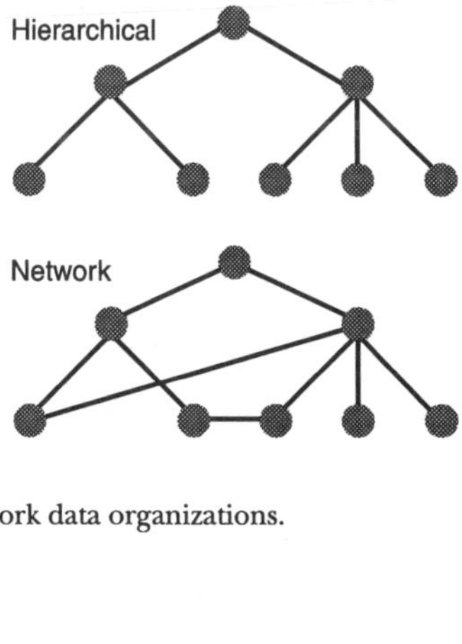

gure 6.2 Hierarchical and network data organizations.

Because of both its theoretical underpinnings (operations on tables are based on set theory, relational algebra, and relational calculus) and the maturity of the available commercial products, the relational model has gained wide acceptance. In contrast to a simple flat file approach to data management, the state-of-the-art RDBMS offers a data-independent interface, considerable software support, and machine portability. Because of their table-based framework, spreadsheet programs are a special form of a RDBMS.

The relational data model exhibits two levels of data independence. First, because the user need not be aware of how the data are actually stored, the model provides *physical data independence.* Second, because all relationships among data are based on the *value of the data* (rather than on explicit pointers), the model provides *logical data independence.*

The relational data model possesses two primary organizational properties. First, to facilitate rapid search, data can be *indexed* with respect to either a particular attribute or a combination of attributes. Regardless of the number of fields that are conjoined, however, indexing effectively generates only *linear* (that is, one-dimensional) search dimensions. These dimensions can be organized alphabetically and/or numerically (last_name + first_name + date_of_birth). Because multiple search dimensions can be required for a single table (name, profession, zip_code, profession + name), indexing, which does not lead to data replication, is generally preferred to sorting of the base table.

Second, elements from two or more separate databases (tables) can be viewed together by *joining* the databases via a unique and static identifier (referred to as a *primary key*) of each row in the first table. For example, in a management information system application, the employee social security number often serves as the primary key permitting personnel and payroll databases to be *joined.* A *table join* represents the logical *union* of the two data sets.

The Structured Query Language (SQL) is the relational database query language standard supported by the American National Standards Institute (ANSI) and the International Standards Organization (ISO). The principal database-templating command (which effectively captures much of the functionality of the relational algebra) is the SQL command

SELECT (attribute) FROM (table name) WHERE (condition clause).

Conventional relational database management systems possess a number of highly desirable properties, including:

1. The relational model subsumes both hierarchical and network models [7];
2. The representation is data-independent;

3. The formal relational model is almost entirely supported by commercial products;
4. The model offers standardized query interface;
5. The system is application-independent;
6. Most commercial RDBMS run on numerous hardware platforms.

However, in addition to their virtues, relational database systems possess a number of significant shortcomings, including:

1. A table is the only representation structure;[4]
2. Complex relational joins can be computationally expensive to generate;
3. Relationships between tables depend on *unique, static,* and *explicit* keys;[5]
4. Since only linear search dimensions are supported, spatial, combined spatial and temporal, and hierarchical data sets may be inefficient to both search and manipulate;
5. For highly dynamic data, considerable re-indexing (which implies a high computational overhead) may be necessary to maintain database search dimensions;
6. A table-based data model can not preserve complex semantic relationships among data.

Therefore, although the relational model provides a powerful framework for access and maintenance of one-dimensional table-based records, the basic relational data model provides limited support to complex, real-time data fusion applications.

In recent years, numerous extensions to the relational model have been proposed. These extensions include the support for abstract data types, spatial queries, procedures, and rules.

.2.3 Geographic Information Systems and Spatial Data

While the relational data model was developed to support the storage and retrieval of *table-based information,* Geographic Information Systems (GIS) systems were developed to support the storage and retrieval of *spatially organized information.*[6] Prior to the emergence of GIS systems in the late 1970s, application-developers were forced to select data representation structures and develop their own spatial search mechanisms and manipulation tools. Not only did such an undertaking involve an exten-

---

4. RDBMS have addressed nonstandard data types by providing a capability to store unstructured "long records" or "binary large objects" (BLOBS).

5. In the relational database literature, a *primary key* is often referred to as being "implicit" (since it is the *value* of the key that establishes the link) to contrast it with joins that are based on the value of an "explicit" pointer.

6. The term "GIS" was actually coined in the 1960s.

sive commitment, the complexity of the task was often underestimated, leading to numerous unsuccessful development efforts.

Today, a large number of commercial GIS systems are available and widely used by city planners, utility companies, commercial freight haulers, resource managers, environmental research programs, and mining companies. A systematic treatment of the wide range of GIS applications in geography and cartography is found in [8]. A tabular summary of the characteristics of a number of prominent GIS systems is provided in Table 6.1; a much more comprehensive listing is published annually by GIS World [9]. As can be observed from this table, GIS systems employ either *vector-based, raster-based,* or both such representations.

Table 6.1
Characteristics of a Number of Widely Distributed Commercial GIS

Product	*Raster*	*Topo Vector*	*Nontopo Vector*	*Date Introduced*
PC ERDAS	•			1979
ARC/INFO	•	•	•	1982
GRASS	•	•		1985
SPANS	•	•	•	1985
GENAMAP/ GENACELL	•	•	•	1986
Windows/on the world	•	•		1986
IDRIS I	•		•	1987
GEO/SQL		•		1987
Terrain/ GEMSTONE	•	•	•	1987
SMALLWORLD GIS	•	•		1988
VISION	•	•	•	1990
GEOLOCATE	•			1991
ERDAS IMAGINE	•	•	•	1992

In general, spatial representations can be either *decomposition-based* or *nondecomposition-based. Vectors* are the most common nondecomposition-based representation. Vector representations store:

1. *Points* as tuples, (x_i, y_i);
2. *Lineal features* (open lines) as the list of endpoint tuples associated with piecewise-linear approximation of the feature;
3. *Region boundaries* (closed lines) as the list of endpoint tuples associated with a piecewise-linear approximation of the boundary.

Vector-based representations can be either *nontopological* or *topological*. Because they do not preserve topological relationships among features, traditional vector representations are nontopological. Topological representations, on the other hand, explicitly preserve logical relationships among features. For example, since it is shared by both regions, the topological representation of the line between a forest and an adjacent meadow need only be represented once. As a result, topological vector-based representations tend to be more memory-efficient than nontopological representations.

Whereas nondecomposition-based representations require no structural decomposition of the data space, decomposition-based approaches rely on such a decomposition. *Decomposition-based representations* can be either *regular* or *nonregular*. A raster representation is a common example of a regular (data independent) decomposition, while the R-tree is an example of a nonregular (data-dependent) decomposition [10].

GIS systems support two important classes of spatial reasoning: (1) *spatial search* and (2) two-dimensional *set operations*.[7] Set operations include *intersection* and *union*. The intersection operation creates the set that is *common* to two or more sets; the union operation creates a composite set that includes all elements contained in *any* of the individual sets.

In general, *spatial search* is a form of database templating where the template represents a search constraint. Search can be considered either *point search* or *range search*. In the former, the goal is to find all objects that contain a point; in the latter, the goal is to find all objects that intersect a region.[8] If the constraint is represented as a set, spatial search can be interpreted as the intersection of the constraint set with the database. Two related operations are *inclusion* and *proximity*. One set is *included* within another set if the intersection between the two yields the first set. *Proximity* involves the evaluation of a distance metric between spatially organized features.

In general, efficient spatial search requires some form of *spatial indexing*. Just as a book index facilitates search for subjects of interest, a spatial index facilitates search for spatial data of interest. Spatial indexing methods typically involve either:

---

7. Although many systems possess other capabilities (such as the representation of triangulated irregular networks (TIN) for representing three-dimensional surfaces), these are the two elements that are of primary interest to our discussion.

8. The former objective is a special case of the latter.

1. The partition the space into disjoint subspaces;
2. The partition of space into nondisjoint subspaces;
3. The transformation of a two-dimensional feature into either a lower or higher dimensional space.

Grid-based indexing schemes employ *disjoint* subspaces. Paper maps often use a uniform labeled grid that facilitates locating a feature that has been found in the map gazateer (for example, a particular city is identified as being in the grid cell associated with row *C* and column 9). The use of a multiple-resolution, grid-based data structure can support multiple-resolution spatial indexing.

To facilitate search and set operations, vector-based representations typically associate a *minimum bounding rectangle* (x_{max}, x_{min}, y_{max}, y_{min}) with each line or region feature. Since a minimum bounding rectangle (or *bounding box*) partitions space into nondisjoint subspaces, bounding boxes represent a *nondisjoint* subspace indexing scheme.

The mapping of data sets into a new space provides a different form of indexing. A two-dimensional data set, for example, can be mapped onto a single-dimensional axis to determine the spatial ordering of features along that direction. Similarly, a data set can be mapped into a higher dimensional space that preserves a particular search dimension. For example, mapping a region's bounding rectangle to a single point in four-dimensional space might facilitate reasoning about the underlying regions.

Many forms of reasoning rely on set operation generation. As we observed in numerous examples in Part I, set operations can be used to:

1. Define constraints (all regions within 10 km of A);
2. Compose separate constraints (slope 10% AND deciduous forest);
3. Build search windows (locate all *B* such that *B* is within *C*);
4. Evaluate templates (intersect a geometrical pattern with an underlying map database).

Although all GIS systems *support* both search and set operations, the potential efficiency and accuracy of the various systems varies considerably, largely due to the characteristics of the underlying data representation and the supporting indexing scheme. For instance, with relatively coarse resolution requirements, raster-based set operations tend to be considerably more efficient than vector-based set operations. On the other hand, if high-resolution products are required, raster-based set operations tend to be inefficient because they can require the manipulation of extremely large, high-resolution raster representations.

Although vector-based representations are highly memory efficient, their use can lead to computationally expensive set operations and spatial search. Both search and set operation efficiency can be significantly enhanced by associating *minimum bounding rectangles* with each feature. However, for highly extended line and region features, the indexing efficiency of bounding rectangles tends to be low. While indexing schemes based on the transformation of spatial data to a higher dimension space may facilitate proximity testing, such schemes generally do not enhance set operation efficiency.

Because most commercial GIS systems are proprietary, comprehensive analytical comparisons of competing systems are generally not available. Although both formal and informal empirical bench-marking studies have been conducted, a considerable effort is required to generate validated comparisons across a representative range of functionality. Since new versions of the systems are regularly released, side-by-side comparisons quickly become dated.

Because of the perceived limitations of current commercial systems, numerous experimental prototype GIS systems are being developed. Table 6.2 lists the characteristics of several of these prototype systems. A number of trade magazines, journals, and symposia focus on geographic information system products and research.[9] Because of their focus on spatial reasoning and their reliance on conventional database management systems to maintain nonspatial (or *aspatial*) data, GIS systems have generally been used to supplement, rather than replace, more conventional database management systems in data fusion applications

Table 6.2
Prototype GIS systems

System	*Reference*	*Comments*
SAND	[11]	• Separate point, line, and region representation • Bidirectional links between spatial and nonspatial data • Spatial data structure serves as spatial index plus supports spatial operations • Optimization based on both merge operations and operation reordering

GIS World is the primary source book reference; the International Journal of Geographic Information Systems, founded in 1987, is the primary journal; and AUTOCARTO is generally considered to be the major symposium.

Table 6.2 (continued)

System	*Reference*	*Comments*
Gral	[12]	• Uses bounding boxes • Full description of spatial objects is defined as tuples • Optimization takes only selectivity into account and not actual cost of the operation
GEOQL	[13]	• SQL extension to support spatial data • Biased toward relational/spatial processing • Optimizes nonspatial only • Each relation is assumed to have only one spatial attribute
Probe	[14]	• Contains general geometric point set (but no point, line, and region objects) • No spatial query language • User-defined specializations of point set data type • Spatial efficiency low
SIRO-DBMS	[15]	
GEOVIEW	[16]	
Geo-Kernal	[17]	• Spatial information stored in text format • Bidirectional links • Uses nested relations • Spatial efficiency low • Bidirectional links between spatial and nonspatial • Query optimizer prefers relational selections before spatial
GEO++	[18]	• Implemented on top of POSTGRES using spatial primitives • Optimizer attempts to map defined operators into existing database methods
QUILT	[19]	• Optimized point, line, and region quadtrees • Efficient set operations

6.2.4 Object-Oriented Databases

Object-oriented (OO) reasoning is a highly intuitive, machine-oriented reasoning paradigm that has become rather popular in recent years. In this paradigm, *objects* encapsulate knowledge, perform abstract functions, maintain data, and communicate with each other. In particular, *objects* perform the following functions:

1. Autonomously reason based on their internal declarative and procedural knowledge base;
2. Inherit properties from parent objects;
3. Communicate with and task other objects;
4. Pass results to other objects.

In object-oriented jargon, *objects* are said to possess a *class*; the set of all classes form a *class hierarchy*. Objects communicate by *messages* that are operated on by *methods*. *Objects* are analogous to *variables*, *classes* are analogous to *type*, *class hierarchies* are analogous to *type hierarchies*, *methods* are analogous to *functions*, and *messages* are analogous to *function calls* in conventional programming approaches.

Consider the object representation of a tank depicted in Figure 6.3. This figure shows:

1. A partial *vehicle-class hierarchy*;
2. A partially expanded representation of a *generic tank object*;
3. A representation of a *specific tank-class object*;
4. A *specific-instance object* of a particular tank.

The generic tank object inherits all the properties of *military, ground-based, tracked vehicle objects*. The M-1 tank object (specific tank-class object), in turn, inherits the properties of the *generic tank object*. Finally, the specific-instance object in Figure 6.3(d), which represents the current knowledge about a particular target, inherits the properties of all *M-1 tank objects*.

Object-oriented reasoning naturally supports multiple level-of-abstraction analysis. For reasoning with respect to "tanks," only the *generic tank object* and the inheritance through its parent node (*tracked, ground-based military vehicles*) are relevant. On the other hand, if reasoning about a specific tank *class* is required, the relevant specific-class object (M-1 tank) would be used. Finally, if analysis with respect to a specific tank target is required, the lowest level of the object hierarchy would be utilized. In each case, the specific knowledge of the particular object of interest is supported by the knowledge and the reasoning power of all of the more general objects on the path to the root of the object hierarchy. Because the manipulation of highly modular objects tends to be quite natural for humans, object-oriented reasoning provides an important conceptual framework for machine-based reasoning.

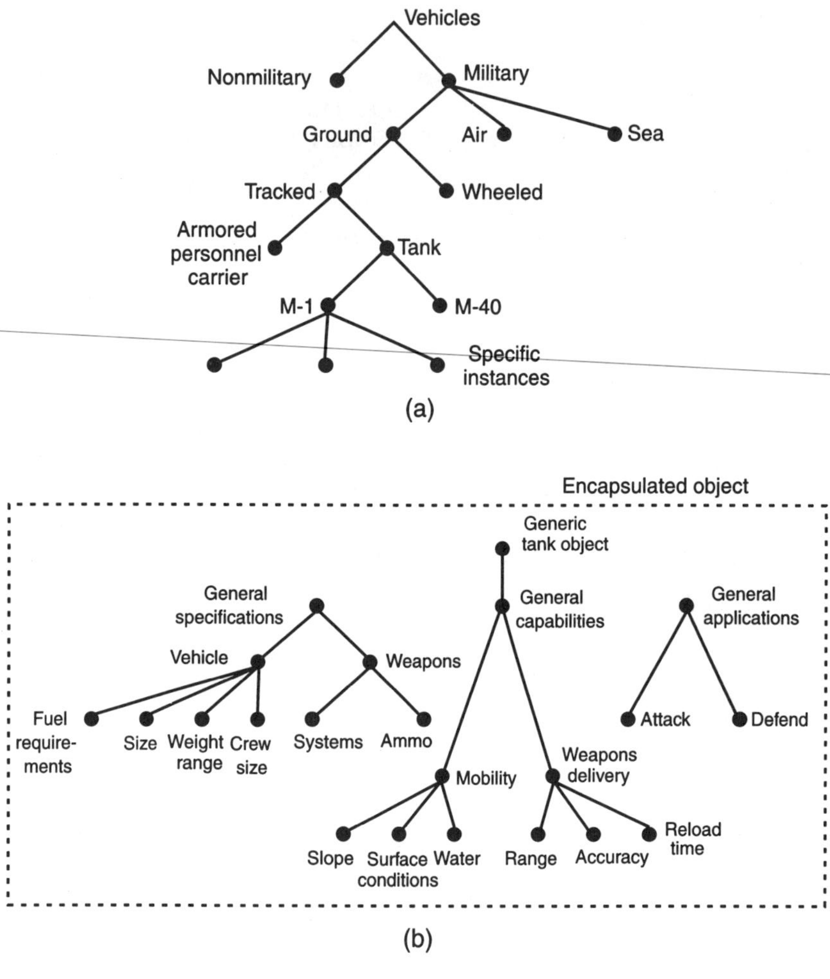

Figure 6.3 (a) Partial object hierarchy, (b) partially expanded generic tank object tree.

An object with little or no internal functionality (no procedural knowledge resembles a *frame*. In fact, frame representations that support "attached procedures can approach the functionality of true object representations. However, sinc frames typically do not possess a formal relational hierarchy, they are a limiting cas of a true object representation.

Considerable interest exists in the development of object-oriented database (OODBs) to support object-oriented reasoning. While object-oriented program ming environments have become relatively mature, persistent object databases ar still fairly recent arrivals. Although a number of commercial and prototype system

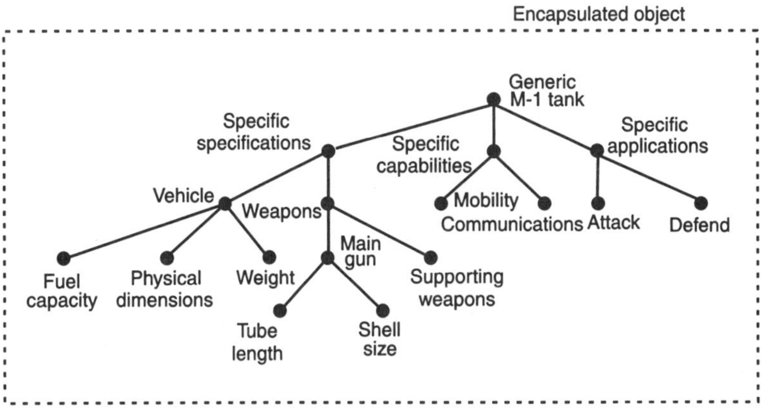

(c)

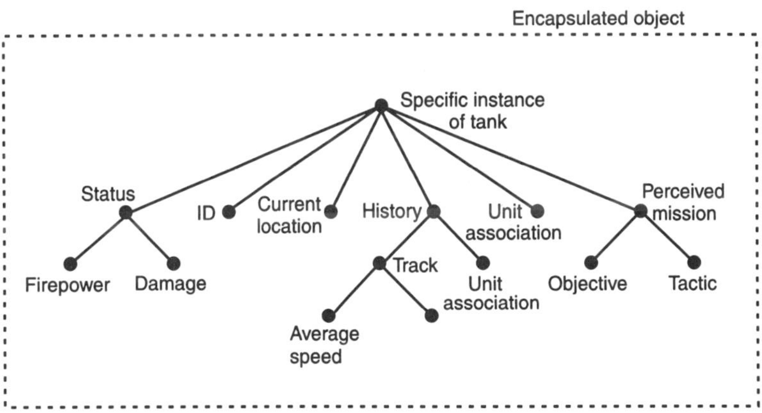

(d)

gure 6.3 (c) M-1 tank object, and (d) specific instance of an M-1 tank object.

ave been built, arguments remain that the defining characteristics of OODBs have ot been completely resolved [20].

Because persistent object databases lack the strong theoretical underpinnings f the relational algebra and the relational calculus, additional research and experi- ıce is required to evaluate the potential of persistent object databases in terms of assical database functionality, such as query optimization, database integrity check- g, and related overhead operations. Table 6.3 lists several commercial and proto- pe OODB systems

Table 6.3
Commercial/Prototype OO DBMS

Product	*Comments*	*Vendor*
Gemstone	Currently available system related to SMALLTALK via OPAL; implemented in C++	Servio Logic Development Corp.
ONTOS	C++	Ontologic, Inc.
Orion	Research prototype; Common LISP	Microelectronics and Computer Technology Corp.
IRIS	Research prototype implemented in C	Hewlett-Packard Laboratory
O_2	Research prototype directed at office automation and management of spatial data	Altair Consortium (France)
OOPS	Research prototype	Germany
TIGRIS	Support for spatial data	Intergraph Corp.
System 9	Support for spatial data	Prime Corp.
SMALLWORLD	Support for spatial data	Smallworld Systems, Ltd.

6.2.5 Multimedia/Hypermedia Databases

Considerable contemporary interest exists in the development of databases that can store and retrieve not only data from tables and maps but one-dimensional signatures, general imagery, graphics, video, and audio as well. Because of the obvious commercial advantages of such systems, multimedia database research is quite active. The interested reader will find many books, journals, and conference proceedings that address this subject. Since a significant portion of the interest in multimedia relates to the storage and retrieval of two-dimensional spatially organized information, extended relational models, GIS systems, and multimedia database system development appear to be on a convergent course. Object-oriented databases provide the foundation for most multiple-media database systems. Although somewhat loosely defined, *hypermedia* is often considered to be *active* or *intelligent* multimedia.

6.3 ALGORITHM REQUIREMENTS

As discussed in Part I, the development of robust, context-sensitive, and efficient data fusion algorithms requires more complex and sophisticated reasoning than ha typically been employed. Because conventional database management systems offer somewhat limited support to the development of such applications, this section reviews algorithm requirements with respect to their database management system implications.

6.3.1 Human Problem-Solving Metaphor

As discussed in Part I, biological systems maintain a dynamic (*temporal reasoning* implied) situation awareness with respect to three-dimensional space (*spatial reasoning* implied) by fusing sensory-derived information with a priori domain knowledge using highly focused, multiple level-of-abstraction analysis (*hierarchical reasoning* implied). The very survival of a species demands an ability to develop and maintain such situation awareness.

Humans learn to identify the spatial direction of a sound based on the amplitude difference at the two ears. The perception of three dimensions is accomplished by stereo processing of two-dimensional visual fields. Motor skills demand global goal planning and dynamic replanning supported by sensory feedback from visual and balance sensors. The fusion of vision and sound over time permits the speed of an oncoming vehicle to be estimated. In addition to the apparent "ease" with which biological systems perform sensory fusion, their analysis is accomplished in "real time."

As discussed in Part I, while most attempts to automate tactical data fusion processes have employed lower order canonical problem-solving forms, human analysts routinely use far more sophisticated, robust, and context-sensitive reasoning. For example, consider the interpretation of the time-stamped location reports of the multiple moving targets shown in Figure 6.4. While a traditional canonical form target tracker might be unable to correctly associate detections arising from closely spaced targets and targets with crossing trajectories, a human is often capable of resolving the associated ambiguity by using powerful reasoning models that are sensitive to a wide range of domain constraints. Rather than simply attempting to "connect the dots" in Figure 6.4, a human analyst could employ sophisticated target behavior models that depend on the underlying domain constraints depicted in Figure 6.5 to help "explain" the observations. In Figure 6.6, when the time-stamped detections are overlaid on a topographic map, tracks 1 to 3 are discovered to be highly consistent with a road-following target model, tracks 4 and 5 are determined to be most consistent with a minimum terrain gradient–following trajectory model, while track 6 is not consistent with *any* ground-based vehicle behavior model.

When interpreted within a more global context (road association with respect to estimated vehicle speeds and target spacings), tracks 1 to 3 can be deduced to be wheeled vehicles traveling in a convoy along a secondary road. Based on their speed and the associated surface materials along their entire trajectory, tracks 4 and 5 can be deduced to be tracked vehicles. Finally, because of its relatively high speed and the rugged terrain in the vicinity, track 6 is most consistent with a low-flying air-

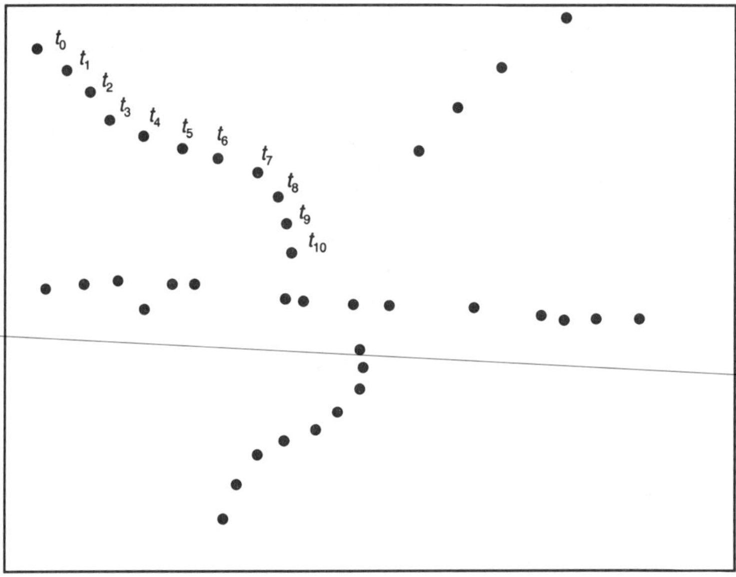

Figure 6.4 Location reports from multiple moving targets (each point is assumed to be time-stamped).

Figure 6.5 Perspective view of the target area.

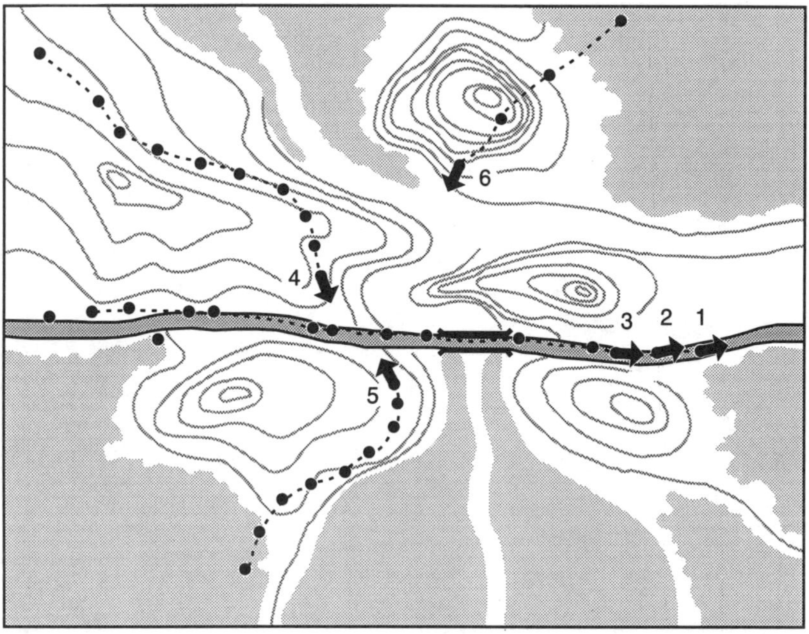

Figure 6.6 Target detections overlaid on terrain and cultural feature maps.

borne target. If the velocity of target 6 is too low to be a fixed-wing aircraft, the target would most likely be a helicopter. Figure 6.7 shows the final product that results from the *synergistic fusion* of both sensor-derived and non-sensor-derived information.

Expectation-based analysis is often handicapped by the limited dimensionality of the available sensor-derived information. However, by supplementing sensor-derived information with supporting non-sensor-derived information, the effective dimensionality of the decision process can be increased. In the previously described examples, a canonical form I tracker that employs *context-free* procedural knowledge might be unable to properly assign individual detections to existing tracks. By contrast, a human (or potentially, a higher order canonical form algorithm) might be capable of producing robust track assignments as well as tentative target classifications.

The difference between the two approaches, of course, is that the human problem-solver effectively reasons in a higher dimensional space using *context-sensitive* procedural knowledge. While *context-free* knowledge simply ignores any possible context sensitivity, *context-general* knowledge can be specialized to the current, relevant context. Thus, a lower order algorithm that relies on context-free knowledge is effectively limited to the dimensionality of the sensor-derived information, while a higher order canonical form paradigm is limited only by the total relevant knowledge that can be brought to bear on the decision. Database support for the instan-

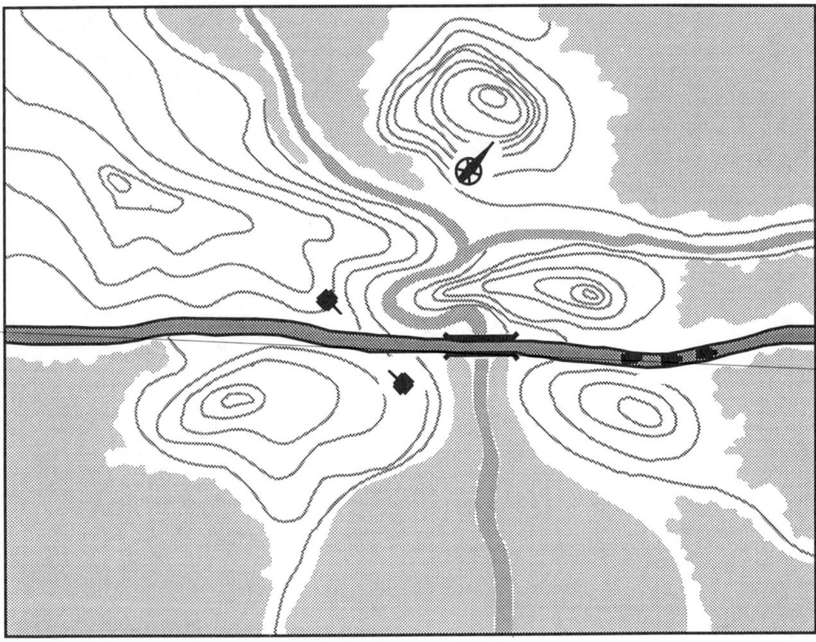

Figure 6.7 Product of context-sensitive target track association and classification process.

tiation of context-sensitive declarative and procedural knowledge is the principal subject of the next several chapters.

The following query provides insight into the power of the human's spatial-reasoning ability:

Find all roads west of River 1, south of Road 10, and not in Forest 11. 6.2

Given the data representation shown in Figure 6.8, a human can readily locate all roads that satisfy this query. However, if the map-based data depicted in Figure 6.8 was provided to the human in the vector-based form shown in Figure 6.9, a human would have considerable difficulty identifying the appropriate roads. Although the two representations might be technically "equivalent," the former representation implicitly depicts all relevant spatial relationships among the data, while the latter representation requires that such relationships be discovered by computational means. In addition to this obvious difference, human perception permits the regions in Figure 6.8 that are depicted as boundary-only representations to be inter

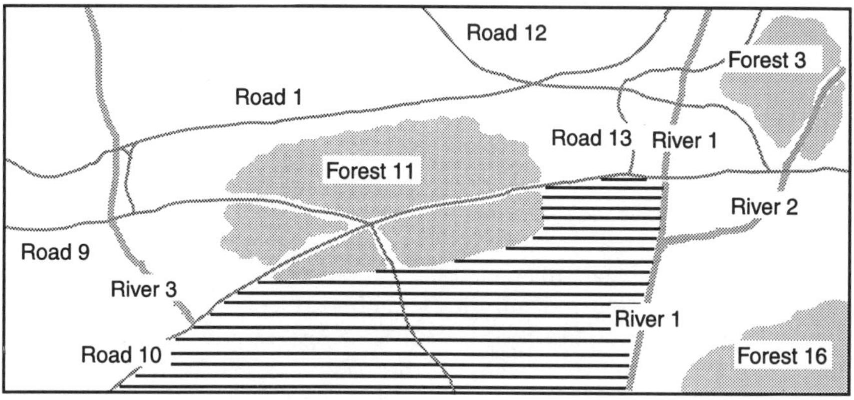

Figure 6.8 Database relevant to query (6.2).

Road 9	(x_1, y_1),	(x_2, y_2),	(x_3, y_3), ...
Road 10	( ),	( ),	( ), ...
Road 11	( ),	. . . .	
Road 12	( ),	( ),	
Road 13	( ),	( ),	
River 1	( ),	( ),	( ), ...
River 2	( ),	( ),	
River 3	( ),	( )	
Forest 3	( ),	( ),	( ), ...
Forest 11	( ),	( ),	( ), ...
Forest 16	( ),	( ),	

igure 6.9 Vector-based representation of map data shown in Figure 6.8.

reted as the representation of the boundary *plus* the enclosed area. Because of this areal-based" perceptual facility, humans perform two-dimensional set operations irtually "by inspection."

For decades, military analysts have capitalized on the ability of humans to erform complex spatial reasoning by providing them with fully registered, two-diiensional topographic maps and acetate overlays. For reasoning that requires conderation of "primary road networks," for instance, only the acetate depicting that ature would be used. Thus, the paradigm avoids the search and analysis of irrelent information. Because newly abstracted spatial features (for example, the reilt of set operations among existing feature sets) are stored in the same uniform, fully registered representation, all resultant products are readily usable in subsequent analysis. Thus, data *representation* plays a critical role in both human perception and reasoning. Figure 6.10 depicts a multiple level-of-abstraction "acetate overlay" view of spatially organized short-, medium-, and long-term declarative knowledge.

6.3.2 Algorithm-Related Issues

In this section, we investigate the database requirements of robust, context-sensitive data fusion algorithms. Since data fusion requires the combination of information from spatially distributed sensors, the entities of interest are spatially distributed, and the domain constraints for analyzing this data possess spatial attributes, *spatial reasoning* necessarily underlies the situation development process.

Because many complex fusion problems exhibit a stratified character or possess a sequential aspect, *hierarchical reasoning* is endemic to many fusion applications. As discussed in Chapter 1, the functional fusion model itself has both a multiple

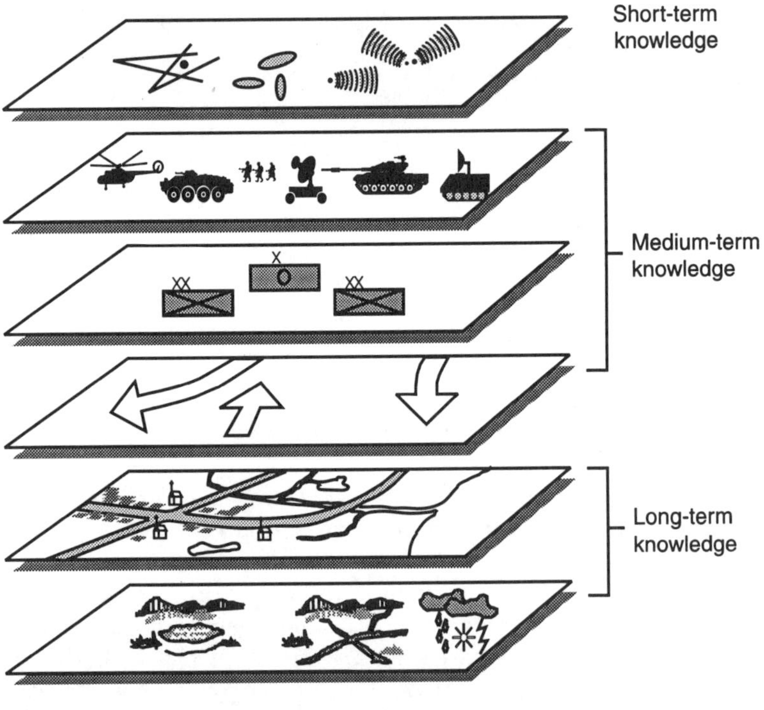

Figure 6.10 Multiple level-of-abstraction, acetate-overlay view of information.

level-of-abstraction/problem decomposition character as well as a sequential process flow character.

Due to the evolving nature of tactical situations (as well as many other problem domains), most data fusion applications require *temporal reasoning.* Targets may be at a particular location at a certain time or during a certain time interval; bursts of radio frequency energy occur as an illuminator sweeps past the observer; and physical phenomenon, such as sunspot effects on communications, all have a time-varying and/or time-sensitive aspect. Both coordinated and independent activities begin, evolve, and terminate. Events often possess a natural ordering that effectively imposes a temporal character on the fusion process. Because sequential phenomena effectively impose an ordering of events, temporal reasoning can be treated as a specialized form of hierarchical reasoning.

There exist two generic classes of temporal reasoning: (1) *implicit* and (2) *explicit. Implicit temporal reasoning* represents those classes of reasoning where time is an implied element of both the data and the analysis process (for example, the filtering of time-series data). Implicit temporal reasoning normally involves a data-driven process flow and a simple control element. For example, radar return provides geopositional and parametric measurements that are an implicit function of time. Similarly, target tracking involves the sequential processing of time-sampled data.

Explicit temporal reasoning, on the other hand, involves more complex reasoning with respect to time, such as time order, duration, or concurrency of events. In a tactical river crossing, for instance, a sequence of events necessarily precede and follow the actual crossing. Low-level units become associated, engage in a range of activities, cross the river, and are subsequently realigned as the river-crossing operation terminates.

Perhaps the most fundamental database search operation required by data fusion algorithms is the retrieval of database objects that are "close" both in time *and* space to a new target report. If both stationary and moving entities are maintained as discrete 3-tuples (x_i, y_i, t_i), the indexing or sorting of such data along any *individual* dimension is straightforward. However, for large data sets, linear indexing of inherently two-dimensional spatial data can lead to inefficient access. Similarly, the retrieval of target data stored in a (linear) temporal history file (organized strictly by its time tag) tends to be ineffective as well.

Consider a simple example that illustrates the problem of retrieving data that is "close" in both time and space to a new target d (where $d = (x_d, y_d, t_d) = (3, 1.5, 5)$) using a data set indexed strictly by time. Suppose that a temporally sorted database currently holds the following target information:

$$(x_a, y_a, t_a) = (1, 1, 1)$$

$$(x_c, y_c, t_c) = (3, 3, 5)$$

$$(x_b, y_b, t_b) = (5, 2, 10)$$

Based on the temporal index dimension, query point *d* appears to be "close" to point **c** (since $t_c = t_d$) and far from either *a* or *b*. In reality, as seen in Figure 6.11, query point *d* is actually "far" from point *c* and very "close" to the trajectory represented by the line *a–b*. Thus, even though moving targets might only be observed at discrete times (for example, at the radar scan rate), the actual trajectory of a physical object is a continuous function in both space and time. Although discrete data sets *can* be analyzed to identify the closest data to line segments that are stored by their endpoint pairs, for large data sets, a high computational overhead can be incurred. Efficient support for the retrieval of database elements that are "close" both in time and space to a query point effectively demands database search along the continuous dimensions of both space and time. Therefore, for a large class of problems, indexing spatially sensitive data strictly by *time* provides ineffective search-space focus.

The explicit representation of both the spatial and temporal history of moving targets as continuous lineal objects (target tracks) provides a better match to the characteristics of physical phenomenon. Although not independently indexed, time is "embedded" in the continuous spatially organized track representation shown in Figure 6.11. Based on the highly localized search of this particular representation, *d* is identified as being spatially near line *a–b*. Next, for all candidate tracks that are "spatially near" *d*, the time attribute can be evaluated by appropriate interpolation between t_a and t_b. Thus, a time-tagged true two-dimensional spatial representation of target tracks provides effective search dimensions as well as potentially preserve spatial relationships between the track and other database elements. Time-tagged point and region data can be handled in a similar fashion.

Figure 6.12 illustrates a simple taxonomy that captures important relationships between spatial, temporal, and hierarchical reasoning. At the highest level in this

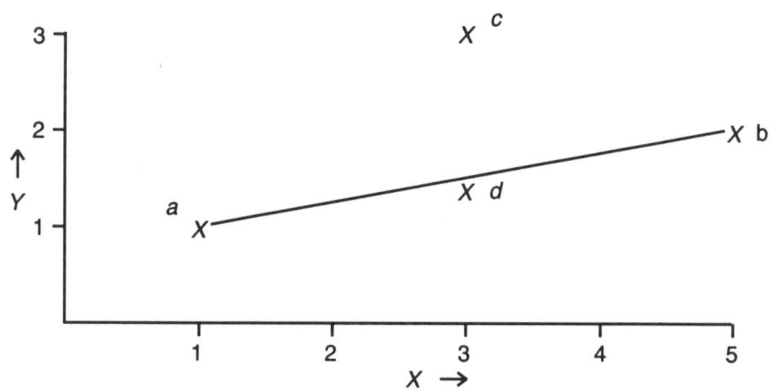

Figure 6.11 Example demonstrating the advantage of a temporally coded true two-dimensional spatial representation over a strictly temporally indexed representation.

taxonomy, reasoning is considered either *spatial* or *nonspatial*. Both reasoning classes can be handled using either *hierarchical* or *nonhierarchical* approaches. Each of these, in turn, may or may not involve temporal reasoning. Since *temporal reasoning* has been effectively subordinated to both *spatial reasoning* and *hierarchical reasoning*, this taxonomy is consistent with the above argument in favor of time-tagged spatial representations and the earlier observation that temporal reasoning is a form of hierarchical reasoning.

From the perspective of this reasoning class taxonomy, there exist four classes of spatial and four classes of nonspatial reasoning. If *temporal reasoning* is referred to as *dynamic reasoning* and *nontemporal reasoning* is referred to as *static reasoning*, four primary classes of both *spatial reasoning* and *nonspatial reasoning* can be defined:

1. Dynamic, hierarchical;
2. Static, hierarchical;
3. Dynamic, nonhierarchical;
4. Static, nonhierarchical.

The supporting database management system must accommodate each of these reasoning classes. Since *hierarchical reasoning* technically subsumes *nonhierarchical reasoning* and since *dynamic reasoning* effectively subsumes *static reasoning*, a database that supports *dynamic hierarchical spatial reasoning* and *dynamic hierarchical nonspatial reasoning* would effectively support all eight reasoning classes. The database kernel developed in Chapter 8 supports both of these representation classes.

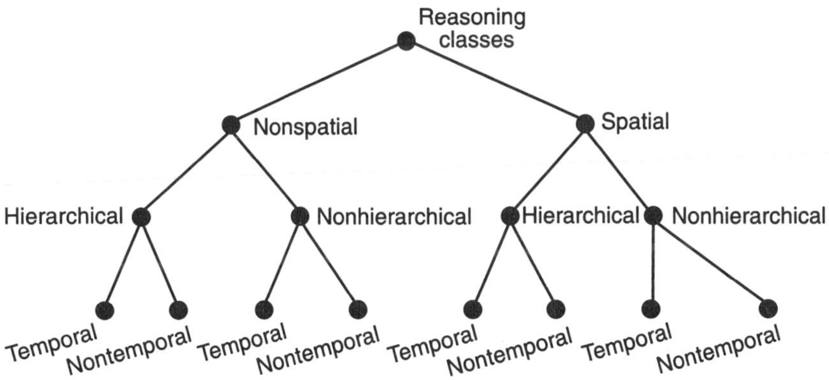

Figure 6.12 Reasoning class taxonomy that results when spatial reasoning is selected as the dominant reasoning form.

6.3.3 Algorithm Requirements Summary

We next investigate the relationship between the various canonical problem-solving forms and the general fusion algorithm requirements. To do this, we first contrast the characteristics of the four lowest and four highest order canonical problem-solving forms and observe that, in general, the more sophisticated the algorithm, the potentially more severe are the database memory, search, and query requirements.

Canonical Fusion Forms I to IV

As discussed in Part I, algorithms that rely on *specific* declarative knowledge and employ *rigid* control often lead to exhaustive generation-and-test or hypothesize-and-test operations. The canonical form I path-planning example from Chapter 3, for example, required the exhaustive generation of all possible paths to the goal state. The speech recognition example required the correlation of an input pattern with all prestored word patterns. The object detection and doctrinal templating examples discussed under canonical form II required the comparison of a data set with an exhaustive set of templates; the composition products were stored in working memory and then rank-ordered using a selected evaluation metric.

While the lower order canonical form approaches can require the storage of a relatively large number of static exemplar patterns, the maintenance of extensive domain databases is generally not required because such algorithms typically rely on context-free knowledge. Thus, in terms of overall database storage requirements, approaches based on canonical forms I to IV generally require relatively small dynamic working memory (*short-term* and *medium-term* declarative knowledge) and only moderately large static exemplar pattern storage (*long-term declarative* knowledge). Due to the relative simplicity of typical canonical forms I to IV algorithms, sophisticated database search is generally not required.

Although most canonical form I to IV approaches impose modest database requirements, exceptions are algorithms that inherently involve extremely large state-space representations (recall the canonical form I path-planning algorithm presented in Chapter 3). Such algorithms can have moderately large memory requirements and generate rather large search spaces.

Canonical Fusion Forms XIII to XVI

Since canonical forms XIII to XVI support *general, context-sensitive* reasoning, such algorithms can require sophisticated, multiple level-of-abstraction search of large domain knowledge bases. Sophisticated database templating can be required to support both their model-based nature and global control character. The canonical form XV path-planning and canonical form XVI doctrinal templating algorithms presented in Chapter 3, for example, both required potentially large spatially organized domain knowledge bases and demanded sophisticated database search and query support.

Therefore, conventional database management systems generally provide adequate support for typical lower order canonical form problem-solving approaches. On the other hand, the highest order algorithmic forms can employ context-sensitive declarative and procedural knowledge, which can lead to the requirement for sophisticated search and manipulation of extremely large multiple-resolution, spatially organized knowledge bases.

Table 6.4 characterizes the potential *short-term, medium-term,* and *long-term* declarative knowledge base size requirements for both the lowest order and the highest order canonical form algorithms. For tactical data fusion applications, spatially organized long-term declarative knowledge is seen to potentially impose the most significant database storage, retrieval, and manipulation requirements.

Table 6.4
Potential Data-Set Size for Both Low-Order and High-Order Canonical Form Algorithms

	Potential Data-Set Size		
Knowledge Class	*Low-Order Forms*	*High-Order Forms*	*Static/Dynamic*
Short term declarative	Small	Small	Highly dynamic
• Sensor measurement			
• Messages			
Medium term declarative	Moderate	Moderate	Moderately dynamic
• Situation description			
Long term declarative			Static
Non-spatial			
–Rigid templates	Moderate	Low	
–General facts	Low	Moderate	
Spatial			
–Map based	Moderate	High	
–Imagery	Moderate	High	
Procedural			Static
Rules	Small	Moderate	
Procedures	Small	Moderate	

Although performance issues associated with the use of conventional database systems are often accepted as temporary problems that will be corrected by the higher performance of next-generation workstations, hardware improvements alone will not overcome performance shortfalls of several orders of magnitude. A more practical solution appears to be the development of database management systems that provide more effective support to the higher order canonical form algorithms. A database system that is capable of meeting the more stringent requirements of the highest order canonical form approaches will readily satisfy the more modest requirements of the lower order problem-solving approaches.

Thus, the effectiveness and efficiency of data fusion systems can be enhanced by the development of highly robust, context-sensitive, and efficient fusion algorithms that, in turn, are supported by database systems that both *facilitate* algorithm development and *enhance* overall algorithm performance efficiency. The objective of the next several chapters is to investigate the data management principles that underlie effective and efficient support to the higher order canonical form algorithms.

6.4 GENERAL LIMITATIONS OF CLASSES OF DATABASE SYSTEMS

The limitations of current database models are briefly reviewed in this section from the perspective of the critical support role of a database management system in the development and performance of sophisticated, higher order canonical form algorithms.

6.4.1 Relational Model

In general, efficient search requires effective database search dimensions. In a table based representation of personnel data, for example, *employees* within a given department of a large firm are relatively "close" along the indexed search dimension (*employee department* + *employee name*). However, index files that facilitate search for one query may be inappropriate for another query. For instance, the close association of *employees* within a specific *department* is lost if the order of indexing is reversed (*employee name* + *employee department*).

A typical RDBMS representation of vector-based spatial data is shown in Figure 6.13. Although highly effective for indexing single-dimensional data, the RDBMS possesses no inherent capability for indexing two-dimensional spatially organized information. In addition, because vector-based representations do not preserve two-dimensional spatial relationships among different data sets, accessing data that is spatially "near" a point, line, or region feature can lead to extensive test and evaluation operations.

When an object moves to a new location, a new object is discovered, or an old object disappears, the dynamic relationships between this object and other objects within the database are likely to change. The explicit maintenance of the spatial relationships among both static and dynamic database features can entail a considerable computational overhead. As a result, spatial search and association operations for data maintained within a RDBMS tend to be highly inefficient.

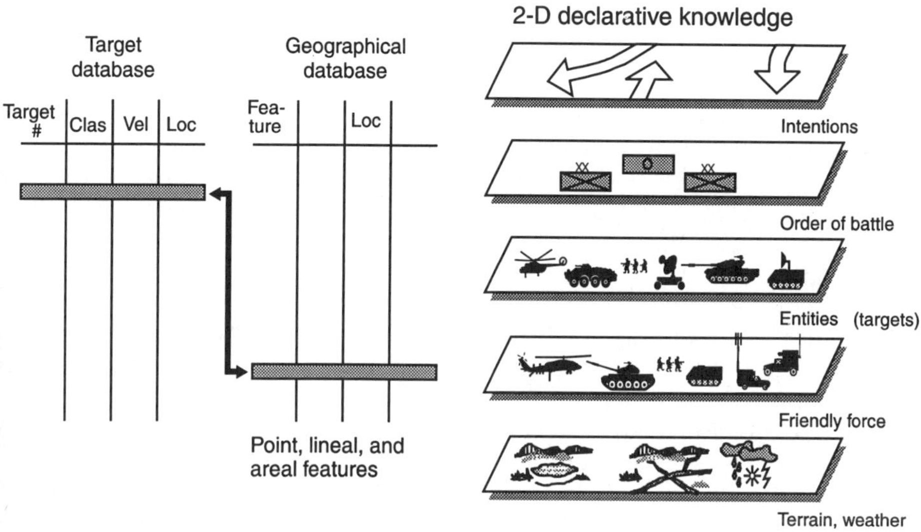

Figure 6.13 Conceptual view of the shortcomings of evaluating spatial relationships using a relational database model.

Rather than attempting to *dynamically maintain explicit* relationships among database elements (for example, as an attribute in a table), a far more efficient approach is to *preserve implicit* relationships among database elements and provide an efficient means of evaluating those relationships, as required. Maps, for instance, naturally preserve both absolute and relative Euclidean distance metrics as well as all *implicit* spatial relationships ("west of," "behind") among spatial features. Distance metrics and spatial relationships are preserved principally because a map maintains spatial data along natural spatial search dimensions.[10]

Thus, at least two critical relational database deficiencies are apparent. The first deficiency is the inability of the relational model to support natural search dimensions for multidimensional data sets (two or three spatial dimensions plus time). A related deficiency is the lack of adequate support for the maintenance of both dynamic and implicit spatial relationships among the data. As a result, the use of a RDBMS to support complex spatial and temporal reasoning can lead to:

1. The generation and maintenance of artificial indexing dimensions;[11]

---

10. Natural spatial search dimensions refer to the use of data structures that mimic conventional analog representations of two- and three-dimensional information. Natural search dimensions are discussed in considerably more detail in the Chapter 7.

11. The creation of a special-purpose linear index dimension such as $(x_i + y_i)$ is an example of an artificial search dimension.

2. Excessive computational requirements to evaluate spatial relationships among both static and dynamic data.

In addition to the above two deficiencies, a number of other limitations of the basic relational model have been widely recognized (for example, [21]). Several relational model extensions have been investigated in an attempt to overcome these shortfalls.

6.4.2 GIS Systems

The two most common spatial representations supported by GIS systems are *raster* and *vector.* The former is a single-resolution true two-dimensional spatial representation, while the latter is a *nontrue two-dimensional* representation.[12] While the former representation tends to facilitate spatial search and preserve spatial relationships, the latter representation does not.

Although commercial GIS products offer sophisticated support for spatial reasoning, current systems do not provide adequate general-purpose database support to replace commercial database management systems. GIS systems that employ a commercial database management system come the closest to providing single-interface support for both spatial and nonspatial reasoning. In general, GIS systems offer only limited support to real-time embedded applications. In addition, current systems have generally focused on the manipulation of static, single level-of-abstraction spatial databases. As a consequence, such systems provide limited support for dynamic data and multiple-resolution reasoning. While GIS systems do overcome many of the deficiencies of the relational model, as shown in Table 6.5, current systems lack one or more of the features that support efficiency across the full spectrum of database requirements.

The critical requirement for database management system support to spatial data is becoming increasingly recognized [22,23]. As a result, there exists considerable interest within the database community to accommodate spatially organized data. The next two chapters focus on some of the key issues associated with the development of more powerful and flexible database models that can overcome some of the deficiencies of both relational database and geographic information systems.

6.4.3 Object-Oriented Representations

While the conventional relational data model supports the representation of only explicit relationships between data (database joins depend on the value of an explicit, static key), object-oriented databases (OODB) can preserve both *explicit* an

---

12. A true two-dimensional representation stores two-dimensional spatial data along natural storag and retrieval dimensions. Thus, a true two-dimensional representation is considered to offer a *na ral representation* for two-dimensional spatially organized data.

Table 6.5
Potential Shortcomings of Current GIS Systems

Deficiency	*Example*
Boolean operations too slow	Intersection among several large, multiply-connected regions
Proximity and inclusion tests are too slow	Nearest road to an arbitrary query point
Do not support efficient spatial search of both classes of features and individual named features	Locating the nearest road may require test of all individual named roads
Implicit spatial relationships among data are not preserved	Spatial search "to the west" of a given line may require exhaustive search and test
Do not handle time-varying data	Spatial data is assumed to be static
Do not support top-down reasoning	Top-down multiple-resolution constraint-satisfaction

implicit relationships among data. As discussed above, due to the underlying hierarchical nature of object representations, a *tank* object inherits generic properties of track-driven military vehicles. In addition, if the tank was known to be located within a particular spatial region during a specific time interval, the tank would inherit the spatial and temporal relationships between the tank and that region. The former inheritance represents *explicit* relationships, while the latter represents *implicit* relationships. In addition, all other targets that were either "within" or "near" this region during the same time interval might be *implicitly* related as well.

Because the spatial relationships among both sensor and non-sensor-derived information must be evaluated before it can be effectively combined, the ability to evaluate *implicit* spatial relationships is a critical database requirement. While an OODB conceptually supports sophisticated higher order canonical form problem-solving approaches, current object databases provide limited direct support for the maintenance, query, or manipulation of spatial objects. In addition, current systems may not provide adequate performance for many time-critical applications.

6.5 CONCLUSIONS

The lack of effective and efficient database support for spatial, temporal, and hierarchical reasoning represents a significant barrier to both the development and performance of sophisticated data fusion algorithms. Because of the inherent linear indexing capability of relational database management systems, such systems provide ineffective search and manipulation of potentially massive spatially organized domain databases. Although GIS systems do provide direct support to spatial reasoning, such systems generally have not been designed to provide high-performance database services. Although interface code can be written to "lash" together a conventional database system for storing and retrieving nonspatial data and a GIS system for storing and retrieving spatial data, the resultant product might exhibit significant performance problems.

Rather than patching together existing systems that were independently optimized for different data classes, the top-down development of a fully integrated spatial and nonspatial database system has a number of significant advantages. The next chapter outlines such a design approach. Based on these design recommendations, Chapter 8 describes the development of a database kernel that can potentially simplify application development as well as enhance fusion system performance.

References

[1] Sedgewick, R., *Algorithms*, Second Edition, Reading, MA: Addison Wesley, 1988.

[2] Lawler, E.L., *The Traveling Salesman Problem: A Guided Tour of Combinatorial Optimization*, New York: John Wiley & Sons, 1987.

[3] Cronin, T.M., "The Voronoi Diagram for the Euclidean Traveling Salesman Problem is Piecewise Quartic and Hyperbolic," *Trans. Eighth Army Conf. on Applied Mathematics and Computing*, Cornell University, Ithaca, NY, June 1990.

[4] Chen, P.P., "The Entity-Relationship Model: Towards a Unified Model of Data," *ACM Trans. on Database Systems*, Vol. 1, No. 1, March 1976.

[5] Sowa, J. (ed.), *Principles of Semantic Networks: Exploration in the Representation of Knowledge*, San Mateo, CA: Morgan-Kaufmann, 1991.

[6] Codd, E.F., "A Relational Model of Data for Large Shared Data Bases," *Comm. ACM*, Vol. 13, No. 6, June 1970, pp. 377–387.

[7] Date, C.J., *Relational Database: Selected Writings*, Reading, MA: Addison-Wesley, 1986.

[8] Tomlin, C.D., *Geographic Information Systems and Cartographic Modeling*, Englewood Cliffs, NJ: Prentice-Hall,1990.

[9] *1993 International GIS Source Book*, GIS World, Inc., Fort Collins, CO, 1993.

[10] Guttman, A., "R-trees: A Dynamic Index Structure for Spatial Searching," *Proc. 1984 SIGMOD Int. Conf. on Management of Data*, Boston, MA, June 1984, pp. 47–57.

[11] Aref., W.G., and H. Samet, "Extending a DBMS with Spatial Operations," *Second Symp. on Advances in Spatial Databases*, SSD'91, O. Gunther and H.J. Schek (eds.), 1991; Lecture Notes in Computer Science, Vol. 525, Berlin: Springer-Verlag, 1991, pp. 299–318.

[12] Guting, R.H., "GRAL: An Extensible Relational System for Geometric Applications," *Proc. Fifteenth Int. Conf. on Very Large Databases (VLDB)*, P.M.G. Apers and G. Wiederhold (eds.), Amersterdam, August 1989, pp. 33–44.

[13] Ooi, B.C., *Efficient Query Processing for Geographic Information Systems*, Ph.D. thesis, Monash University, Victoria, Australia, 1988.

[14] Orenstein, J.A., and F.A. Manola, "PROBE Spatial Data Modeling and Query Processing in an Image Database Application," *IEEE Trans. Software Engineering*, Vol. 14, No. 5, May 1988, pp. 611–629.

[15] Aref, W.G., and H. Samet, "Extending a DBMS with Spatial Operations," *Advances in Spatial Databases*, O. Gunther and H.-J. Schek (eds.), 1991, pp. 299–318.

[16] Waugh, T.C., and R.G. Healy, "The GEOVIEW Design: A Relational Data Base Approach to Geographical Data Handling," *Int. J. Geographic Information Systems*, Vol. 1, No. 2 , April–June 1987, pp. 101–118.

[17] Wolf, A., "The DASDBS GEO-Kernel: Concepts, Experiences and the Second Step," *Design and Implementation of Large Spatial Databases*, A. Buchmann, O. Günther, T.R. Smith, and Y.-F. Wang (eds.), Lecture Notes in Computer Science, Vol. 409, Berlin: Springer-Verlag, 1990, pp. 67–88.

[18] Vijlbrief, T., and P. van Oosterom, "The GEO++ system: an extensible GIS", *Proc. Fourth Int. Symp. on Spatial Data Handling*, Volume 1, Charleston, SC, August 1992, pp.40–50.

[19] Shafer, C.A., H. Samet, and R.C. Nelson, "QUILT: A Geographic Information System Based on Quadtrees," *Int. J. Geographical Information Systems*, Vol. 4, No. 2, April–June 1990 , pp. 103–131.

[20] Unland, R., and G. Schlageter, "Object-Oriented Database Systems," *Expert Database Systems*, K. Jeffery (ed.), New York: Academic Press, 1992.

[21] Codd, E.F., "Extending the Database Relational Model to Capture More Meaning," *ACM Trans. on Database Systems*, Vol. 4, No. 4, December 1979.

[22] Healey, R.G., "Database Management Systems," *Geographical Information Systems: Principles and Applications*, D.J. Maguire, M.F. Goodchild, and D. Rhind (eds.), Essex: Longman Scientific and Technical and New York: Wiley & Sons, Vol. 1, 1991, pp. 251–267.

[23] Frank, A., "Requirements for a Database Management System for a GIS," *Photogrammetric Engineering and Remote Sensing*, Vol. 54, 1988, pp. 1,557–1,564.

CHAPTER 7

Database Design Considerations

7.1 INTRODUCTION

Human analysts routinely fuse sensor-derived information using sophisticated reasoning models that are dynamically conditioned on relevant context. A radar operator analyzing detections from a target that appears to be following a road would naturally expect the target to continue to exhibit a road-following behavior. However, if the target subsequently appeared to be moving away from the road, the analyst might switch behavior models to a possible "minimum mobility resistance" or "goal-seeking strategy" track extension model. If rain began falling at 2 inches per hour in the vicinity of the off-road target, the analyst would once again adapt his expectations of future target behavior based on the local soil type, surface conditions, foliage, and slope of the region. Valid hypotheses would be expected to be supported by subsequent sensor measurements, while invalid hypotheses would tend to be either unsupported or refuted.

Although humans are adept at reasoning in context, automated data fusion algorithms have historically relied on relatively rigid, context-free algorithms. As a result, data fusion algorithms have not been nearly as flexible, adaptable, or context-sensitive as manual analysis approaches. Yet many fusion tasks demand such a capability. For example, determining the "closest" river-crossing site from a given location on a map is inherently highly context-sensitive. Figure 7.1 shows the two "closest" river-crossing sites from point *X*. As can be observed, the location that is the *furthest* Euclidean distance from *X* requires the *shortest* travel distance. In addition, candidate locations may need to satisfy numerous other mobility criterion, such as the ease of ingress and egress from the river bank, recent rainfall amount, season, river-flow rate, and bank height. Thus, the development of a river-crossing site-selection algorithm can require the evaluation of multiple candidate plans under a wide range of both static and dynamic problem-solving constraints.

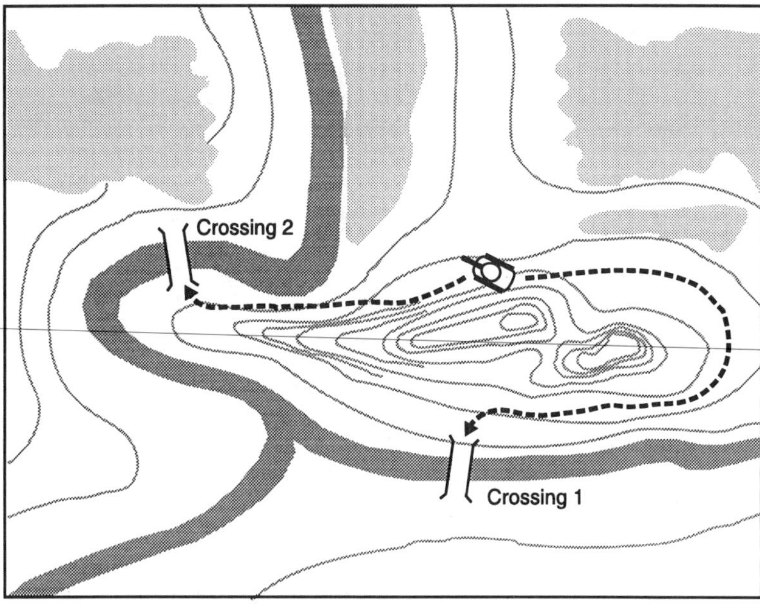

Figure 7.1 Example showing that bridge-crossing location 2 entails a shorter travel distance than the "closest" (Euclidean distance) bridge-crossing location 1.

The fact that both the target-tracking and path-planning algorithms just sketched are both *intuitive and robust* is not really coincidental. The algorithms are robust because they are based on sophisticated, yet highly intuitive, behavioral models that are sensitive to relevant domain characteristics and constraints. In fact highly intuitive algorithms and robust performance tend to go hand-in-hand.

As we have seen, data fusion algorithms employ one or more of the four classes of knowledge defined in Chapter 2: *short-term declarative, medium-term declarative, long-term declarative* , and *long-term procedural* knowledge. Both *short-term* and *medium-term* knowledge are inherently context-sensitive. As an example of context sensitive reasoning, consider the decision process used by a state trooper at a highway speed-trap. In deciding whether to cite a particular driver that is exceeding the posted speed limit, the officer is likely to consider a wide range of general contextual information, including the current weather conditions and the general traffic-flow rate. In addition to general contextual knowledge, specific contextual information might be considered as well. A speeding car that was recklessly changing lanes, for instance, would be more likely to be stopped than one that was merely keeping up with the traffic flow. On the other hand, if the officer was aware of a nearby multiple-alarm fire and the apparently reckless driver had volunteer fire department plates, the officer might choose to escort the vehicle to the fire. Thus, con textual information can be critical to the analysis of sensor-derived information. While a police officer can readily observe and analyze the relevant contextual knowledge in the speed-trap scenario, in remote sensing applications, context must be established using a combination of sensors (meteorological, imagery) and non-sensor-derived information sources (geographic and cultural feature databases).

The proper interpretation of context-sensitive short-term and medium-term knowledge virtually demands the use of context-sensitive algorithms. Context-sensitive algorithms can employ either *context-specific* or *context-general* knowledge. For relatively simple problems, *context-specific* algorithms can be entirely adequate. However, as the problem complexity increases, dependence on context-specific (or, equivalently, context-embedded) algorithms leads to the development of a combinatoric number of algorithm variants in an attempt to accommodate all possible context states.

The development of a set of mobility maps, for example, that represent a spectrum of specific context conditions (rain/no rain, summer/winter, below freezing/above freezing, wheeled/tracked vehicle) can lead to massive storage requirements. Since context-specific mobility maps depict only nominal conditions, algorithms based on such knowledge bases tend to be inherently brittle. Because context-specific knowledge is analogous to a set of rigid templates, the same arguments offered in Part I against dependence on rigid templating paradigms applies to the use of *context-specific* reasoning paradigms. In addition, because they cannot be readily adapted to domain changes or to support entirely new applications, context-specific algorithms tend to be inherently nonextensible.

Rather than attempting to construct a set of algorithms that apply to a range of narrowly defined context conditions, a much more powerful approach is to develop *context-general* algorithms that are implicitly sensitive to the current context conditions. In addition to enhanced robustness, such algorithms tend to be much more extensible. The taxonomy of knowledge developed in Chapter 2 (Figure 2.7) can be extended to include the four classes of context sensitivity, as shown in Figure 7.2.

Database maintenance includes those operations that are required in order to maintain the database over time. In general, database maintenance is enhanced by the logical separation of the four classes of context sensitivity within the knowledge base. Since *context-free* knowledge is static and insensitive to domain context, no maintenance is required as a result of changes in the domain. *Generic knowledge* is highly stable knowledge that rarely changes. *Context-general* knowledge is generally easy to extend and requires little maintenance. Thus, three of the four classes tend to be rather stable forms of knowledge that do not impose significant database maintenance requirements. *Context-specific* knowledge, on the other hand, may require extensive modifications to accommodate time-varying behavior, changes in the problem domain, or to support entirely new context conditions. Table 7.1 summarizes the major *algorithm extensibility* and *database maintainability* issues associated with the four classes of context-sensitive knowledge.

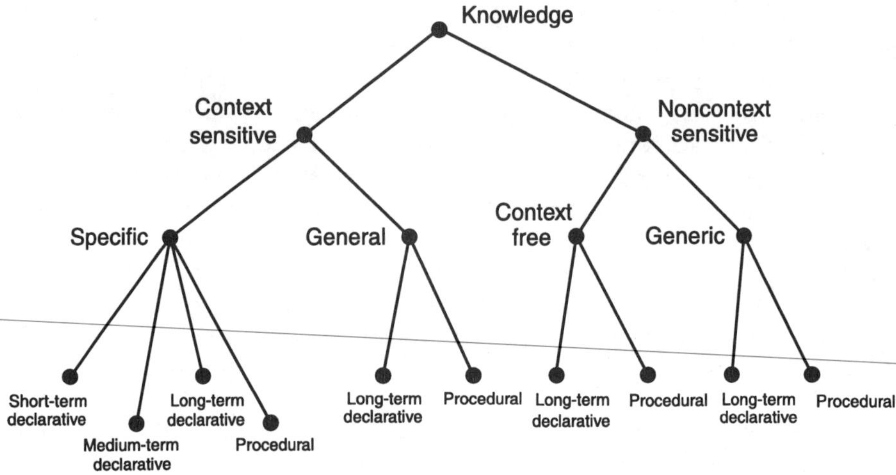

Figure 7.2 Taxonomy of knowledge context sensitivity.

Table 7.1
Algorithm Extensibility and Database Maintainability Issues associated with Four Classes of Context Sensitivity

Knowledge	*Algorithm Extensibility Issues*	*Database Maintainability Issues*
Context-specific	Knowledge must be changed to account for different contexts	Many variants of same basic knowledge may be required
Context-general	Relatively easy to extend	Minimal maintenance requirements
Context-free	May not extend well	Minimal maintenance requirements
Generic	No extensibility issues	Large storage requirements; no maintainability issues

As has been demonstrated, context-sensitive reasoning can require the maintenance and search of massive domain knowledge bases. With relatively small data sets and state-of-the-art hardware, performance problems associated with potential inefficient algorithms and inefficient algorithm/database interaction often go unnoticed. However, as indicated during the discussion of the traveling salesman problem in the last chapter, as problem sizes are scaled upward, even seemingly trivial tasks can generate potentially enormous search and problem-solving spaces. Thus, for real-world problems that demand both sophisticated algorithms and large, spatially organized domain databases, control of the computational requirements and search-space size is of considerable importance.

7.2 FUNDAMENTAL APPROACHES TO DATABASE DESIGN

The goal of this chapter is to identify appropriate design criteria for developing a maintainable database kernel that is capable of supporting the higher order canonical form algorithms. In this section, three basic database design approaches are discussed as shown in Figure 7.3. A *query-product* database form maintains pre-established answers to all permitted queries. A *single-representation, on-line query* database form evaluates queries on-line using data stored within a single data structure. Finally, a *multiple-representation, on-line query* database form stores data using multiple data structures and answers queries by on-line processing.

The advantages and disadvantages of each database form are illustrated by the following database query:

$$\text{Find the closest } \textit{feature}\,1 \text{ to } \textit{feature}\,2. \qquad 7.1$$

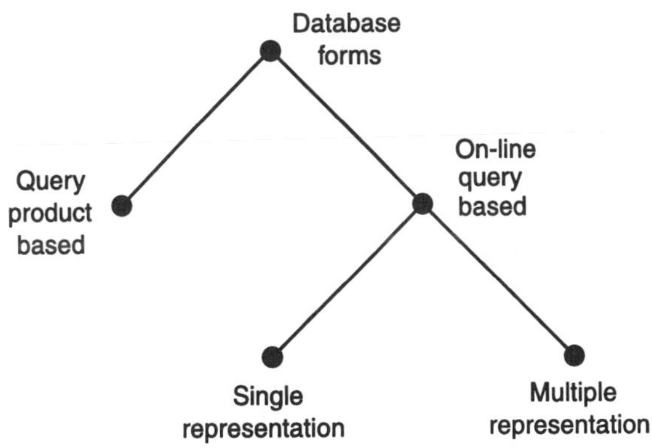

Figure 7.3 Three basic database forms.

In a *query-product*-based database, a table (or equivalent representation) developed during preprocessing maintains the answer to all queries. Table 7.2 offers a simple example of a *query-product* database. Since insertions, deletions, and changes to the database can require extensive table rebuilding, "table-lookup"-based approaches tend to be most appropriate for relatively small, static databases that service specific, well-defined queries.

Table 7.2
Example of Query-Product-Based Database Organization

	Closest City	*Closest Airport*	*Closest Reservoir*
City 1	City 10	Airport 103	Lake 62
City 2	City 38	Airport 31	Lake 905
City 3	City 19	Airport 11	Lake 92
•	•	•	•

In a *single-representation, on-line query* database, rather than maintaining all precomputed answers to queries, only the underlying data sets are maintained. Queries are evaluated on-line using sorted data sets or index files. Since only the affected data sets require resorting or reindexing, this database form is more appropriate for dynamic data than a *query-product*-based database. Table 7.3 depicts the *single-representation, on-line query* database representation of the data shown in Table 7.2.

Table 7.3
Example of a Single-Representation, On-Line Query–Based Database Organization

Cities	*Location*	*Airports*	*Location*	*Reservoirs*	*Location*
City 1	(x_1, y_1)	Airport 1	(x_{100}, y_{100})	Lake 1	(x_{200}, y_{200})
City 2	(x_2, y_2)	Airport 2	(x_{101}, y_{101})	Lake 2	(x_{201}, y_{201})
City 3	(x_3, y_3)	Airport 3	(x_{102}, y_{103})	Lake 3	(x_{202}, y_{203})
•	•	•	•	•	•

A *query-product* database form requires exhaustive preprocessing but offers rapid query response for highly circumscripted queries. A *single-representation, on-line query* database supports more flexible queries and requires less preprocessing but can require extensive search and computation during query evaluation. Although this latter database form can support dynamic data, extensive reindexing operations can be required for certain classes of data (for example, spatial data). The memory requirements for a *query-product* database and the computational requirements for a *single-representation, on-line query* database approach tend to increase exponentially with the data-set size.

As opposed to the single-representation approaches of the previous two database forms, a *multiple-representation, on-line query* database form permits the use of a wide range of data storage structures that can be matched to the various data types. For instance, table-based information could be stored in tables, hierarchical data in tree-based representations, two-dimensional spatial data in inherently two-dimensional data structures, video images as sequences of two-dimensional images, audio as compressed time-series signals, and surfaces in inherently three-dimensional data structures.

As mentioned earlier, data structures that inherently match the character of the information to be represented are referred to as *natural data representations.* A representation that is "natural" for one class of data might be ill-suited for another class of data. For personnel records, inventory records, and Tables of Equipment, a table-based data organization is quite natural. On the other hand, as discussed in the last chapter, a table-based data model exhibits a number of deficiencies for two- and three-dimensional data. In the case of personnel data, the database fields represent relatively *independent* attributes, whereas in the case of two-dimensional spatial data, the two-dimensional attributes (x and y) are highly *dependent.* Figure 7.4 depicts a spatially organized representation of the sample database. Just as in the second database form, the data are stored directly and queries are evaluated on-line.

Although the first two database forms are well supported by a RDBMS, the relational model provides inadequate support for the third database form. Object-oriented databases, on the other hand, directly support this latter database organization. A *multiple-representation, on-line query*–based database can potentially maintain natural search dimensions as well as preserve important relationships among data sets. A *single-representation, on-line query*–based database form subsumes a *query-product*-based form. A *multiple-representation, on-line query*–based database form, in turn, subsumes a *single-representation, on-line query*–based form.

7.3 DESIGN ISSUES

Although there exist numerous design criteria and functional requirements (integrity checking, lock-out) that must be considered when developing a database management system, we restrict our attention to five relatively high-level issues that are directly related to algorithm development and algorithm performance. In general, data fusion automation benefits from a database that:

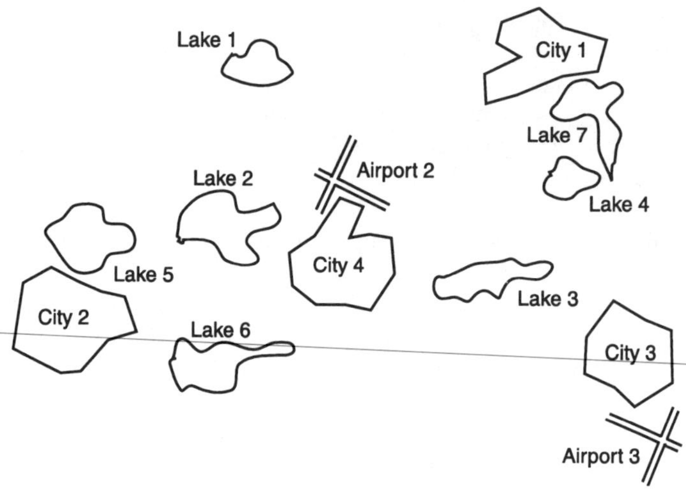

Figure 7.4 Spatially organized representation of airports, cities, and water reservoirs.

1. Employs natural data representations to facilitate the development of highly intuitive algorithms that achieve robust, context-sensitive performance;
2. Employs representations that enhance algorithm efficiency;
3. Provides high-accuracy data representations;
4. Ensures efficient data storage;
5. Achieves efficient performance.

Due to the importance and potentially massive size of spatially organized knowledge, this chapter focuses on database support to spatial representation and reasoning. Each of the above issues are now discussed.

Support for Intuitive Algorithm Development. The use of *natural* data representations can support the development of highly intuitive algorithms. A table-of-equipment database is naturally stored in a table-based data structure. An order-of-battle database is naturally stored in a hierarchical data structure. A true two-dimensional representation of images supports the development of much more intuitive image processing than do list-oriented spatial data structures. Because fusion algorithms can require the use of a wide range of data classes, only a *multiple-representation, on line query* database provides adequate support.

In addition to data representations that are matched to data types, algorithm development can be enhanced by a database that provides for the full integration among the various representations. Since a *river* consists of both semantic attributes (*class, width, flow rate*) *and* spatial attributes (*lineal* and/or *region representations*), object representations that fully integrate both semantic and spatial representations further facilitate the development of highly intuitive algorithms.

Support for High-Efficiency Algorithms. Data representation can have a significant impact on problem-solving efficiency. While a signal-analysis algorithm that directly operates on a time-series signal might require computationally demanding matrix inversions, an algorithm that operates on the "dual" frequency domain representation may require much more computationally efficient algorithms. An algorithm that must locate "all roads that have an east-west orientation" and that are "within 5 km of a particular city" would be potentially more efficient if the road database employed a true two-dimensional spatial representation rather than a vector-based representation.

Support for Data Representation Accuracy. In general, data *representation accuracy* must be adequate to support the full range of data fusion applications. For finite-resolution representations, accuracy is related to the form of data *sampling* that is employed. In general, sampling can be either uniform or nonuniform. Nonuniformly sampled numerical data are normally represented as floating point numbers, while uniformly sampled data are normally represented as integers. For a fixed memory size, *nonuniformly sampled* representations normally provide higher accuracy than *uniformly sampled* representations. A vector-boundary representation of a region, for instance, is a nonuniformly sampled representation that maintains the vertices of the piecewise linear boundary as a floating point list. A pixel-based representation of a region boundary, on the other hand, represents a uniform, integer-based sampling. For a given region, a vector-boundary representation offers both potentially higher accuracy and lower memory requirements than a pixel-boundary representation of the same region.

Data Storage Efficiency. The storage requirements of alternative spatial data representations can vary considerably. *Memory efficiency* requires either inherently compact representations or the use of some form of data compression. As shown in Figure 7.5, the storage requirements for a simply connected region (a closed region that contains no holes) can range from rather small for a boundary-only polygon representation to rather large for a high-resolution raster-based representation that maintains the boundary plus the enclosed area. In addition to the memory requirements for the data itself, pointers, index files, and other overhead support structures can be required that increase the overall memory requirements. The primary memory efficiency considerations are summarized in Figure 7.6

Database Performance Efficiency. In order to provide effective support to sophisticated, context-sensitive, real-time algorithms, adequate *database performance efficiency* is required. Ooi [1] recommends that a database intended to support spatial reasoning should possess the following three primary characteristics:

1. Spatial indexing that provides efficient access based on proximity;
2. Spatial indexing that permits efficient evaluation of spatial constraints;

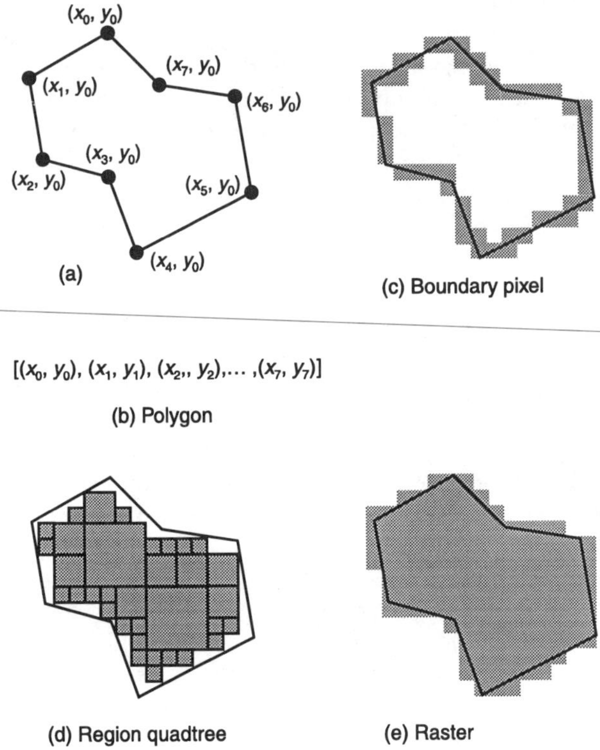

Figure 7.5 Alternative representations of the singly connected region in (a).

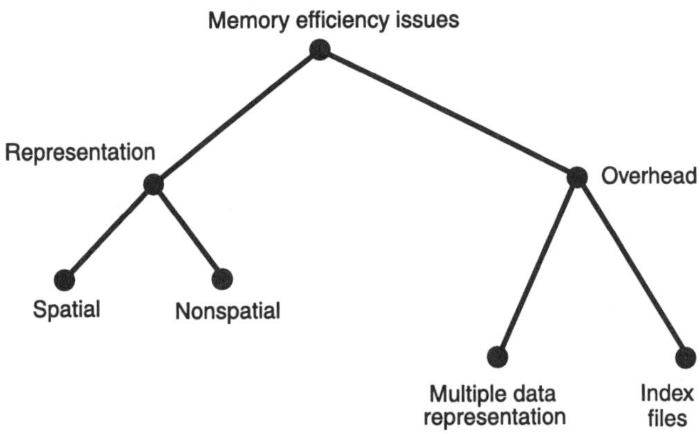

Figure 7.6 Memory efficiency issues.

3. Efficient optimization on both *spatial* and *aspatial* selections.

These considerations are addressed with respect to the five broad efficiency classes that are now discussed.

Overhead Efficiency

Overhead efficiency includes both *indexing efficiency* and *data maintenance efficiency.* Indexing efficiency is associated with operations that are required to both create and preserve required logical search dimensions (reindexing, reorganization of pointers), while data maintenance efficiency includes physical data structure–dependent operations that can be required following insertions, deletions, and updates (tree rebalancing, table reconstruction, node splitting, and node collapsing). Many forms of data possess natural representations that, if preserved, eliminate the need for explicit indexing structures. When spatial data are maintained in a true two-dimensional spatial representation, for example, the representation inherently preserves the most natural data-search dimensions. As a direct consequence, such representations do not require extensive overhead operations following database updates. Although overhead efficiency is a relatively minor issue for static databases, it can be a major issue for dynamic data sets. The primary overhead efficiency considerations are summarized in Figure 7.7.

Search Efficiency

Search efficiency requires effective control of the search-space size for the full range of typical database queries. A key aspect of search efficiency is the storage of

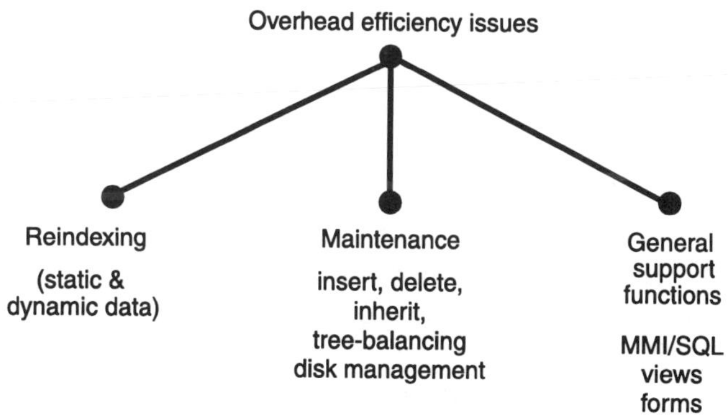

Figure 7.7 Overhead efficiency issues.

data along effective search dimensions. For a relational database, index files facilitate access to data that are organized along an arbitrary linear search dimension. Many kinds of data possess a natural representation that, if preserved, facilitates search. True two-dimensional spatial representations, for instance, support highly focused spatial search. Multiple-resolution representations can further improve search efficiency by permitting efficient tree-oriented search. The primary search efficiency considerations are summarized in Figure 7.8.

Association Efficiency

Association efficiency requires the efficient development of relationships among data sets. Although a relational database can *join* two tables that are related through a static explicit key (for example, employee social security number), the relational model does not inherently support joins among data sets that are implicitly related by their combined spatial and temporal proximity, for example. Natural two-dimensional spatial data representations tend to be more efficient in assessing spatial proximity or inclusion than representations that do not preserve the inherent organizational properties of the data. The primary association efficiency considerations are summarized in Figure 7.9.

Query Efficiency

Query efficiency requires both *set operation efficiency* as well as *complex clause evaluation efficiency*. For the applications of principal interest in this book, *set operation efficiency*

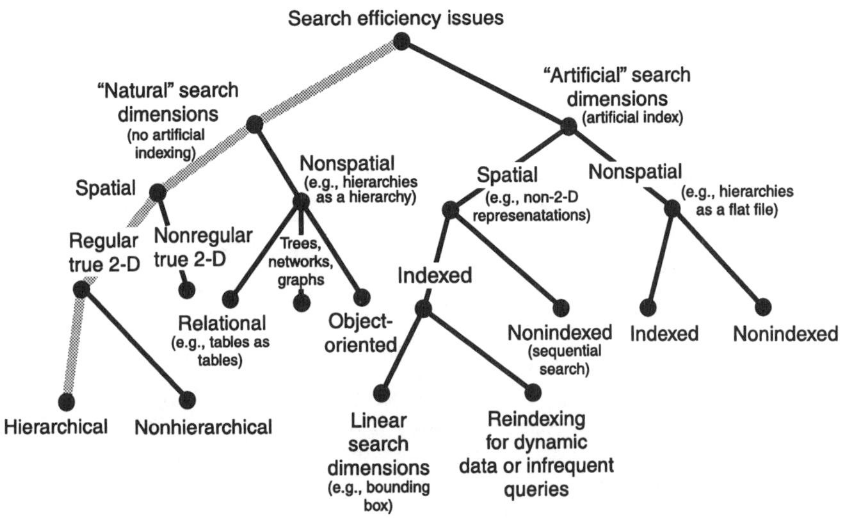

Figure 7.8 Search efficiency issues.

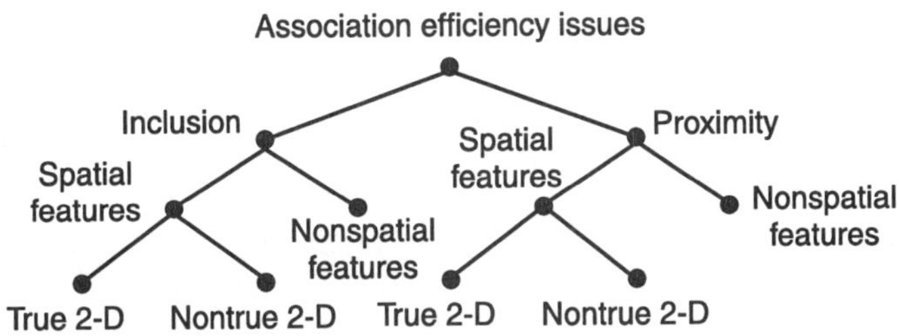

Figure 7.9 Association efficiency issues.

demands efficient Boolean and fuzzy set operations among spatial point, line, and region features. *Complex clause evaluation efficiency* requires query optimization for compound query clauses that may include mixed spatial and semantic constraints. The primary query efficiency considerations are summarized in Figure 7.10.

Implementation Efficiency

Implementation efficiency is enhanced by employing a system architecture and data structures that support the effective distribution of data, processing, *and* control.

Because significant dependencies exist among the above efficiency issues, no single spatial representation is equally effective across all the above database design requirements. For example, while boundary-only representations of regions are

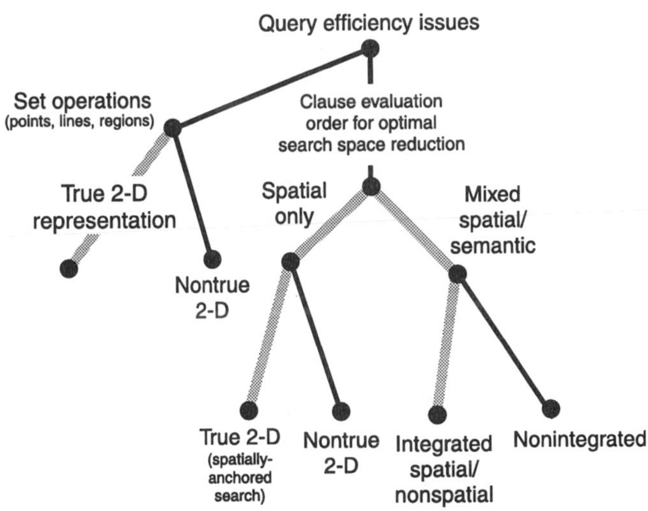

Figure 7.10 Complex query efficiency issues.

highly memory-efficient, set operations among multiply-connected regions tend to be computationally expensive. Therefore, the development of an "optimal" database system requires an effective *compromise* among the above design objectives. Figure 7.11 provides a high-level view of these design considerations.

Under restricted conditions, a *query-product-based* database supports *search efficiency* and a *single-resolution, on-line query* database supports *memory efficiency*. However, only a *multiple-representation, on-line query* database permits an effective compromise across all the issues for all relevant data types. Therefore, as opposed to simply supporting multiple data types, an "optimal" database design must effectively integrate the multiple data types. In the next section, *data representation* is shown to be the principal thread that ties these database design considerations together.

7.4 DATABASE REPRESENTATION ISSUES

Because *object-oriented reasoning* supports the development of complex information processing systems, *object-oriented databases* can be expected to offer significant support to the development of complex applications. In order to optimize an object database with respect to the design issues just presented, the notion of an abstract object must be "fleshed out." This section identifies specific data representation characteristics that achieve the overall database design objectives.

In general, since no single data representation is equally effective for all data classes (as is required by the relational model, for instance), each distinct data class should be permitted to possess a specialized object representation. Because data fu-

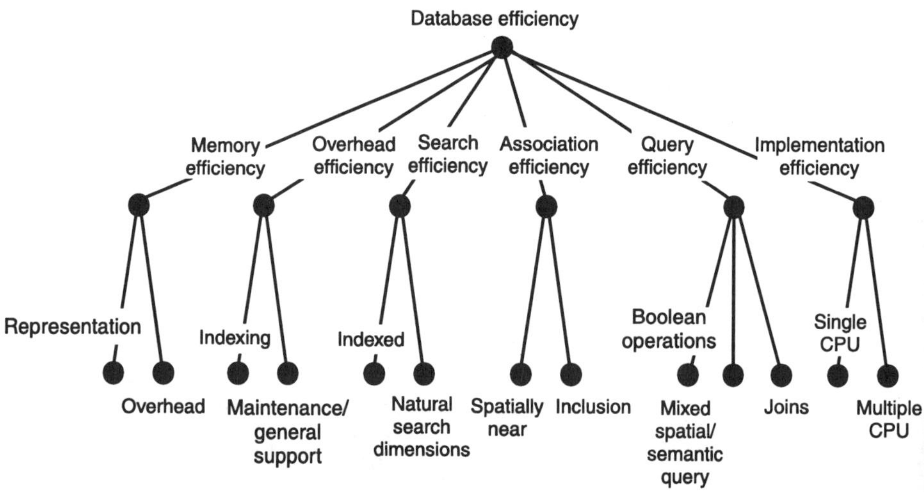

Figure 7.11 Top-down database design considerations.

ion algorithms rely on spatial, hierarchical, and temporal reasoning, the efficiency of specific algorithms can be enhanced by data representations (or object representations) that facilitate storage of and access to spatially, hierarchically, and temporally organized information. Properly designed spatial data structures tend to enhance the access to and manipulation of spatially organized data. Hierarchical representations tend to be both natural and highly efficient for complex, multiple level-of-abstraction problems that possess extensive parent/child relationships or that lend themselves to global problem-solving approaches. As argued previously, temporal reasoning can be supported by maintaining appropriate time-stamps on objects and their attributes.

.4.1 Nonspatial Declarative Data

Nonspatial declarative knowledge can be represented as *n*-tuples, arrays, tables, transfer functions, frames, trees, and graphs. As suggested in the last chapter, although nonoptimal for hierarchical data, the relational model adequately supports the representation of many classes of nonspatial data. Conventional object-oriented representations, likewise, offer excellent support to nonspatially organized information.

.4.2 Spatial Declarative Data

Whether performed manually or automated, data fusion requires the analysis of many classes of data and knowledge. However, none is as pervasive or as extensive as spatially organized information. In addition, none can be as critical to the development of robust and context-sensitive data fusion algorithms. Thus, *data storage, search, overhead, association, query,* and *implementation efficiency* issues for spatial data must be critical elements of any database design.

In this section, spatial data structures are discussed with respect to the spatial representation taxonomy shown in Figure 7.12. Although alternative taxonomical representations can be developed, this particular form reveals important relationships between data representation characteristics and the aforementioned database design issues. At the highest level of abstraction, a spatial representation is considered to be either *finite resolution* (*discrete*) or *infinite resolution* (*continuous*). Functional representations, analog images, and other similar classes of spatial information are continuous resolution representations. Only finite resolution (or sampled) representations of spatial data are considered in the following discussion.

Sampled representations of two-dimensional space can be based on either regular or nonregular decompositions. *Regular* decompositions provide a uniform, data-independent representation, while *nonregular* decompositions provide a nonuniform, *data-dependent* representation. At a low resolution, spatial features tend to be relatively uniformly distributed, while at a high resolution, spatial data tends to be nonuniformly distributed. Thus, *regular* decompositions are most appropriate for maintaining relatively low-resolution spatial representations. Because of their uni-

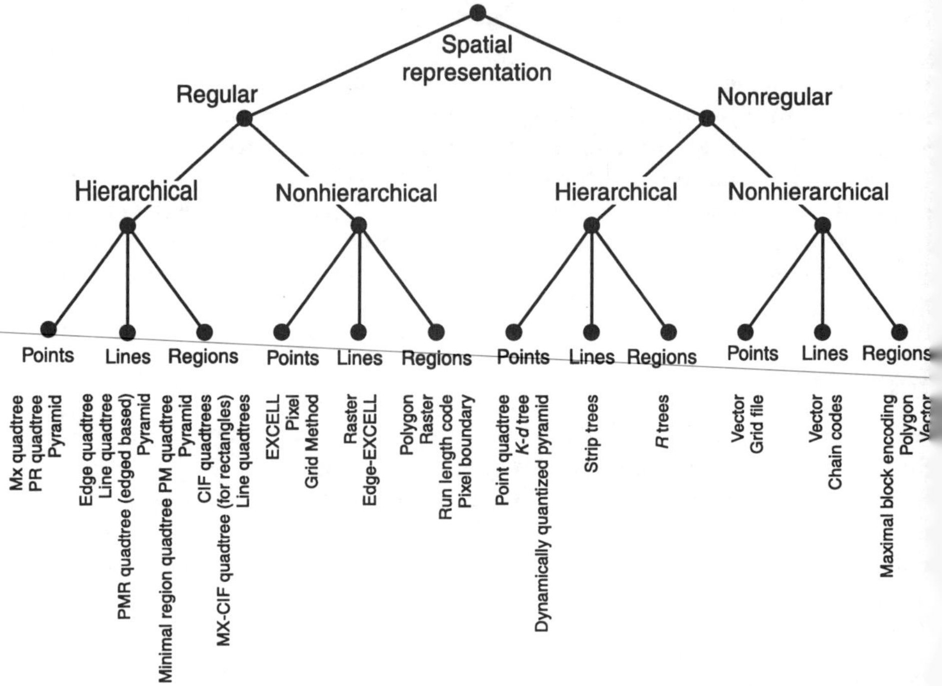

Figure 7.12 Taxonomy of representative spatial data structures.

formly sampled character, regular decompositions naturally support registratio among data sets. In addition, as we will see, regular representations can support ef cient spatial indexing. *Nonregular,* data-dependent decompositions (nonuniform sampled representations), on the other hand, offer much more memory-efficie representations for nonuniformly distributed high-resolution spatial data.

Hierarchical decompositions of two-dimensional spatial data support the conti uum from highly global to highly local spatial problem solving. For instanc supported by a multiple-resolution spatial representation, the closest road to a pa ticular (x, y) coordinate can be efficiently located by region-growing the point $(x,$ in large stepsizes until an intersection occurs and then refining the intersection ι ing progressively higher resolution region contraction or growth.

Areal-based representations store two-dimensional regions by their area (the boundary *plus* interior area), while nonareal-based representations explicitly sto only region boundaries. Because set operations among two-dimensional spatially (ganized data sets inherently involve both the region interior, as well as its bounda areal-based representations tend to support efficient set operations among classes of spatial features. Efficiency is especially important for set operations amo large multiply-connected and multiply-embedded regions.

Explicit data representations literally depict a set of data, while *implicit* data rep-esentations store information that permits the reconstruction of the data set. For nstance, while an implicit data representation of a lineal feature might maintain only the endpoint pairs of individual line segments (vector representation), an ex-plicit data representation would store the associated line segments (raster-based lin-eal representation). In general, implicit representations tend to be considerably more memory-efficient than explicit representations.

Specific feature–based spatial representations maintain individual point, line, and egion features, while *composite feature–based* spatial representations store the pres-ence or absence of multiple features. As an example, a raster-based representation hat stores the value of all features in a given spatial cell is a *composite spatial repre-entation,* while a pixel boundary list for a particular region is a *specific spatial repre-entation.* A specific spatial representation tends to be most effective for performing operations with respect to individual features, while a composite representation ends to be more effective for supporting spatial search with respect to classes of eatures.

Finally, *two-dimensional representations* are those data decompositions that pre-erve the inherent two-dimensional character of spatial data. Such data structures rovide natural spatial-search dimensions that can eliminate the need for artificial ndexing. In addition, two-dimensional representations preserve Euclidean distance netrics and other spatial relationships. A raster representation is a *two-dimensional patial representation,* while list-oriented representations of lineal features and regions re inherently *non-two-dimensional spatial representations.*

Data representations that are both *regular and hierarchical* support efficient ee-oriented search. Since regular decompositions have a fixed grid size at each evel of the decomposition and a fixed resolution relationship between levels, the esulting fully registered data sets can be readily associated at all resolution levels.

Fully registered data representations that are both *regular* and *areal-based* sup-ort efficient set operations among all classes of two-dimensional spatial features. ecause raster representation is both regular and areal-based, it supports efficient t-operation generation.

Spatial representations that are both *regular* and *two-dimensional* preserve spa-al adjacency among all classes of two-dimensional spatial features. Regular, two-di-ensional representations do not require either (1) extensive database-rebuilding perations following insertions and deletions (because they use a data-independent, gular decomposition), or (2) reindexing (because no artificial indexing is re-uired). Thus, regular, two-dimensional representations support efficient database aintenance operations. Because they do not require extensive maintenance opera-ons, spatial decompositions that are both *regular* and *two-dimensional* tend to be in-erently *dynamic data structures.* Since extensive modifications can be required llowing spatially local changes, representations that are either nonregular or non-o-dimensional, or both nonregular and non-two-dimensional are relatively *static ta structures.*

Although it might seem appropriate to store static data using inherently static data structures and dynamic data using inherently dynamic data structures, because data fusion requires efficient composition among both dynamic (short-term knowledge, medium-term knowledge) and static (long-term knowledge) data sets, the efficiency of such composition operations is enhanced by employing compatible data representation forms. For instance, if the static road network is maintained in a nonregular representation and the dynamic target track files are maintained in a regular two-dimensional data structure, association efficiency would be degraded due to the disparate nature of the representations. The use of relatively dynamic spatial data structures for both static and dynamic information can enhance both *database association efficiency* and *database maintainability*.

Spatial representations that are *explicit, regular, areal, and two-dimensional* support efficient set operations, offer natural search dimensions among fully registered point, line, and region data, as well as represent relatively dynamic data structures. Because they possess the key characteristics of analog two-dimensional spatial representations, such as paper maps or photographs, such representations are defined to be *true two-dimensional spatial representations. Nontrue two-dimensional spatial representations,* on the other hand, possess one or more of the following properties:

1. Spatial data is not stored along natural search dimensions (for example, list- or table-based representations).
2. Regions are maintained in nonareal-based representations (for example, as bounding polygons).
3. Nonregular decompositions are employed.

True two-dimensional spatial representations tend to support both intuitive algorithm development and efficient algorithm performance. For *nontrue two-dimensional* spatial representations, set operations among complex, multiply-connected regions can require extensive search-and-test of large portions of the database. Although the association of bounding rectangles with all lineal and region features can significantly improve search efficiency, their use offers ineffective search-space control for multiply-connected regions and extended lineal features, such as roads, rivers, and elevated or buried utility lines.

Specific spatial representations that are *explicit, regular, areal, two-dimensional, and hierarchical* (or *hierarchical true two-dimensional*) support highly efficient top-down spatial search and top-down areal-based set operations among *specific* spatial features. In a nonhierarchically organized true two-dimensional spatial database, on the other hand, the search-space size tends to be exponentially related to the resolution size. Finally, *composite spatial representations* that are *explicit, regular, areal, two-dimensional, and hierarchical* support efficient top-down spatial search among *classes* of spatial features.

A number of generalizations can be offered based on the relationships between the database efficiency issues and the classes of spatial data decomposition.

1. For problems that require relatively low-resolution spatial reasoning, *algorithm efficiency* tends to be enhanced by hierarchical, true two-dimensional spatial representations.
2. For reasoning with respect to specific spatial features, *specific spatial representations* are preferable; conversely, for reasoning with respect to classes of spatial features, *composite spatial representations* are preferable. The determination of the closest road to a given point would be potentially more efficient using a true two-dimensional *composite* road network representation than if spatial search of all individual road representations was required, for example.
3. Finite resolution data decompositions that are implicit, nonregular, nonhierarchical, nonareal, and non-two-dimensional tend to be highly *memory-efficient.*
4. *Indexing efficiency* is enhanced by hierarchical, true two-dimensional spatial representations that support natural, top-down search dimensions.
5. Since spatially local changes require only highly locally changes to the underlying representations, database *maintenance efficiency* is enhanced by regular, dynamic, two-dimensional spatial representations.
6. For *individual spatial features,* database *search efficiency* is enhanced by spatial representations that are specific, hierarchical, true two-dimensional and that support distributed search.
7. Search efficiency for *classes of spatial features* is enhanced by *composite hierarchical spatial feature* representations.
8. Since all spatial relationships among data sets are preserved, database *association efficiency* for specific spatial features is enhanced by hierarchical true two-dimensional spatial representations for individual spatial features; for classes of spatial features, association efficiency is enhanced by hierarchical, true two-dimensional composite feature representations.
9. For specific spatial features, *complex query efficiency* can be enhanced by hierarchical, true two-dimensional representations that support distribution of data; for spatial feature classes, complex query efficiency can be enhanced by composite spatial representations.
10. *Database implementation efficiency* is supported by data structures that support the distribution of data, processing, and control.
11. Finally, *representation accuracy* is supported by implicit, nonregular spatial representations.

These general conclusions are summarized in Table 7.4.

In many data fusion applications, spatially organized data is temporally sensitive. Although a temporal history file supports search along the time dimension, it offers limited support to numerous data fusion applications. A time-tagged spatial representation, on the other hand, was recommended in the last chapter for data that is dependent on both spatial and temporal variables. Rather than an exception

Table 7.4
Relationship Between Spatial Database Requirements and Spatial Representation Characteristics

Spatial Database Requirements	*Spatial Representation Attributes*															
	Regular	*Non-regular*	*Hier-archical*	*Non-hierarchical*	*Areal*	*Nonareal*	*2-D*	*Non-2-D*	*Static Structures*	*Dynamic Static Structures*	*Distrib-uted*	*Non-distributed*	*Explicit Feature Repre-sentation*	*Implicit Feature Representation*	*Specific Feature-Based*	*Composite Feature-Based*
Spatial reasoning algorithm efficiency																
•Individual features	●		●	•			●						●		●	
•Feature classes	●		●		●		●						●			●
DB memory efficiency		●		•		●		●						●		
DB overhead efficiency																
•Indexing	●		•		●		●						●		●	
•Maintenance	●						●			●			●		●	
DB search efficiency																
•Individual features	●		•		●		●				●		●		●	
•Feature classes	●		●		●		●				●		●			●
DB association efficiency																
•Individual features	●		•		●		●				●		●		●	
•Feature classes	●		●		●		●				●		●			●

	Spatial Representation Attributes															
Spatial Database Requirements	*Regular*	*Non-regular*	*Hier-archical*	*Non-hierarchical*	*Areal*	*Nonareal*	*2-D*	*Non-2-D*	*Static Structures*	*Dynamic Static Structures*	*Distrib-uted*	*Non-distributed*	*Explicit Feature Repre-sentation*	*Implicit Feature Representation*	*Specific Feature-Based*	*Composite Feature-Based*
DB query efficiency																
•Individual features	•		•		•		•				•		•		•	
•Feature classes	•		•		•		•				•		•			•
DB imple-mentation efficiency											•					
Representation accuracy		•												•		

to the principle that "data should be stored along its most natural search dimension," *time* must be viewed as an *attribute* of data that *is* maintained along its natural spatial search dimensions. In this case, the subordination of certain attributes to another natural search dimension can dramatically enhance overall database performance efficiency.

7.5 DATABASE DESIGN COMPROMISE

As discussed at the beginning of Section 7.3, the design of a database system to support data fusion automation necessarily requires consideration of five interrelated issues:

1. Support for intuitive algorithm development;
2. Facilitation of algorithm efficiency;
3. Provision of accurate data representation;
4. Minimization of data storage requirements;
5. Enhancement of database performance efficiency.

With a single-representation spatial database, typical spatial database designs are selected between variants of a raster-based representation or variants of a vector-based representation. As suggested previously, the degrees of freedom that are required to achieve an effective compromise among the five competing requirements is too high to be satisfied by either a raster-based representation or a vector-based representation.

On the other hand, a multiple-representation-based database architecture permits additional degrees of freedom that can yield a much more effective compromise among the design criteria. In the next chapter, an integrated spatial and semantic object database kernel design is discussed that achieves an effective compromise among both the algorithm and database requirements.

7.6 SUMMARY AND CONCLUSIONS

This chapter discussed a wide range of issues associated with the development of a database management system capable of supporting the construction of robust, context-sensitive, and efficient data fusion algorithms. As observed in the numerous examples, many of the shortcomings of the relational database model with respect to data fusion applications relate directly or indirectly to the inability of table-based data structures to support efficient search and manipulation of two-dimensional spatially organized data. Because of the potential size of the spatial databases required to support the higher order data fusion algorithms, spatial database design can be a critical element of data fusion automation.

The goal of this chapter was to identify critical database requirements and establish a top-down methodology for the design of an effective and efficient database system to support the higher order canonical form algorithms. The spatial database design must achieve an effective compromise with respect to the following five primary requirements:

1. Support for intuitive algorithm development;
2. Support for highly efficient fusion algorithm performance;
3. Provision of high-precision data representations;
4. Minimization of storage requirements;
5. Maximization of database performance efficiency.

Database performance efficiency was discussed in terms of the following five criteria:

1. *Overhead* efficiency:
   a. Search dimension maintenance;
   b. Underlying data structure maintenance;
2. *Search* efficiency;
3. *Association* efficiency;
4. *Complex query* efficiency:
   a. Set operations;
   b. Concatenated query clauses with mixed spatial and nonspatial constraints;
5. *Implementation* efficiency.

The following *spatial representation issues* were shown to be relevantto achieving an effective design compromise:

1. *Regular* versus *nonregular* representation;
2. *Hierarchical* versus *nonhierarchical* representation;
3. *Areal* versus *non-areal-based* representation;
4. *Two-dimensional* versus *non-two-dimensional* representation;
5. *Static* versus *dynamic* representation;
6. *Monolithic* versus *distributed* implementation potential;
7. *Explicit* versus *implicit* feature representation;
8. *Single* versus multiple (or *hybrid*) spatial representation.

Table 7.4 summarizes the relationships between these goals and the design considerations. Algorithm efficiency, database search efficiency, overhead efficiency, association efficiency, query efficiency, and implementation efficiency are all seen to be enhanced by hierarchical, true two-dimensional spatial representations, while memory efficiency and representation precision are enhanced by nonhierarchical, nontrue two-dimensional spatial representations. Although no single spatial representation achieves all the design objectives, a hybrid spatial representation that combines a hierarchical true two-dimensional low-resolution spatial representation and a memory-efficient high-precision spatial representation can potentially provide an effective compromise among the conflicting representation requirements.

Not only does Table 7.4 identify required representation attributes, it suggests representation attributes that are *not* required as well. For instance, there appears to be no pressing requirement for a database to support:

1. Hierarchical, nonregular representations;
2. Fundamentally static data representations;
3. Nondisjoint data structures that do not lend themselves to data distribution.

In the next chapter, a database kernel is developed that integrates a hybrid representation of specific spatial features, an object representation of two-dimensional space, and a conventional semantic object representation.

References

[1] Ooi, B.C., *Efficient Query Processing in Geographical Information Systems*, Lecture Notes in Computer Science, Vol. 471, Heidelberg, Germany: Springer-Verlag, 1990.

CHAPTER 8

Integrated Semantic/ Spatial Database Kernel

8.1 INTRODUCTION

The development of robust, context-sensitive, and efficient fusion algorithms has been seen to require the use of the higher order canonical form problem-solving approaches that, in turn, can require extensive search and manipulation of large spatially organized databases. Chapter 6 offered a critical look at the strengths and weaknesses of both contemporary database management systems and current geographic information systems. In terms of their support to tactical data fusion automation, both classes of systems were found to exhibit a number of fundamental deficiencies.

For example, although a relational database adequately supports the maintenance and search of *static, explicit* relationships among tabular data sets (*near* along an indexed linear search dimension or through an explicit *join* of two tables), tactical data fusion algorithms often demand the search and manipulation of *implicit* spatial, temporal, and hierarchical relationships among data sets. Examples of these types of relationships include:

1. Boolean and fuzzy measures of proximity relative to point, line, and region objects (spatial);
2. Reasoning with respect to time-sequenced events (temporal);
3. Changes to one object can affect subordinate and superordinate objects (hierarchical).

To achieve adequate performance, system developers may be forced to either abandon commercial database systems in favor of a custom file-oriented approach or to expend significant resources to "try to work around" the most serious bottlenecks.

In addition to performance considerations, conventional database systems can represent a barrier to the development of higher order canonical form algorithms. Consider the highly intuitive, top-down canonical form XV path-planning algorithm described in Chapter 3. Because the algorithm relied on multiple-resolution, true two-dimensional spatial search, a database that stored the natural and cultural features using vector-based data structures would provide inadequate support. As a consequence, the proposed algorithm would be *straightforward to develop* as well as *performance-efficient* only if the supporting database management system accommodated multiple-resolution, true two-dimensional spatial search. Thus, data representation issues can influence both the ease of algorithm development and the efficiency of many forms of automated reasoning.

The widespread reliance on highly combinatoric problem-solving paradigms (rigid templating, exhaustive compute-and-test) can be attributed, at least in part, to the lack of database support that facilitates the development of more intuitive, higher order canonical form algorithms. Reliance on nontrue two-dimensional spatial representations can lead not only to nonintuitive spatial-reasoning algorithms but to highly inefficient algorithm performance as well.

Chapter 7 outlined a formal set of requirements for both evaluating existing spatial database systems as well as designing new systems. The top-down design approach identified key tradeoffs that were required to achieve an effective compromise across a wide range of conflicting algorithm and database requirements. This chapter outlines the development of a powerful integrated semantic and spatial object database kernel based on these design considerations. The following chapters will demonstrate the advantages of the database kernel in terms of its support to the development of more sophisticated and more efficient data fusion algorithms.

The construction of a large building is metaphorically similar to the construction of a complex data fusion system. A project that begins with basic raw materials (rough-sawed lumber, hand-forged nails) and relies on a rather primitive tools (handsaws, hammers) would be expected to yield a much more modest structure than one that employs higher order building materials (prefabricated walls, steel I-beams, prehung doors) and more powerful tools (radial-arm saws, nailguns). Although both buildings might have the same "cost" in terms of labor hours or physical effort expended, the second building is likely to be both larger and more in-line with contemporary building standards. Even if all "cost" metrics are ignored, it is unlikely that a construction approach using primitive handtools and basic raw materials would be able to produce a building that was comparable to one produced using more modern construction techniques.

Analogously, with basic raw materials (relatively low-level programming languages) and primitive tools (relational database management systems), the "cost" of automating complex data fusion applications can be very high. However, given higher order building materials (high-level languages, object-oriented programming environments) and more powerful tools (advanced database management systems), algorithm development costs can be significantly reduced. Just as a radial-arm saw can make straighter, more accurate cuts than a handsaw, a database system that has been designed to support more complex applications can be expected to be more effective than a DBMS that was optimized for storing and retrieving table-based information. In addition to the quality of the woodworking that can be achieved using a radial-arm saw, a handsaw cannot approach the efficiency of its powered cousin. Analogously, a database that is optimized to support complex multiple level-of-abstraction spatial, temporal, and hierarchical reasoning tasks would be expected to outperform a system that was designed to support table-based storage and retrieval.

The broad goal of this chapter is to outline the key elements of a database management system kernel that can both *facilitate the development* of more sophisticated algorithms as well as *provide efficient operational support* to those algorithms. By treating physical entities, events, and domain knowledge (targets, river-crossing operations, and geographic and cultural features, respectively) as abstract objects, object-oriented reasoning offers a powerful conceptual framework for developing sophisticated data fusion algorithms. Object-oriented reasoning, in turn, relies quite naturally on object-oriented representations. This chapter discusses a high-level object-oriented database architecture that supports the development of both *highly intuitive* and *highly efficient algorithms* as well as meets the *representation accuracy* and *database efficiency* requirements discussed in the last chapter.

8.2 HIGH-LEVEL DATABASE KERNEL ORGANIZATION

The database kernel developed in this chapter consists of two principal object classes: *semantic objects* and *spatial objects.* Since all object databases support general semantic object representations, the chapter focuses on the development of the spatial object representation. In general, spatial objects consist of two subordinate object classes: an *object representation of two-dimensional space* and an *object representation of individual point, line, and region features.* The former representation supports top-down, multiple level-of-abstraction spatial reasoning with respect to *classes* of spatial features as well as permits spatial indexing of specific spatial features. The latter representation supports efficient multiple-resolution spatial reasoning with respect to *individual* spatial features. Figure 8.1 depicts a high-level view of such a database organization.

8.2.1 Object-Oriented Representation of Two-Dimensional Space

Consider the query: *Identify the class* 1 *road that is closest to query point* (x_1, y_1). If the database maintains only individual spatial representations for each point, line, and region object, the above *feature-class query* could potentially require interrogation of all

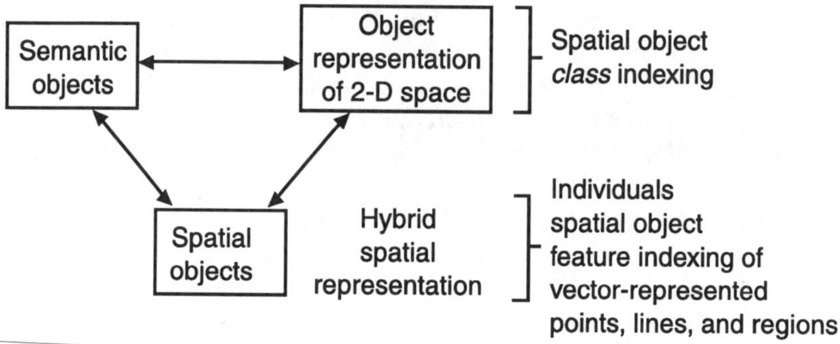

Figure 8.1 High-level view of the composite object database kernel.

class 1 road representations. Just as a conventional semantic object hierarchy permits efficient query with respect to classes of semantic objects, a hierarchical representation of two-dimensional space can support efficient query with respect to classes of spatial objects. If the spatial object and semantic object representations are fully integrated, the composite database can potentially support both efficient multiple level-of-abstraction spatial reasoning as well as more conventional multiple level-of-abstraction semantic reasoning.

At the highest level of abstraction, the *object representation of two-dimensional space* consists of a single spatial object that characterizes the entire area of interest (a map sheet, a Division Area of Influence, Central Europe).[1] At each successively lower level in the spatial object hierarchy, two-dimensional space would be decomposed into progressively smaller regions. Just as higher order semantic objects possess more general properties than their offspring, higher order spatial objects effectively characterize the general properties of their offspring nodes. Thus, the object representation of space establishes a *spatial object hierarchy* that is analogous to a conventional *semantic object hierarchy.*

Based on the general principles outlined in Chapter 7, an *object representation of two-dimensional space* should possess the following properties:

1. Finite resolution;
2. Regular decomposition;
3. Areal-based;
4. Two-dimensional;
5. Explicit;
6. Hierarchical;

1. Since there may be no advantage in summarizing all of two-dimensional space at the coarseness associated with the highest level of abstraction, a truncated hierarchy may be of more practical interest.

7. Composite feature-based;
8. Relatively dynamic data structure;
9. Distributed data representation potential.

A representation that possesses properties 1 to 5 has been previously defined as a *true two-dimensional spatial representation.* With the addition of property 6, the spatial object hierarchy would provide a uniform hierarchical spatial representation of classes of point, line, and region features. From Figure 7.12, a pyramid representation is seen to fully satisfy properties 1 to 6. Since a pyramid can maintain multiple classes of spatial objects (roads, waterways) within each cell, the pyramid is a natural candidate for a *composite feature representation.* Because a pyramid is a true two-dimensional spatial representation, it is a relatively *dynamic data structure.* In particular:

1. Spatially local changes require only relatively local changes to the data structure;
2. No reindexing is required following updates;
3. Because the pyramid is a data-independent representation, extensive tree-balancing operations are not required following insertions or deletions.

Finally, due to both the hierarchical and grid-based character of the pyramid, a pyramid data structure readily accommodates the *distribution of data.* Therefore, a pyramid data structure satisfies all nine requirements for an *object representation of two-dimensional space.*

8.2.2 Hybrid Spatial Object Representation

Rather than selecting a single spatial data structure for individual point, line, and region features, a multiple-representation database supports the use of a hybrid spatial representation. An effective hybrid representation can be developed that consists of a multiple-resolution, true two-dimensional low-resolution spatial representation combined with a memory-efficient, high-resolution vector-based representation.

Low-Resolution Spatial Representation

Based on the database principles developed in Chapter 7, the low-resolution spatial representation should possess the following properties:

1. Finite resolution;
2. Regular decomposition;
3. Areal-based;
4. Two-dimensional data structures;

5. Explicit;
6. Hierarchical;
7. Specific feature–based;
8. Dynamic data structure;
9. Distributed data representation potential.

With the exception of property 7, the *low-resolution spatial representation* requirements are identical to those of the *object representation of two-dimensional space.* Whereas a *composite feature–based representation* supports efficient spatial search with respect to classes of spatial features, a *specific feature–based representation* supports effective search and manipulation among specific point, line, and region features. As will be discussed in detail in Section 8.3, the *region quadtree* satisfies all of the aforementioned nine representation criteria.

High-Resolution Spatial Representation

High-resolution spatial reasoning demands access to high-resolution spatial data representations. Since regular decompositions become less efficient as the resolution is increased, a multiple-resolution spatial decomposition tends to be highly inefficient for maintaining high-resolution spatial features. Based on the design recommendations from Chapter 7, the high-resolution spatial representation should possess the following ten properties:

1. Finite resolution;
2. Nonregular;
3. Nonhierarchical;
4. Non-areal-based;
5. Non-two-dimensional;
6. Implicit;
7. Specific feature–based;
8. Dynamic data structure;
9. Distributed data representation potential;
10. High precision.

As seen in Figure 7.12, traditional vector-based spatial representations clearly meet the first five criteria. In addition, because they describe a *specific feature* using an *implicit* piecewise-linear representation of lines and region boundaries (line segments are represented by their endpoints), vector representations satisfy properties 6 and 7. Because changes to a feature only require modification of the property list of a single feature, vector representations are inherently *dynamic data structures.* Since the representation of individual features is self-contained, vector representations can be readily *distributed.* Finally, vector-based representations inherently offer a *high precision.* Thus, vector-based representations satisfy all the requirements of the high-resolution spatial representation.

8.2.3 Composite Database Representation

As mentioned in Chapter 6, the *acetate-overlay paradigm* is a well established and highly intuitive spatial-reasoning methodology. In this paradigm, paper maps maintain relatively static spatially organized information while acetates can hold both static and dynamic information. As a consequence, the *acetate-overlay paradigm* accommodates highly intuitive true two-dimensional spatial reasoning among both static and dynamic data sets. With the addition of multiple-resolution acetates and paper maps, the paradigm readily supports highly intuitive multiple-resolution spatial reasoning.

The emulation of both the problem-solving style and the database form of the acetate-overlay paradigm provides a powerful model for automated spatial reasoning. Although the resolution and accuracy of the manual analysis products are generally relatively low, spatial-reasoning algorithms that are based on this metaphor are not constrained by many of the shortcomings of manual approaches, such as the thickness of grease pencil lines, registration of among the acetates, and map reading and placement errors.

Because the acetate-overlay reasoning paradigm employs both static and dynamic spatially organized information that is naturally distributed among multiple maps and overlays (for example, *mobility resistance* overlays are maintained separately from *concealment* overlays), *both* the pyramid and low-resolution spatial representations support reasoning that is metaphorically similar to the acetate-overlay paradigm. Just as maps and acetate overlays can maintain either *individual features* or *classes of features,* the low-resolution *representation of individual spatial features* and the *object representation of two-dimensional space* support the maintenance of individual spatial features and classes of spatial features, respectively.

In addition to spatially organized information, military analysts use a wide range of table-based (Tables of Equipment, logistics databases), hierarchical (military organization charts), and text-based (doctrinal rules, free-form messages, operational orders) information. In order to facilitate reasoning across all classes of knowledge, the spatial representation must be integrated with the representations of these other data-object types.

Figure 8.2 depicts a high-level architecture of a simple *multiple-representation, on-line query*–based database consisting of the three object classes:

1. General semantic objects;
2. Object representation of two-dimensional space;
3. Hybrid spatial representation of individual point, line, and region features.

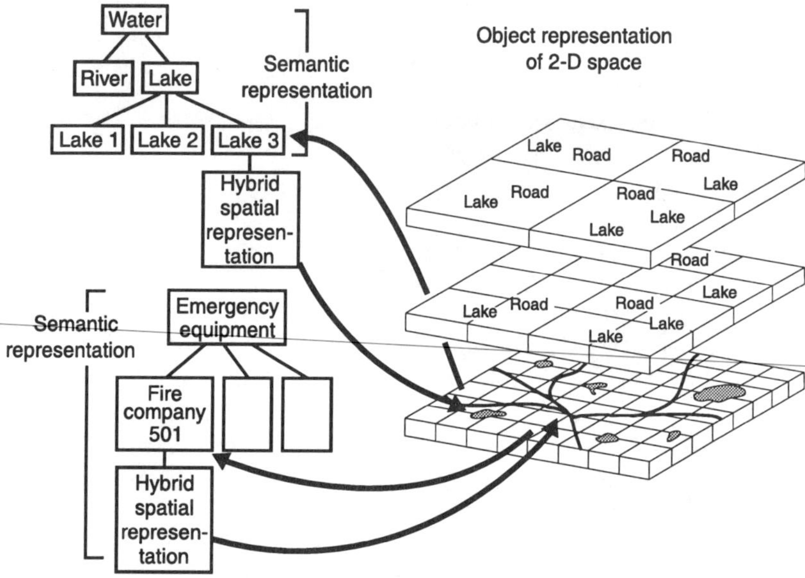

Figure 8.2 Simple example illustrating the high-level relationship between spatial and semantic objects.

To investigate the potential utility of such an organization, consider the following query:

Find the nearest *lake* to *fire company* 501. (8.1)

The objective of this query might be to locate the closest water source for refilling the tanker truck. Top-down search of the *semantic object* hierarchy can rapidly locate the specific object of interest (*emergency services>fire company>fire company* 501) from all other semantic objects in the database. The location attribute maintained by the object *fire company* 501 is used to initiate the spatial search for lakes using the *object representation of two-dimensional space.* Top-down spatial search of the pyramid representation in the vicinity of the current location of this rural fire company identifies grid cells that contain lakes. The precise location of the closest lake requires evaluation of the *hybrid representations of individual* lakes that are identified to exist within the identified grid cells. Since the current location of the tanker truck could be the company fire house, its location en route to a fire, at the scene of an alarm, or at the Fireman's Parade, the query product can be both dynamic and context-sensitive.

Because data are stored along highly "natural" search dimensions, the data base depicted in Figure 8.2 possesses a number of highly desirable properties:

1. Supports true two-dimensional spatially anchored search;[2]
2. Preserves spatial relationships among data (since no data transformation is required);[3]
3. Does not require extensive reindexing following insertions, deletions, or changes;[4]
4. Facilitates mixed semantic and spatial searches;
5. Supports efficient areal-based set operations.

Because true two-dimensional spatial representations preserve natural spatial-search dimensions, they support highly focused spatial search. Objects that are "spatially near" in the physical world are also "near" along two-dimensional spatial search dimensions in the data representation. An object that is "to the left of" another object in the physical world retains this same property within the spatial representation. The preservation of natural search dimensions can eliminate both preprocessing and reindexing requirements. The areal-based character of the low-resolution spatial representation supports highly efficient set operations among extended spatial features (lines and regions). Top-down spatial search further minimizes both the search-space size, as well as the overall computational requirements for many classes of spatial reasoning.

8.3 THE QUADTREE DATA REPRESENTATION

Because the *region quadtree* is a candidate data structure for *both* the low-resolution component of the hybrid spatial representation of *individual spatial features* and the *object representation of two-dimensional space* (as we will see shortly, a pyramid is just a special form of region quadtree), this section offers a high-level overview of some of the prominent characteristics of the region quadtree data structure.

As its name implies, a quadtree is a recursive hierarchical decomposition of a node into four offspring nodes. Just as a binary tree supports efficient tree-oriented search of an ordered one-dimensional list, a quadtree supports efficient tree-oriented search of two-dimensional spatially organized data. From this perspective, a quadtree is simply the logical extension of a binary tree to two dimensions.[5]

---

2. Spatially anchored search is defined as true two-dimensional spatial search relative to a point, line, or region feature. This capability is discussed in considerable detail in Chapter 12.

3. For instance, when a region is stored as its bounding polygon, the region interior is effectively eroded and the remaining boundary is represented by an ordered set of piecewise-linear line segments.

4. Although the underlying physical data representations might change, no logical-level reorganization is required.

5. Technically, the *point quadtree* is the direct extension of the binary tree to two dimensions.

Section 8.3.1 provides a brief background of the region quadtree data structure. For most readers, this introduction will suffice. However, since the balance of Section 8.3 discusses important properties of the region quadtree that justify its foundational role in the integrated semantic/spatial object database kernel, for those interested in implementation details, a complete reading is recommended. Although much of the original interest in quadtrees was related to its data-compression potential for binary images, in this book, the quadtree is treated as a general hierarchical data structure that can support the storage, search, and manipulation of two-dimensional, spatially organized information.

8.3.1 Introduction

The quadtree is sometimes treated by its advocates as if it possesses mystical powers and by its detractors as a costly, cumbersome, and unnecessarily complex representation. However, as we will demonstrate, the quadtree is neither savior nor demon. Rather, the quadtree is a very natural two-dimensional spatial data structure that has a number of useful properties as well as a number of significant shortcomings.

Although the term "quadtree" is associated with a large number of hierarchically organized two-dimensional data representations, our discussion will focus on the *region quadtree* representation. For readers interested in a more complete treatment of the field, Samet's two books [1,2] present a comprehensive description of the design and application issues associated with numerous hierarchical spatial data structures; in addition, the books provide over 800 references.

A region quadtree is a hierarchical areal-based decomposition of two dimensional space. A data representation is referred to as regular if the decomposition is independent of the data; conversely, a representation is non-regular if the decomposition is data dependent. Thus, a *regular region quadtree* [3] is an areal-based decomposition of data into equal-sized quadrants. The regular region quadtree supports highly efficient top-down search and manipulation, requiring only integer arithmetic and simple table-lookup operations.

The regular region quadtree representation of a binary image (black image against a white background) is generated by recursively decomposing all cells that are *not* completely black into four cells (or quadrants). The decomposition terminates when either no further decomposition is required, or some specified maximal resolution block size is reached. The *minimal region quadtree* representation of a binary image is described by the *leaf nodes* (or terminal nodes) of the resultant tree. Thus, the minimal region quadtree is a variable-resolution representation that contains the minimum number of nodes (or alternatively, blocks or cells) required to represent a region at a given resolution.[6] Specification of the maximum quadtree

6. Since, as will be discussed below, the regular region quadtree is a fixed-origin representation, lower node-count may exist for a different coordinate system origin.

node resolution size (or equivalently the tree depth) permits a tradeoff between memory requirements and representation accuracy.

A *true two-dimensional* spatial representation stores points as points, lines as lines, and regions as the boundary *plus* the enclosed area. In addition, the representation preserves all inherent two-dimensional spatial relationships among the data sets.[7] If points are maintained as a single node, lines as a set of spatially adjacent nodes through which the line passes, and regions as a set of hierarchical nodes that completely *covers* (analogous to the *minimal covering set* under the specific decomposition constraints) a region, a regular region quadtree representation of these three spatial features would exhibit all the required properties of a *hierarchical true two-dimensional spatial representation.* Because the minimal region quadtree recursively decomposes homogeneous regions into maximal quadtree nodes, the bulk of the interior of a large region is represented by relatively few large nodes while areas closer to the boundaries are represented by progressively smaller-sized nodes.

Although a regular region quadtree decomposition produces a representation that is independent of the order in which the data is installed, the decomposition itself is highly translation-dependent.[8] For a given region, a change in the coordinate system origin can lead to the generation of a qualitatively different quadtree representation. However, as long as a common decomposition grid is employed for all data sets, such translation-sensitivity is generally not significant. Although the decomposition origin can be adjusted to optimize storage requirements, the modest compression improvement does not normally justify the effort.

The physical representation of a quadtree can be either *pointer-based* or *non-pointer-based.* In pointer-based representations, all quadtree leaf nodes are linked by pointers. The *linear quadtree* [4] is a non-pointer-based hierarchical indexing scheme that facilitates both quadtree representation and tree traversal.[9] In the following discussion, we restrict our attention to non-pointer-based representations.

In a pointerless representation, individual quadtree nodes are generally stored as a string $X_0X_1.. X_l$ that provides both a unique identification of a particular quadtree node as well as defines a unique tree traversal path from the tree root to that node. Each term in the string X_i defines a specific quadrant X at tree depth i, where X is a four-valued variable that will be defined as A, northeast quadrant; B, northwest quadrant; C, southwest quadrant; and D, southeast quadrant.[10]

In the interest of clarity of the concepts, the discussion of spatial representation and reasoning has been restricted to two dimensions. However, the extensions to a third dimension are reasonably straightforward. For instance, the *oct tree* is the natural extension of the region quadtree to three dimensions. Since the *regular volume oct tree* preserves all relationships among 3-D objects, it would be considered to be a true 3-D spatial representation.

All regular, grid-based decompositions exhibit such translation sensitivity.

The linear quadtree representation presented here is generally referred to as the *FD linear quadtree* in the literature.

. The notation used here (that is, for example, A is assigned to the first quadrant, and B is assigned

Since each term in the string takes on only four values, two bits are required to store each value of X_i. For all k, $0 < k < j$, $(X_0X_1 \ldots X_j) = (X_0X_1 \ldots X_{j-k}) + (X_{j-k+1} \ldots X_j)$ where + can be interpreted as either addition or concatenation. Thus, the quadtree and its associated string notation is a relative, incremental coding scheme that represents a specialized two-dimensional run-length code.[11] The radix 4 node-string coding scheme exhibits Abelian group properties with respect to addition and subtraction. A 32-bit word can index $2^{16} \times 2^{16} = 4 \times 10^9$ maximum resolution cells. The regular region quadtree spatial decomposition is illustrated in Figure 8.3. The representation of a specific point feature is depicted in Figure 8.4 and an example of a regular, minimal region quadtree decomposition of a multiply-connected region is illustrated in Figure 8.5.

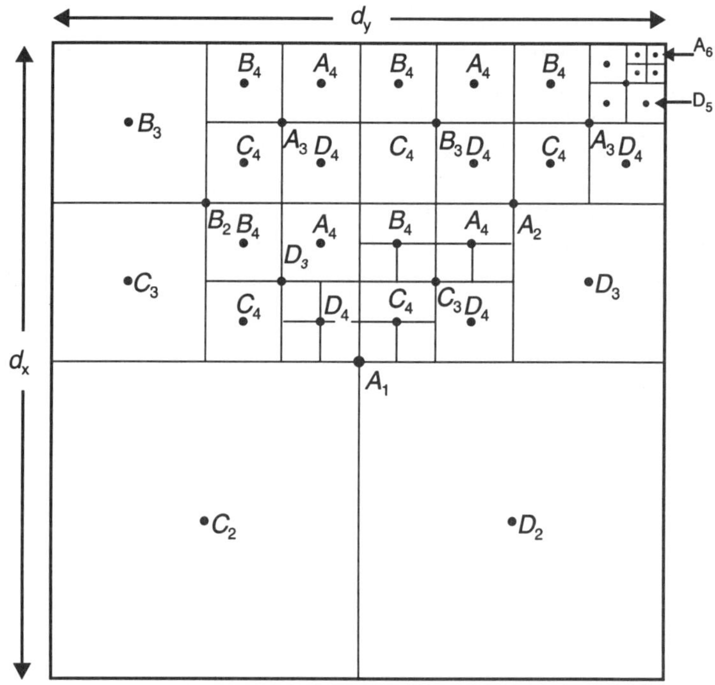

Figure 8.3 Node-string convention for a regular quadtree decomposition.

to the second quadrant) is nonstandard in the representation literature. However, the use of a single-letter assignment rather than a mnemonic ("NW") or a binary code ("00") facilitates a conceptual understanding of many of the algorithms presented.

11. A run-length coding scheme uses the length of a "run" (the number of consecutive replications) c a given symbol to compress a sequence of such symbols.

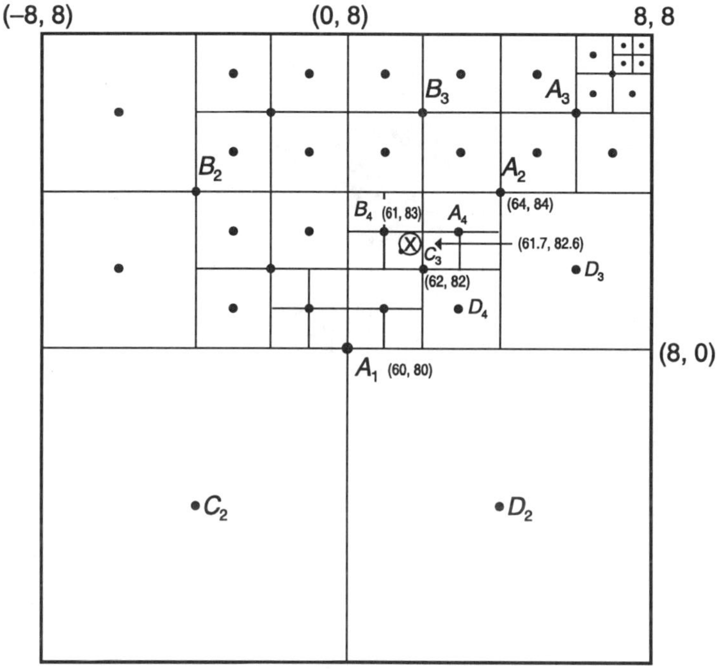

igure 8.4 Example of a quadtree decomposition for the point (61.7, 82.6).

The cardinal and noncardinal spatial directions can be treated as func-onal operators on individual quadtree nodes. For example, east($A_1B_2A_3C_4$) = $A_1B_2A_3D_4$). Since the functional operators are commutative, concatenated opera-›rs can be applied sequentially and in any order (east(south()) = south(east())). rom symmetry, concatenated opposing operators cancel each other (east(west()) =) while nonopposing cardinal operators produce the noncardinal operators east(north()) = northeast()).

.3.2 Move Rule

ovement within the quadtree data structure at a single-resolution level relies on e nearest neighbor relationships among the nodes. Movement to higher or lower solution nodes can involve both nearest neighbor and hierarchical relationships nong nodes. A simple, noncomputational neighbor-finding algorithm [5,6] is dis-ssed next. Unlike approaches that require specialized functions for different move-ent classes [7,8], this algorithm uses a single uniform approach for all single-step ovements within the tree. When applied recursively, the nearest neighbor algorithm rmits general, spatially based movement within a linear quadtree data structure.

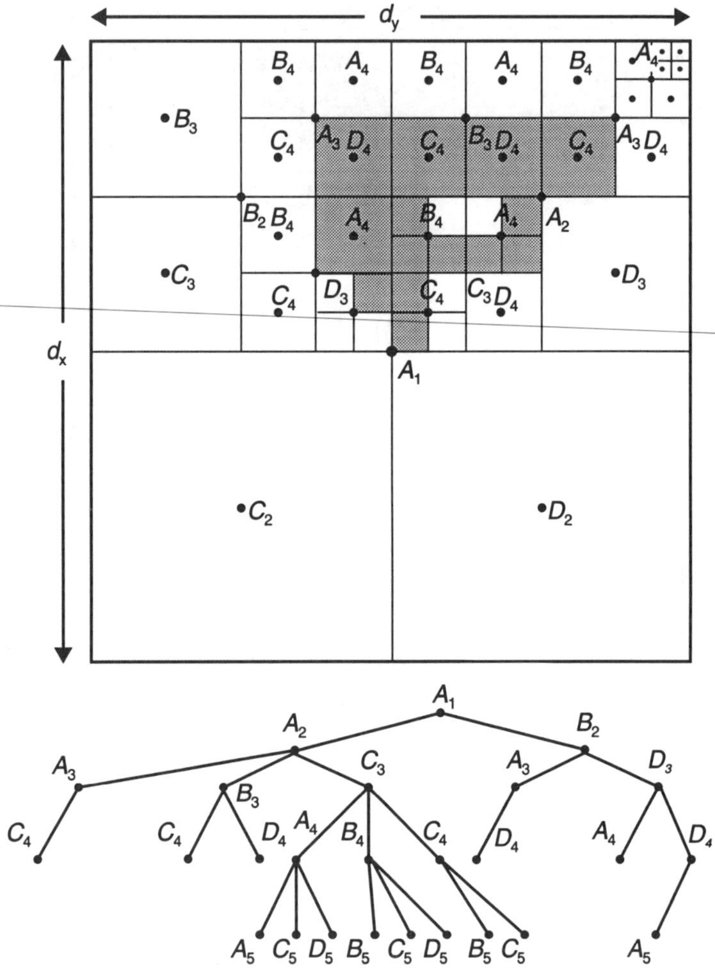

Figure 8.5 Regular region quadtree decomposition of a region with an included hole.

As observed in Figure 8.3, a quadtree node has eight nearest neighbors. Mov ment to a neighbor node requires the identification of the lowest order node in th linear quadtree data structure that is common to both the current node *and* the s lected neighbor node. If the current node and its neighbor have the same pare node, the move is considered to be "local." On the other hand, when a neighb node does not have the same parent as the query node, a move to the neighb node is outside the query node's immediate "family" (it "penetrates" the bounda of the query node's parent). In this case, the grandparent of the query node mu be interrogated to determine if the desired move remains within the grandparen family. If the neighbor node is within this extended family, the move is "local" at the grandparent level. However, if the move penetrates the boundary of the grandparent node, the desired move is again nonfamilial and the great-grandparent node must be checked. This bottom-up process continues recursively until the patriarch of the family that includes both the original node and desired neighbor node is discovered. Due to the hierarchical symmetry of the regular region quadtree, the identical move rules apply at all resolution levels.

Figure 8.6 shows all possible cardinal and noncardinal moves from positions *A*, B, C, and *D*. Table 8.1 summarizes the eight legal move directions *F* from a given cell and classifies the "direction" of the parent-node boundary "penetration" *P*.

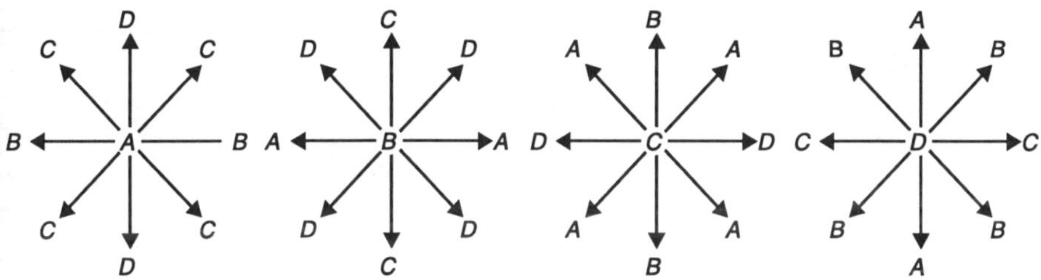

Figure 8.6 Nearest neighbor quadtree rewrite rule set.

Table 8.1
Nearest Neighbor Movement Rule Boundary Penetration Classification

Move Direction	*Type of Parent Node Boundary Penetration P*			
F	*A*	*B*	*C*	*D*
North	North	North	Null	Null
South	Null	Null	South	South
East	East	Null	Null	East
West	Null	West	West	Null
NE	NE	North	Null	East
SE	East	Null	South	SE
SW	Null	West	SW	South
NW	North	NW	West	Null

The following pseudocode computes the nearest neighbor nod $(Y_{i-k} \dots Y_{i-1} Y_i)$ of node $(X_{i-k} \dots X_{i-1} X_i)$ in the direction F:

$Y_i = F(X_i)$
$F = P(X_i)$
DO FOR $j = 1, k$
IF $(F$ *.NE. Null*$)$ *THEN*
$Y_{i-j} = F(X_{i-j})$ node transformation rule (Figure 8.6)
$F = P(X_{i-j})$ parent-node penetration direction (Table 8.1)
OTHERWISE
$Y_{i-j} = X_{i-j}$
RETURN
ENDDO

Each application of the above algorithm recursively transforms the string not tion of the original node (starting with the least significant character) into the dest nation node, one string character at a time. As soon as the common ancestor of th current node and the desired move location is discovered, the process terminat and the resultant transformed string represents the desired neighbor node. No bac tracking operations are required once the common ancestor node is discovered.

Several examples of this neighbor-finding algorithm are presented to demo strate the simplicity of the recursive transformation process. First, we locate the *we* neighbor of $A_1B_2A_3A_4$. From Figure 8.6, a west move from an A node yields a node. Thus, the west neighbor of A_4 is B_4. Since, as observed in Table 8.1, a we move from an A node yields a B node with the same parent, west$(A_1B_2A_3A_4)$ $A_1B_2A_3B_4$.

Consider an *east* move from the original location. An east move from a level A node results in a level-4 B node and causes an east boundary penetration of th level-3 parent node. An east move from a level-3 A node results in a level-3 B no and causes a level-2 east boundary penetration. Since an east move from a level-2 node produces a level-2 A node that does not penetrate the level-l parent node, th recursion ceases and the destination east$(A_1B_2A_3A_4) = A_1A_2B_3B_4$.

Next, consider a *southeast* move from the original location. A southeast mo from a level-4 A node results in a level-4 C node. A southeast move from a level-4 node causes a level-3 east boundary penetration. An east move from A_3 yields B_3 a results in a level-2 east boundary penetration. An east move from B_2 yields A_2 wi no resulting level-l boundary penetration. Thus, the final destination node $A_1A_2B_3C_4$.

Finally, consider a *south* move from $A_1B_2A_3D_4$. A south move from a level-4 node results in a level-4 A-node with a level-3 south boundary penetration. A sou move from A_3 yields D_3, which is a local move. Thus, south$(A_1B_2A_3D_4) = A_1B_2D_3A$

Two important characteristics of the algorithm should be noted. First, if t desired move is, in fact, beyond the data-set boundary, the algorithm will disco that the neighbor node does not have the same root node as the current data s ither another data set that does contain the desired common ancestor must be upplied, or the algorithm can terminate with an "end of image" flag set.

Second, the nearest neighbor move algorithm identifies the "inherent" equal-zed neighbor node. If an equal-sized neighbor node does not actually exist within ne data set, the parent of the neighbor node can be checked for membership ithin the database (the parent of any node can be determined by simply deleting ne rightmost string character of the linear quadtree representation). If the parent ode is not discovered, the parent of the parent (that is, the grandparent node) is necked. This recursive process can be repeated until a node in the data set is dis-overed. When they exist, smaller-sized neighbors can be determined by recursively sting progressively smaller neighbor nodes. For instance, $A_1C_2B_3C_4$ is an eastern eighbor of B_1D_2 because:

1. East$(B_1D_2) = A_1C_2$; *and*
2. $A_1C_2B_3C_4$ is on the western edge of A_1C_2.

This characteristic of the neighbor-finding algorithm, which produces the ring representation of an equal-sized neighbor node (even if such a node does not tually exist in the database), can often be exploited by windowing and search gorithms.

3.3 Important Regular Region Quadtree Properties

general, the regular region quadtree representation:

1. Provides a relatively compact representation of two-dimensional regions;
2. Supports efficient set-operation generation;
3. Permits top-down multiple-resolution reasoning;
4. Preserves Euclidean metrics among data sets;
5. Supports efficient mixed top-down and bottom-up spatially focused search;
6. Accommodates simple transformation, scaling, and rotation operations.

Each of these properties will be discussed in some detail.

ximum Storage Requirement

mentioned earlier, initial interest in the quadtree data structure focused on its plication to image compression. The image compression potential of any form of -length encoding depends on the likelihood that adjacent pixels possess the ne value (color, gray-scale value), which, in turn, depends on both the number of el quantization levels and the local correlations present in the particular data set. the number of quantization levels increase, the run-lengths can be expected to go down. Thus, run-lengths would be considerably longer with a 1-bit quantization than an 8-bit quantization. Data sets that possess large homogeneous regions increase the potential for long run-lengths. Binary images and certain terrain features may possess large homogeneous regions, while high-resolution imagery tends to be nonhomogeneous. Thus, although a significant compression ratio might be achievable for a binary map that distinguishes trafficable regions from those that do not support mobility, the region quadtree offers little compression potential for high-resolution, 24-bit color photographs.[12]

Because of its "maximal block size" encoding character, the regular region quadtree requires considerably less storage for large homogeneous regions than comparable resolution, raster-based representations. However, due in part to its translation sensitivity, the storage requirements of the regular region quadtree cannot be directly compared to the storage requirements of nontrue two-dimensional region representations, such as one-dimensional run-length coding or pixel boundary-list representations. In this section, "worst case" bounds on the storage requirements of the minimal region quadtree are established that permit a comparison with the memory requirements of competing representations.

In general, since areal-based representations maintain both the boundary plus the interior of all regions, regions with large area-to-perimeter ratios tend to have larger storage requirements than those with a small area-to-perimeter ratios. Since for a given perimeter length, a disk possesses the maximal interior area of any two-dimensional region, a disk is the worst case region in terms of relative storage requirements. Because a square is the four-sided, rectilinear analog of a disk, a square is a natural choice for computing the approximate worst case storage requirements for large regions.[13]

Hunter and Steiglitz [7] proved a number of fundamental theorems on the space requirements of binary images represented by quadtrees that have held up to empirical evaluation. For instance, the maximum number of quadtree nodes required to represent a polygon with perimeter p embedded in a $2^n \times 2^n$ image is $16(n+p) - 11$. Dyer [10] showed that an average of $0(2^{m+2} + n - m)$ quadtree nodes is required to represent a square of size $2^m \times 2^m$ at any position in a $2^n \times 2^n$ image.

Whereas the above storage bounds depend on the image size, in a general database application the worst case node-count for a particular region is effectively independent of the overall database size. A simple constructive approach that is not a function of the overall data-set size can be used to compute the worst case number of linear quadtree nodes required to represent both the interior and the boundary

12. This is not to suggest that there do not exist methods for compressing high-resolution images (for example, first transform the image into an optimal 8-bit version and then apply run-length coding [9]).

13. For a quadtree-based decomposition, a square region is much more easily analyzed than a disk.

of a square region. As will be shown, the largest node-count (worst case) occurs when the number of highest resolution leaf nodes is maximized.

A simple two-stage algorithm can be used to compute the node-count for the worst case positioning of a square with side length m. In stage 1, the number of highest resolution nodes in the perimeter of the square r_1 is determined. For odd-value-sided squares, the worst case alignment between the quadtree decomposition grid and the square occurs when the highest resolution nodes along two sides of the square must be represented. The worst-case alignment for even-valued m requires representation of the maximum resolution nodes along the entire perimeter of the square. Thus, in stage 1, if m is odd, $r_1 = (2m - 1)$ nodes must be represented; if m is even, $r_1 = (4m - 4)$ nodes must be represented. Figure 8.7 shows the worse-case relationship between a square of size $m = 15$ and the quadtree decomposition grid. In this case, two sides of the square possess maximum resolution nodes, which implies that $r_1 = 29$.

In stage 2, the node-count required to represent the interior nodes and the remaining border nodes (in the case of m odd) is computed. The remaining square has sides of length m_1 where

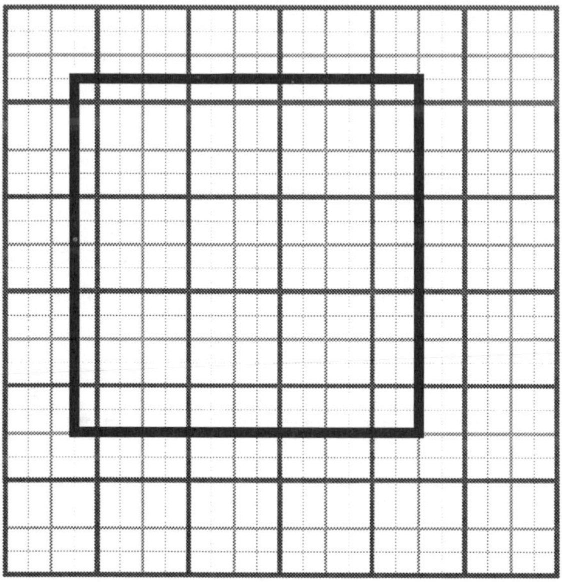

Figure 8.7 Worst-case alignment between a 15-node × 15-node square and the quadtree decomposition grid.

$$m_1 = m - 1 \quad \text{if } m \text{ is odd}$$

$$m_1 = m - 2 \quad \text{if } m \text{ is even}$$

Note that regardless of whether m is odd or even, m_1 is always even. The relationship between the original square of length m and the quadtree grid ensures that the square of side length m_1 will be "optimally aligned" with the quadtree grid. The *binary code* for a square of side length m_1 provides the required optimal decomposition strategy. Since m_1 is always even, the value of the 2^0 bit is never equal to 1. In the above example, m_1 = 14 which has the binary representation 1110. Therefore the decomposition consists of quadtree nodes of size 2^3, 2^2, 2^1 corresponding to all the 1-bits in the binary representation of 14. As observed in Figure 8.8, the quadtree decomposition consists of one node of size 2^3, five nodes of size 2^2, and thirteen nodes of size 2^1.

In general, the node-count r_2 for the optimally aligned square of size m_1 can be computed as

$$r_2 = \sum (2m_{i+1} - 1) \qquad 8.$$

$$\text{over } set\ k = [k_1, k_2, \ldots] \text{ where i} = 1, 2, \ldots$$

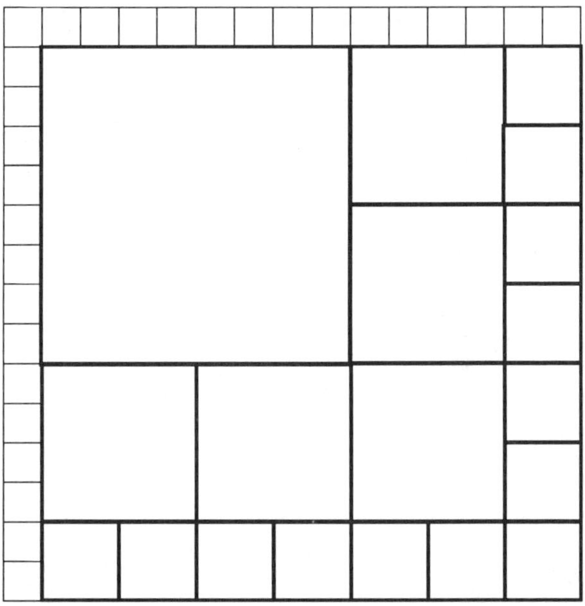

Figure 8.8 Worst-case quadtree node assignment for a 15-node × 15-node square.

where $m_{i+2} = \dfrac{m_{i+1} - 1}{2^{(k_{i+1} - k_i)}}$, $m_2 = {}^{m_1}\!/_{2^{k_i}}$ and the set k is the set of binary exponents of all 1-bits in the binary expression of m_1.

In the above example, since the set $k = (1, 2, 3)$,

$$r_2 = (2)(7) - 1 + 2(3) - 1 + 2(1) - 1 = 19 \quad 8.3$$

Thus, for $m = 15$, the worst-case node-count is

$$r_1 + r_2 = 29 + 19 = 48 \quad 8.4$$

Since a pixel-boundary representation of this region would require $(4m - 4) = 56$ nodes, the worst-case quadtree node-count for storing the boundary *plus* the interior area is less than the pixel boundary-only representation.

The original claim regarding the worst-case orientation of the square with respect to the coordinate system can be verified by comparing the node-counts r_1 and r_2. The minimum value of r_1 is

$$\begin{aligned} MIN(r_1) &= 2m - 1 \\ &= 2(m_1 + 1) - 1 \\ &= 2m_1 + 1 \end{aligned} \quad 8.5$$

while, since $m_1 \geq m_i$ and $\sum (1/2^k)$ is always less than unity, the maximum value of r_2 is[14]

$$MAX(r_2) = \sum [2m_i/2^k - 1] < 2m_1 \quad 8.6$$

Thus, the largest possible value of r_2 is strictly less than the smallest node-count of the perimeter cells removed in stage 1. Since any other translation of the original square would require less (or at most the same) minimum resolution nodes, the overall node-count is maximized.

Figure 8.9 shows r_1, r_2, and the sum $(r_1 + r_2)$ as a function of side length, m. Figure 8.10 compares (1) the pixel area (m^2), (2) the pixel node-count of the pe-

---

14. In fact, unless the binary code for r_2 has a bit in the 2^1 position, $\Sigma\ (1/2^k) < 1/2$. If no 1-bit occurs in either the 2^1 or 2^2 position, $\Sigma\ (1/2^k) < 1/4$, and so forth.

rimeter $(4m - 4)$, and (3) the worst-case quadtree node-count, $(r_1 + r_2)$ for the repre sentation of a square region. As observed, r_1 monotonically increases (at separate rates for odd- and even-sided squares) with the size of the square, while r_2 fluctuate due to the multiple-resolution nature of the quadtree decomposition. The uppe bound for the quadtree node-count is thus

$$(4m - 4) + (2m - 2) = 6m - 6 \quad \text{for } m \text{ even} \qquad 8.$$

$$(2m - 1) + (2m - 2) = 4m - 3 \quad \text{for } m \text{ odd}$$

For large regions, the worst case quadtree node-count is strictly less than the pixel-boundary node-count for odd-sized squares and less than 50% more than the pixel-boundary node-count for even-sided sized squares. These general results ar consistent with the actual worst case node-count results shown in (8.4) for $m = 15$.

Since an optimally aligned square whose side length is an integer power of can be represented by a single linear quadtree node, the *lower bound* on the storag requirements for a square is a single node while the *upper bound* is about 1.5 time the pixel-boundary node-count. Therefore, the non-pointer-based, minimal regio quadtree representation (which stores the boundary *plus* the interior area of square) compares favorably with that of the pixel boundary-only representation. A the area-to-perimeter ratio is reduced (small regions or moderate-sized regions wit convoluted boundaries), the relative cost of the explicit areal-based representatio (the regular region quadtree) goes down relative to a nonareal representation. I the limit as the area of the region approaches the pixel-boundary area, regio

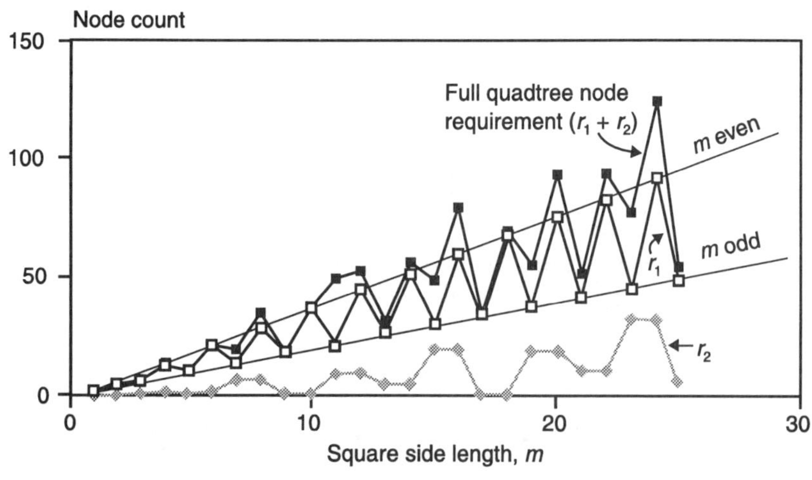

Figure 8.9 Worst-case quadtree node requirements showing both r_1 and r_2 components.

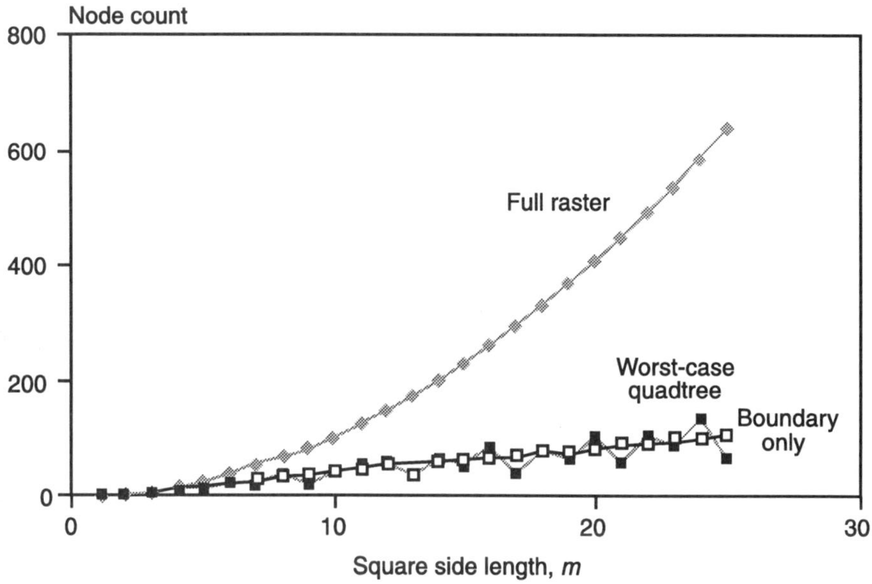

igure 8.10 Comparison of the worst-case quadtree-node requirements with full raster and pixel boundary-only representations.

uadtree and boundary-only pixel representations would require the same number f nodes (or cells).

real-Based Boolean Set Operations

nce set operations among extended line and region features fundamentally involve areal-based operations, areal-based data representations tend to support more ficient set operations than nonareal-based representations.[15] Intersection and unn operations among fully registered (areal-based) raster representations can be enerated using simple pixel-by-pixel AND and OR operations, respectively. Since e region quadtree provides an efficient hierarchical representation of raster-based formation, the region quadtree representation facilitates top-down areal-based set erations. Such top-down analysis can be considerably more efficient than comparable single-resolution raster-based approaches. In addition, with areal-based representations, operations on sets that include multiply-embedded or nested holes (a

. Although vector-based representations do not typically support efficient set-operation computation, a highly parallel vector-based set-operation algorithm has been developed [11].

hole within a forest within a hole within the forest) tend to be highly efficient because the representation explicitly depicts only the area that is actually "within" an object.

The potential efficiency of quadtree-based set operations will be demonstrated using a simple example. The following sorted linear quadtree node-strings for region 1 and region 2 are illustrated in Figure 8.11(a) and 8.11(b), respectively:

Region 1	Region 2
$A_0B_1C_2C_3$	$A_0B_1C_2$
$A_0B_1D_2B_3$	$A_0B_1D_2C_3B_4$
$A_0B_1D_2C_3$	$A_0B_1D_2C_3C_4$
$A_0B_1D_2D_3C_4$	
A_0C_1	

Figure 8.12 shows the tree representation of the two regions with all shared nodes circled. Table 8.2 summarizes the results of the intersection and union operations. For both classes of set operations, the string representation of each quadtree terminal (or leaf) node in region 1 is compared to the corresponding string in region 2. An *intersection* between the two regions occurs if:

1. There exists an identical node in both regions; or

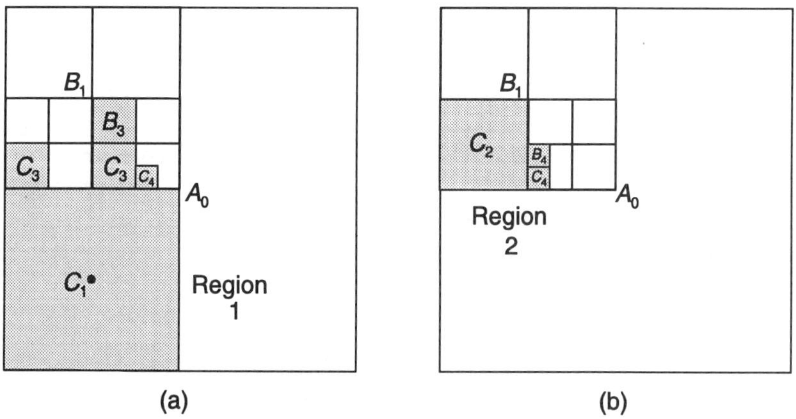

Figure 8.11 Quadtree-based representation for region 1 and region 2.

2. A node-string of length k in one region matches the first k characters of a node-string in the other region. The intersection product is the longer of the two strings (the physically smaller of the two nodes).

If no string match occurs (the node $A_0B_1D_2B_3$ in region 1 does not match any string in region 2), the nodes do not intersect.

Table 8.2
Set-Operation Example

Region 1	*Region 2*	*Intersection*	*Union*
$A_0B_1C_2D_3$	$A_0B_1C_2$	$A_0B_1C_2D_3$	$A_0B_1C_2$
$A_0B_1D_2B_3$			$A_0B_1D_2B_3$
$A_0B_1D_2C_3$	$A_0B_1D_2C_3B_4$ $A_0B_1D_2C_3C_4$	$A_0B_1D_2C_3B_4$ $A_0B_1D_2C_3C_4$	$A_0B_1D_2C_3$
$A_0B_1D_2D_3C_4$			$A_0B_1D_2D_3C_4$
A_0C_1			A_0C_1

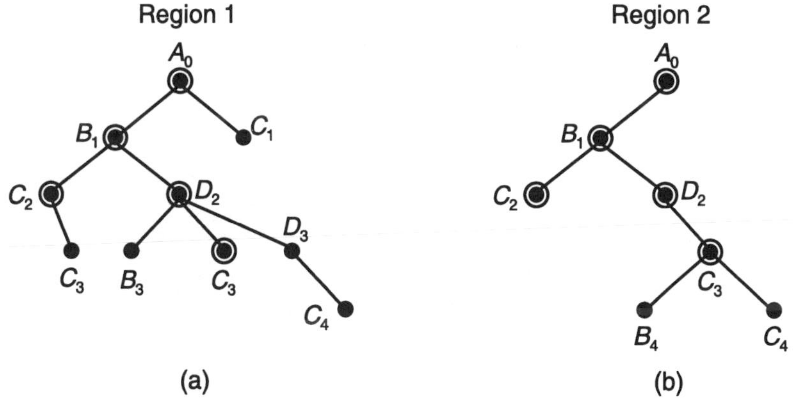

igure 8.12 Tree representation of the two regions shown in Figure 8.11 with all shared nodes circled.

The *union* of the two regions is constructed by following one of the following ules:

1. Add all node-strings in region 2 that are not in region 1 to the nodes of Region 1;
2. If a node-string of length *k* in one region matches the first *k* characters of a node-string in the other region, the union of the two nodes is the shorter of the two strings (the physically larger of the two nodes).

Figure 8.13 demonstrates the result of the intersection and union of the two regions. In terms of data compression, the regular region quadtree provides a compromise between the relatively high-compression ratio offered by pure run-length coding schemes that require decompression prior to performing set operations and a raster representation that offers no compression but supports direct set-operation generation.

In terms of the computational efficiency of manipulating such data structures the minimal region quadtree tends to be more efficient than either conventional run-length coding or raster-based representations. Although only a very simple set operation example was illustrated here, these same basic principles apply equally well to complex data sets. Chapter 9 explores set-operation development in considerably more detail.

Hierarchical Bounding Box

Because it employs blocks of varying spatial resolution, the region quadtree is a *variable-resolution* representation. However, since a *leaf node* in the linear quadtree notation implicitly identifies the set of nodes along an entire *branch* of the representation, the linear quadtree is effectively a true *multiple-resolution* data struc

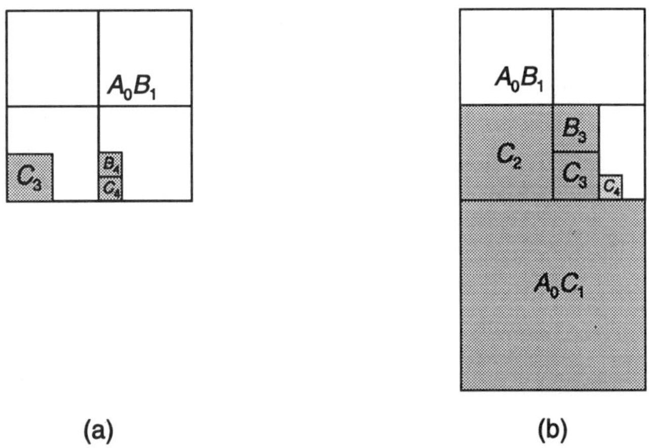

Figure 8.13 (a) Intersection, and (b) union of the two regions shown in Figure 8.12.

ture. For example, the explicit representation of the leaf node $A_0B_1C_2D_3$ implicitly identifies the nodes $A_0B_1C_2$, A_0B_1, and A_0.

For a given spatial feature, the smallest region quadtree node that *completely contains* a feature is the *minimum bounding node,* which is analogous to the *bounding box* traditionally associated with vector-based spatial representations. Each successive decomposition of this minimum bounding node effectively provides a multiple grid-based bounding box approximation of the object.[16] Figure 8.14 compares the conventional single-resolution minimum bounding rectangle with the "recursive approximation" representation afforded by the linear region quadtree. Although he linear quadtree representation of a region *explicitly* identifies only the associated eaf nodes, as just discussed, the minimum bounding node as well as the higher order root nodes are embedded in the locational codes of all leaf nodes. In the present example, the highest order root node shown represents the bounding box designated as square 1, while the next lower order node refers to the bounding box designated as 2 in Figure 8.14(c). Since all leaf nodes of the region are found within 2 and no lower order node has this property, A_0A_1 is the *minimum bounding node.* The minimum quadtree-bounding node will be referred to as the *object root node.*

Because all leaf nodes of a particular region possess the same *object root node,* he replication of this node-string in each leaf node is highly redundant. Therefore, torage efficiency is enhanced if leaf nodes are represented as node-string offsets elative to this minimum bounding quadtree node. Figure 8.14(e) suggests a simple, ixed-field representation for the region shown in Figure 8.14(a).[17] In this representation, each data element has:

1. A name;
2. A feature class;
3. A set of offsets that permit the root node of the representation to be referenced within both a global and local coordinate system;
4. The root node of the particular region;
5. The list of associated leaf node offsets.

Using this convention, bounding box tests for intersection and inclusion reuce to highly efficient bit-string operations on feature root nodes. Since the miniıum bounding node of region 1 (that is, A_0) intersects the minimum bounding ode of region 2 (that is, A_0B_1), more detailed analysis is warranted; otherwise the ıtersection product is null.

---

16. As will become clear, this representation can be viewed as a multiple-resolution spatial index.

17. In an actual implementation, the leaf nodes would be represented in a tree (*for example,* B-tree). Although the tree depth can be embedded in the string notation (as implied by the string subscripts), there are significant advantages to handling the decomposition depth separately so that only 2 bits are required to code each level.

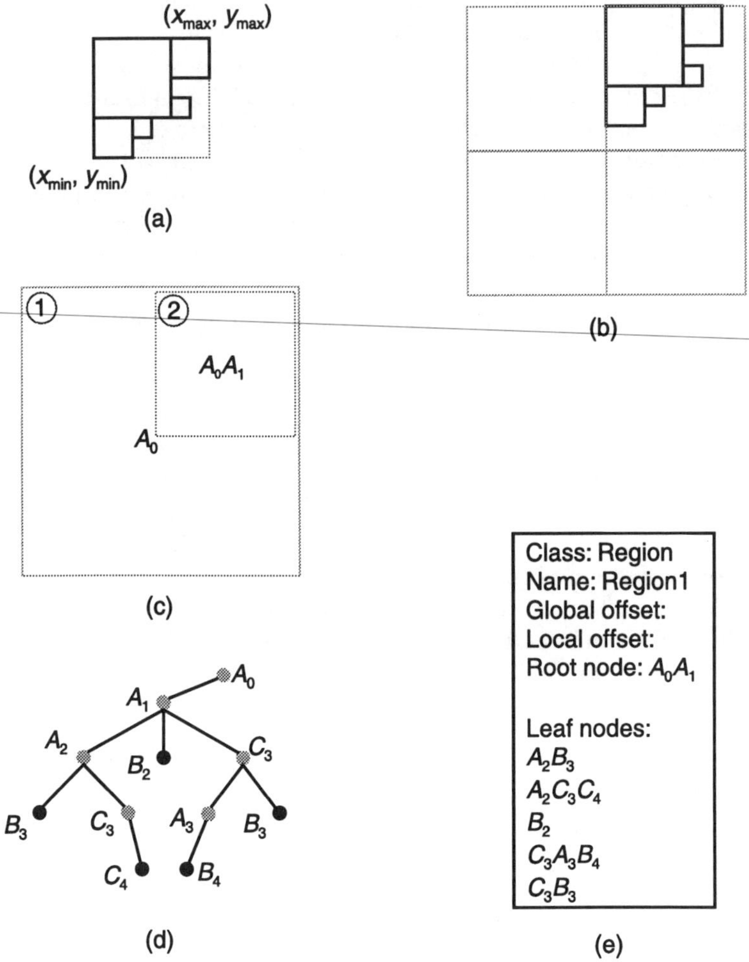

Figure 8.14 (a) Conventional bounding box representation of a region, (b) region quadtree-based rep resentation of the same region, (c) bounding squares 1 and 2, (d) quadtree representatio of region (leaf nodes are black), and (e) table-based representation that explicitly stores th root node and all leaf nodes as linear quadtree offsets.

The comparison between a conventional bounding box and a linear quadtre based representation for a vector-based lineal feature is shown in Figure 8.15. Figur 8.15(a) shows the implicit vector-based bounding box associated with two adjacer endpoint pairs. Figure 8.15(b) shows the vector-based bounding box for an e tended piecewise-linear feature. Finally, Figure 8.15(c) depicts a multiple-node e pansion of the root node of the linear quadtree representation of the same line feature. Higher resolution depictions of the lineal feature merely require consideration of lower level nodes in the linear quadtree representation. For instance, if a road enters a map at the southwest corner and leaves at the northeast corner, the conventional bounding box would include the entire map, which would provide no search-space size reduction potential. Thus, for highly extended lineal features, the use of multiple bounding nodes can permit a significantly reduced search space as compared to the conventional single bounding box approach associated with vector-based representations.

Distance Metrics

Because the locational code is a relative coding scheme (all nodes represent offsets relative to a root node), quadtree-based representations facilitate the computation of both absolute and relative distance metrics. The Euclidean directed-distance between two nodes can be computed as the term-by-term subtraction of their node-strings. The *x*- and *y*-components of all possible resulting difference terms are summarized in Table 8.3. The magnitude of the relative distance between any two nodes is the square root of the sum of the squares of the respective x and y-compo-

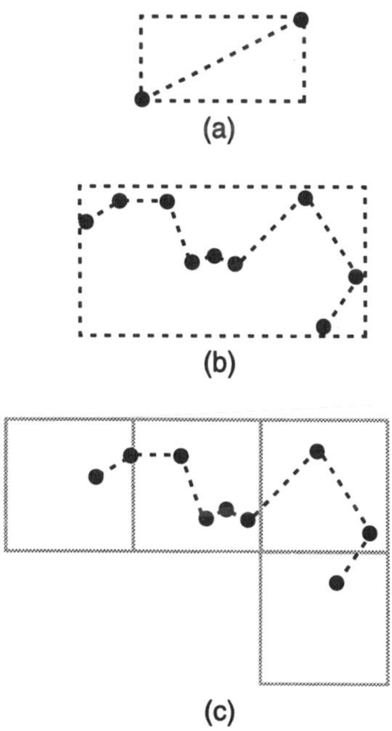

Figure 8.15 (a) Bounding box implied by an implicit endpoint pair representation, (b) an explicit non-regular bounding box representation of a piecewise-linear feature, and (c) a multiple regular bounding box representation of the feature in (b).

nents. As demonstrated in Figure 8.16, the relative distance depends only on the node size and the move class (either cardinal or noncardinal). By accumulating the *x*- and *y*-components of a series of nearest neighbor moves, arbitrary path lengths can be computed. Absolute distance is developed by merely scaling the relative metrics by the physical spatial dimensions associated with a node.

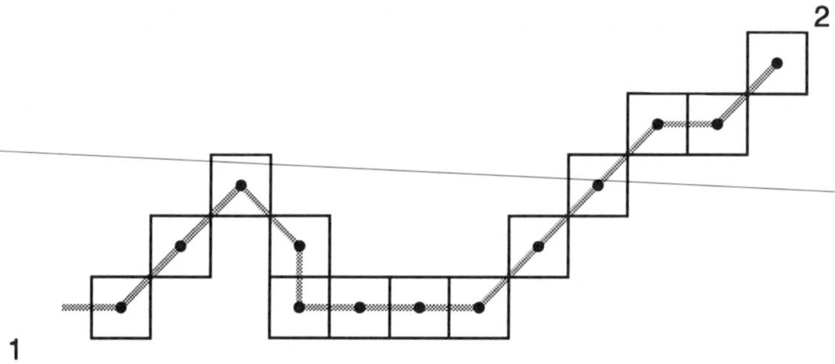

Figure 8.16 Path length between node 1 and node 2 is the sum of the cardinal nearest neighbor distances (that is, $(6)(2^{-i})$) and the noncardinal nearest neighbor distances (that is, $(7\sqrt{(2)}(2^{-i}))$.

Table 8.3
Distance Metrics associated with String Operations

String Term	*x-component*	*y-component*
A_i	2^{-i}	2^{-i}
B_i	-2^{-i}	2^{-i}
C_i	-2^{-i}	-2^{-i}
D_i	2^{-i}	-2^{-i}
$A_i - B_i$	2^{-i+1}	0
$A_i - C_i$	2^{-i+1}	2^{-i+1}
$A_i - D_i$	0	2^{-i+1}
$B_i - C_i$	0	2^{-i+1}
$B_i - D_i$	-2^{-i+1}	2^{-i+1}
$C_i - D_i$	-2^{-i+1}	0

Note: $(U_i - V_i) = (V_i - U_i)$

General two-dimensional distance metrics among combinations of point, line, and region features can be evaluated by region growing, followed by intersection testing. Figure 8.17 shows a coarse-resolution growth of a region that was generated by appending the neighbors of all boundary nodes of the original region. If the new region (the original region plus the boundary dilation) intersects a feature of interest, the original region is within the region growth size of the feature. If an intersection does not occur, the algorithm can be reapplied until the approximate distance to the feature of interest has been discovered. Because of the multiple-resolution character of the linear quadtree, the algorithm efficiency can be improved by beginning with very coarse resolution region growth and intersection, followed by top-down resolution refinement.

Spatial Search

Spatial windows define the search-space limits for spatially organized data. Spatial windows in conventional database systems are often restricted to being rectangular. The region quadtree representation inherently supports the use of highly efficient geometric (rectangular, sector, segment-based) windows and arbitrarily shaped spatial regions [12].

The spatial window size can be scaled by a factor of 2^n (where n is an arbitrary positive or negative integer) by merely incrementing or decrementing the linear quadtree-node subscripts that define the decomposition depth. Consider the spatial

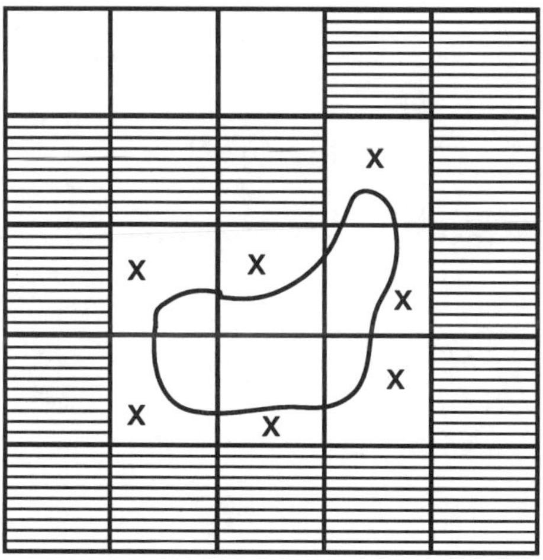

Figure 8.17 Distance computation via single-resolution region growth on the boundary nodes of a region.

window depicted in Figure 8.18. When this region is scaled by a factor of 2, the resultant window is shown in Figure 8.19.

A spatial window can be translated to an arbitrary position by either of the following methods:

1. Applying the appropriate set of x and y offsets to all nodes within the window;
2. Rebuilding the window at a new location based on the nodal relationships within the original window.

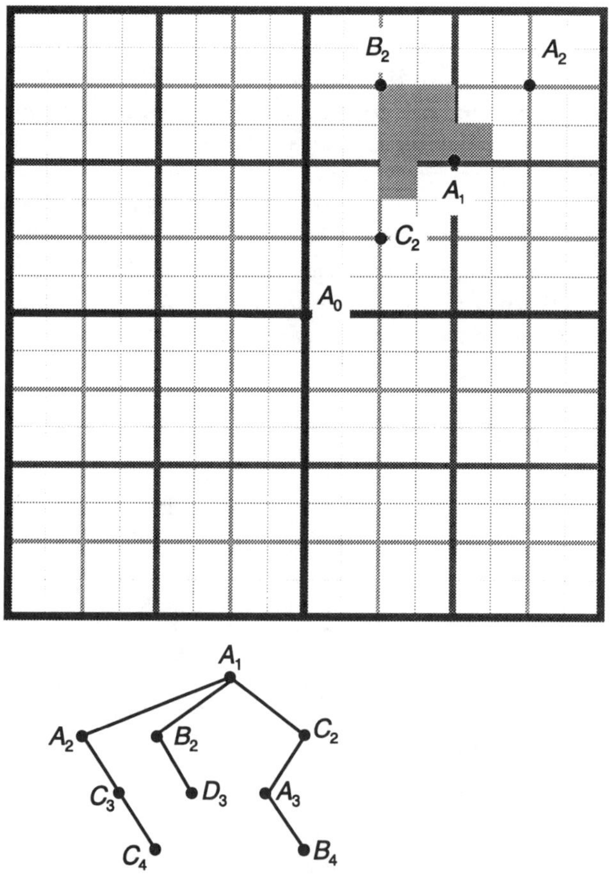

Figure 8.18 Region representation example.

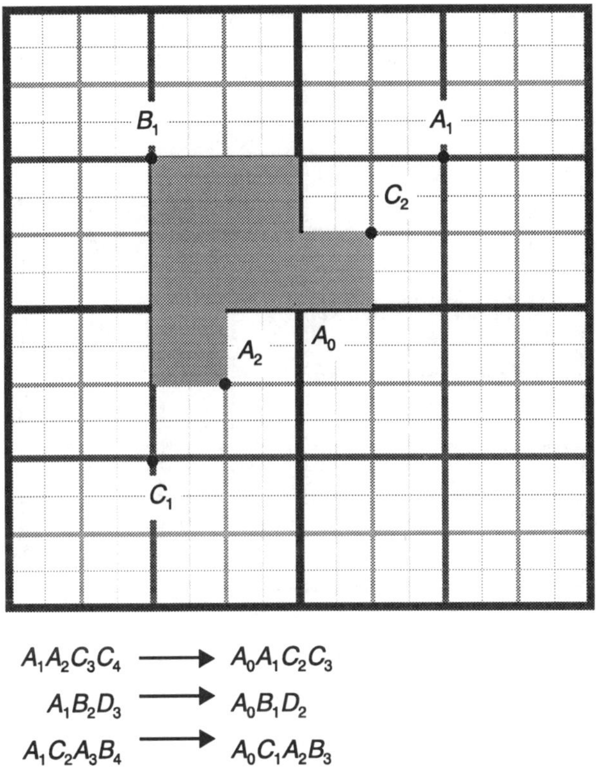

igure 8.19 Scaled region shown in Figure 8.18 generated by decrementing the node-level designations of the original region.

Figure 8.20 shows an example of the translation of the region in Figure 8.18. If he upper-left node of the noncompressed (or nonminimal) quadtree repreentation of the original window $A_1B_2D_3B_4$ is to be translated to node $A_1C_2A_3D_4$, the lirected distance between the original and final position is computed (using Table 3.3) as the term-by-term subtraction $(A_1C_2A_3D_4) - (A_1B_2D_3B_4) = (1/8, -3/8)$.

Using the first window translation approach, each quadtree node of the origiıal window is translated by $(1/8, -3/8)$, which is equivalent to a south move at ıode level 3 (that is, 2^{-2}) and a southeast move at node level 4 (that is, 2^{-3}). In the econd approach, the window is constructed about the translated window uppereft corner node based on the nearest neighbor relationships among the nodes f the original window (remembering that $A_1B_2D_3$ is equivalent to the four nodes $A_1B_2D_3A_4$, $A_1B_2D_3B_4$, $A_1B_2D_3C_4$, and $A_1B_2D_3D_4$).

Due to the recursive symmetry of the quadtree representation, regions can be asily rotated by multiples of 90 degrees. Table 8.4 summarizes the node-string ransformation rule for rotations 90, 180, and 270 degrees. The 90 degree counter-

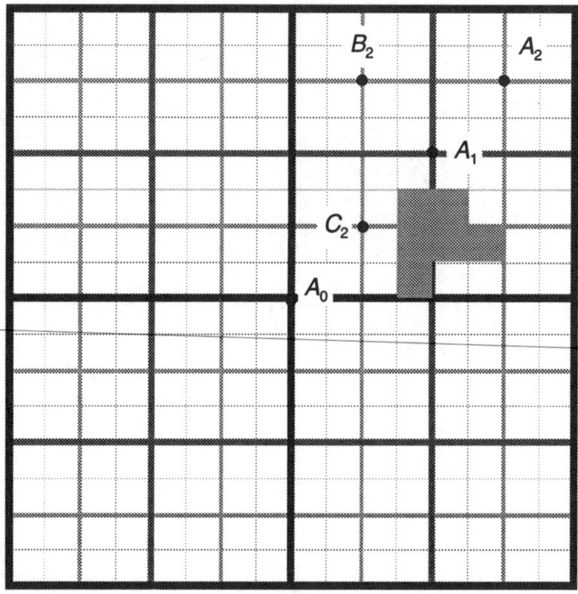

$A_1A_2C_3C_4$ $\quad$ $A_1C_2A_3D_4$

$A_1B_2D_3\,(S_3,\ SE_4)$ $\quad$ $A_1C_2D_3A_4$

$A_1C_2A_3B_4$ $\quad$ $A_1C_2D_3D_4$

$A_1D_2B_3C_4$

$A_1D_2C_3A_4$

$A_1D_2C_3B_4$

Figure 8.20 Translation of the region shown in Figure 8.18.

clockwise rotation of a quadtree-based region is shown in Figure 8.21. The minimal region quadtree leaf nodes of the original region are shown to the left, and the rotated (and transformed) nodes are shown to the right:

$$
\begin{aligned}
D_0 &\rightarrow A_0\\
A_0C_1 &\rightarrow B_0D_1\\
A_0D_1B_2 &\rightarrow B_0A_1C_2\\
A_0D_1A_2C_3 &\rightarrow B_0A_1B_2D_3\\
A_0D_1C_2A_3 &\rightarrow B_0A_1D_2B_3
\end{aligned}
$$

Table 8.4
Quadtree Node-String Operation for Rotation by 90, 180, and 270 Degrees

90°	*180°*	*270°*
A→B	A→C	A→D
B→C	B→D	B→A
C→D	C→A	C→B
D→A	D→B	D→C

8.4 DETAILED DATABASE DESIGN

Based on the database organizational principles outlined in Section 8.2 and the quadtree data structure background just provided, this section outlines a framework for a database kernel that supports highly efficient mixed spatial and semantic reasoning. As discussed in Section 8.2, the database consists of the following three major components:

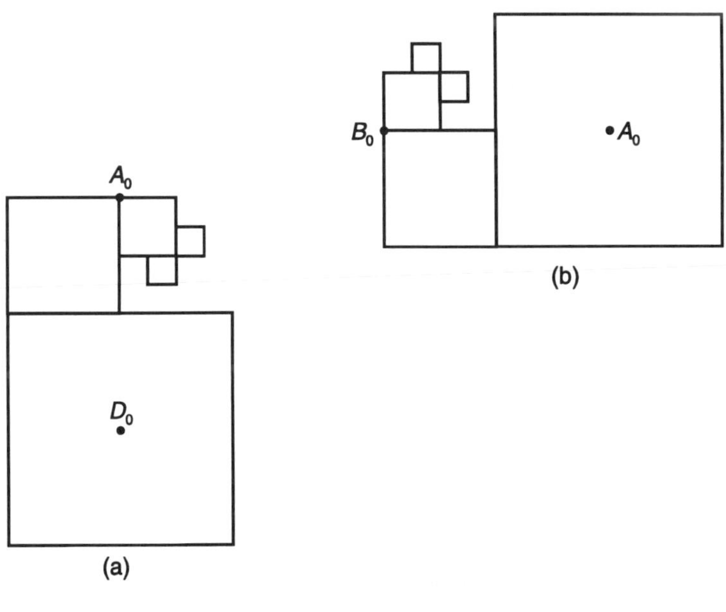

Figure 8.21 (a) Quadtree represented region, and (b) the 90 degree counterclockwise rotation of (a).

1. A general semantic object representation;
2. An object representation of two-dimensional space;
3. A hybrid spatial representation of *individual* point, line, and region objects.

Object-oriented representations facilitate property inheritance and permit the representation of complex relationships among sets of hierarchically organized entities. Semantic reasoning is facilitated by database representations that support search and reasoning with respect to individual semantic objects as well as classes of objects. Similarly, spatial reasoning is facilitated by representations that permit efficient search of both *individual* spatial features and *classes* of spatial features. Thus, component 1 supports object-oriented semantic reasoning and components 2 and 3 support object-oriented spatial reasoning.

8.4.1 Individual Spatial Feature Representation

As discussed previously, the hybrid representation of individual spatial features consists of a hierarchical low-resolution and a nonhierarchical high-resolution representation. The minimal region quadtree is a natural candidate for the low-resolution spatial representation, and a vector representation is a natural candidate for the high-resolution spatial representation. Table 8.5 shows a simple implementation of such a hybrid data structure.

Table 8.5
Summary of Hybrid Point, Line, and Region Spatial Representations

Feature	*Hybrid Representation*	*Example*
Point	Node location and within-node offset	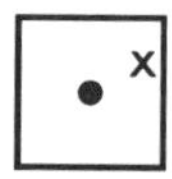
Line	Node location of each associated node and list of vertices (node-offset form) of piecewise line segments within each node	
Region	List of minimal region quadtree nodes; line feature representation within all boundary leaf nodes (assumed counterclockwise order to boundary list)	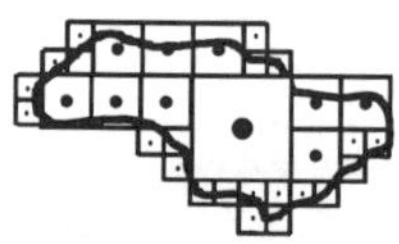

As observed in this table, the spatial representation of specific *point* features consists of a single linear quadtree node (for example, a maximum-resolution pyramid node) plus the *x* and *y* offsets of the point with respect to the center of the node.[18] The spatial representation of *lines* consists of the set of linear quadtree nodes that cover the line feature plus a list of the vertices (maintained in node-offset form) of a piecewise-linear approximation of the line segments associated with each quadtree node.[19] Finally, *regions* use a minimal region quadtree representation that covers both the region interior plus its boundary, coupled with a vector representation of the part of the boundary that is associated with each boundary node.

The region quadtree-based component of the hybrid spatial representation can be treated as a hierarchical spatial index to an associated *bucket* or *bin* that stores the detailed vector data.[20] Since all representations share an identical, hierarchical true two-dimensional representation, the spatial indexing of point, line, and region features is fully registered. Because the locational code associated with individual nodes preserves the hierarchical structure of a particular branch of a tree, the hybrid spatial representation effectively provides both a low-resolution feature representation as well as an embedded search structure that supports highly efficient search and set operations among all spatial feature classes.

Hybrid Point Representation

With its *regular low-resolution* and *nonregular high-resolution* representation, the *hybrid point feature* spatial representation offers advantages over:

1. *Hierarchical regular* point representations, such as the *MX quadtree* [13] where the points are discrete and the non-bucket-based *PR quadtree* [14], where the tree depth is related to the distance between the closest data points;
2. *Nonhierarchical regular* point representations, such as a fixed grid representation [15];
3. *Hierarchical nonregular* point representations, such as the *point quadtree* [16] and the $k-d$ *tree* [17] that does not support highly efficient set operations among general point, line, and region features;
4. *Nonhierarchical nonregular* point representations, such as the inverted list method [15] that requires exhaustive search.

---

18. Alternatively, the reference location could be one of the node corners.

19. The node-offset form expresses **x** and **y** position offsets relative to the center of the respective quadtree node. For a 1 km minimum resolution pyramid cell, 1 m accuracy can be achieved by using 10-bit position offsets ($2^{10} = 1024$). Because they eliminate the replication of redundant root-node substrings, node-offset representations enhance memory efficiency.

20. A number of bucketing methods are associated with spatial data structures. In disk-oriented databases, by selecting a bucket size that is related to the disk page size, page faults can be minimized.

The hybrid point representation can be implemented as a *PR quadtree* that possesses an arbitrary bucket capacity [18]. Due to the location uncertainty and spatial extent of physical objects associated with data fusion applications, many "point" features might be more appropriately represented as regions.

Hybrid Lineal Representation

Because it supports both efficient spatial search and set operations, the *hybrid lineal* feature representation offers significant advantages over:

1. *Hierarchical regular* line representations, such as the *edge quadtree* [19] where each node contains only a single straight line segment and the *line quadtree* [20] where lines must be rectilinear and lie on the border of pixels;
2. *Nonhierarchical regular* line representations, such as the *edge-EXCELL* [21];
3. *Hierarchical nonregular* line representations, such as the *strip tree* [22];
4. *Nonhierarchical nonregular* line representations, such as *chain codes.*

The hybrid lineal representation can be implemented using a *PMR quadtree* [23], where the bucket maintains the list of all attributed line segments, as shown in Figure 8.22(b).

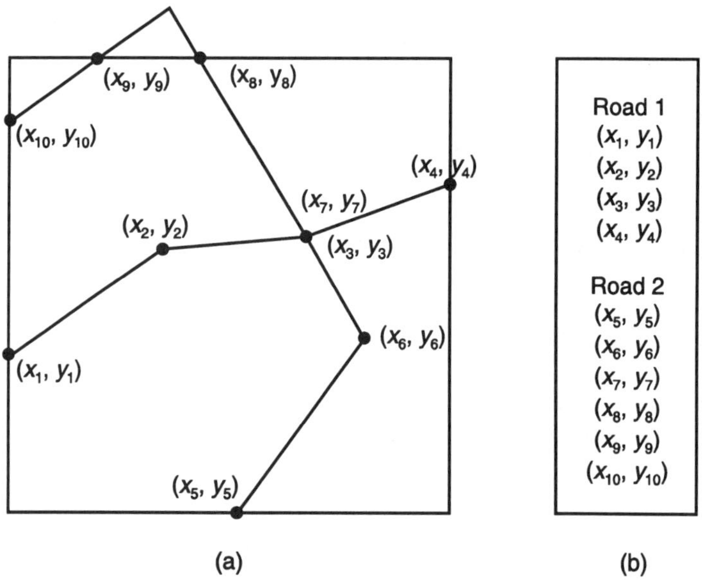

Figure 8.22 (a) Line segments within a maximum resolution quadtree node, and (b) the ordered vertex list of associated line segments.

Hybrid Region Representation

Because it supports both efficient and high-precision search and set operations, the hybrid region representation offers advantages over strictly:

1. *Hierarchical regular* region representations, such as the *regular region quadtree* and the *PM quadtree* [24] that require large tree depths to provide high definition region boundary representation;
2. *Nonhierarchical regular* region representations, such as *raster* and *pixel boundary* that do not facilitate top-down search, association, and Boolean operations, as well as require high resolution representations to provide high-definition region boundary representation;
3. *Hierarchical nonregular* region representations, such as the *R-tree* [25] that does not support highly efficient set operations;
4. *Nonhierarchical nonregular* region representations, such as conventional *polygon* and *pixel boundary-list* that do not facilitate highly efficient search, association, or set operations among spatial features.

The *hybrid region representation* capitalizes on the advantages of both hierarchical grid-based areal representations and the compactness and accuracy of vector-based spatial representations. Since the region boundary representation is maintained as a set of tuples stored in node-offset form, the high-resolution region representation can be implemented using a *PMR quadtree* as shown in Figure 8.23.

Interior and boundary nodes in the region representation can be differentiated by a 1-bit field or by the existence of a pointer to a vector-boundary representation. The "interior" of the vector-boundary representation can be determined by imposing an order to the vertex lists. For example, the interior would be to the left of a counterclockwise-represented boundary. In the case of regions that have embedded holes, the vertex ordering of the region's interior boundaries must be reversed so that the region interior remains consistent with the vertex ordering of the exterior boundary.

Traditional vector-based representations associate a minimum bounding box with all extended objects to enhance both search and set-operation efficiency. In the case of region intersection, if the bounding boxes of the two regions overlap, the regions potentially intersect; on the other hand, if the bounding boxes do not intersect, then the regions cannot intersect. The regular region quadtree component of the hybrid spatial representations can be considered to be a hierarchical bounding box. For example, since all region 1 and region 2 nodes in Figure 8.12 share the parent node A_0B_1, their root nodes intersect. Further analysis is required to verify that the regions themselves actually intersect. However, unlike conventional bounding box–based tests, the hybrid spatial representation permits the de-

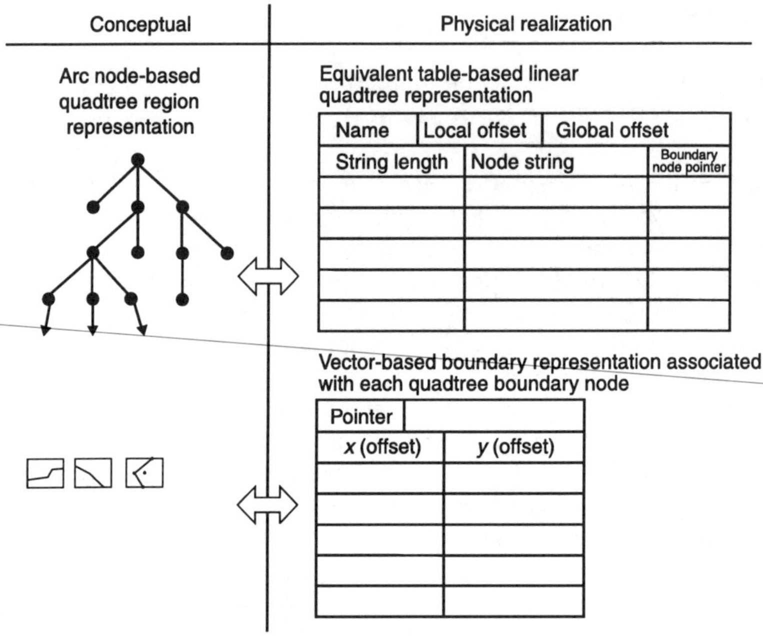

Figure 8.23 Candidate implementation of hybrid region representation for variable node-string representations.

tailed set-operation products to be developed using highly efficient top-down AND and OR operations among quadtree nodes. If the resolution of the resulting product is sufficient, no further processing is required. When a higher resolution product is required, the associated vector representations can be utilized.

Thus, the hybrid true two-dimensional region representation supports high-efficiency spatial search and set operations, while, at the same time, providing high-definition region boundary representation and manipulation. Rather than two separate and distinct spatial representations, the low-resolution spatial representation can be treated as a hierarchical spatial index to memory-efficient, high-resolution representations of specific points, lines, and regions. The hierarchical, true two-dimensional character of the spatial index permits highly efficient tree-oriented search of the high-resolution spatial database. In much the same way as an index file in a RDBMS offers efficient one-dimensional indexing, the linear quadtree-based low-resolution representation provides an efficient hierarchical two-dimensional index to higher resolution spatial data. Because all data sets are fully registered, the hybrid representation supports efficient search among all classes of spatial features.

8.4.2 Spatial Object Class Indexing

While the hybrid spatial representation supports efficient true two-dimensional search and manipulation of *specific* spatial objects, the representation does not directly support spatial search with respect to *classes* of spatial features. With no spatial feature class representation, the test for the existence of a paved road within 1 km of an arbitrary query point could require search of the spatial representations of *all paved* roads or worse yet, *all* roads. With an object representation of two-dimensional space, on the other hand, highly effective spatial search is possible.

The pyramid offers an ideal data structure for a hierarchical, true two-dimensional representation of spatial feature classes. The apex of a pyramid would represent the entire spatial domain of interest; each successively lower level node would provide twice the resolution of the preceding level. Assuming each cell has equal storage requirements, the total storage requirements of a pyramid is 1.33 times the storage requirement of its highest resolution layer.[21] If each pyramid node maintains a semantic description of its associated geographic region, the pyramid could provide a multiple-resolution, spatially organized semantic description of two-dimensional space (in effect, an object-oriented representation of space).

Direct Verses Indirect Indexing

Two related pyramid-indexing schemes are of interest. In *indirect indexing,* each pyramid node maintains a bit-coded representation of all feature *classes* resident in that cell. For example, if bit 10 was set to 1, at least one four-lane highway would be known to exist within the spatial region ascribed to that cell; if a 10 bit was set to 0, no four-lane highway exists within the cell. Because of its multiple-resolution character, the pyramid supports efficient top-down spatial indexing of *classes* of spatial features. When refinement below the resolution provided by the pyramid is required, the *semantic object representations* of *all* feature classes of interest would be searched using the associated hybrid spatial features. Because pyramid cells index classes of features, the scheme provides only an *indirect spatial index* to the specific spatial feature database.

In *direct indexing,* nonterminal pyramid nodes would maintain a bit-coded feature-class representation (just as in the indirect scheme), but the terminal nodes would point to a namelist of the specific point, line, and region features that are associated with that cell. Thus, direct pyramid indexing provides a *pointer* to *specific* semantic object representations, which, in turn, point directly to hybrid spatial representations of *specific* spatial features. Although an indirect pyramid index has fewer storage and (potentially) lower maintenance requirements than the direct in-

---

21. In a simple pyramid, each layer requires only one-fourth the storage requirements of the layer below. Normalizing the sum of the storage requirements from each layer results in the series (1 + 1/4 + 1/16 + 1/64 +...) which converges to 1.33.

dexing scheme, the search efficiency is poorer than the direct index approach.[22] Figure 8.24 shows the general relationships between the various database representations for both direct and indirect pyramid-indexing schemes.

In addition to the top-down indexing of the hybrid spatial representation (via a specific named semantic object), an inverse link exists between the hybrid spatial representations and the pyramid. For *point features* there exists a *one-to-one* relationship between an individual spatial representation and the pyramid (that is, a point feature is associated with only a single pyramid node). However, since lines and regions are extended objects, there can exist a *one-to-many* relationship between a hybrid spatial representation and the pyramid (a given feature may extend across multiple pyramid nodes).

Because the pyramid primarily serves an indexing, rather than a data representation role, the resolution of the pyramid base cells can be considerably lower than the resolution of the individual spatial feature representations. For example, the root node of the quadtree-based spatial representation of a specific feature might be on the order of 100m, while the highest resolution pyramid cell might be on the order of 800m. Since both representations employ fully registered quadtree-based decompositions, the 800-m × 800-m pyramid base node would effectively "point to" sixty-four 100-m × 100-m cells. Thus, even the *direct pyramid-indexing scheme* can represent a "one-to-many" pointer.

Because a pyramid is a *complete* (rather than a *minimal*) regular region quadtree, all quadtree nodes in the hybrid spatial representation are either explicitly represented within the pyramid or are direct offspring of such nodes. Since both the *pyramid base resolution size* and the *minimum resolution* of the individual spatial features are design parameters, a tradeoff exists between the desired *search space focus* and resultant *indexing overhead.*

Pyramid Depth

With a direct indexing scheme, the base nodes of the pyramid maintain the name of a variable number of features.[23] When the number of features in a particular node increases above some threshold, a base node is said to "overflow." This overflow is normally handled by simply decomposing the node into four offspring nodes. Similarly, if the number of features within four sibling leaf nodes falls below some threshold, the data in the leaf nodes can be merged into the parent node and the sibling nodes can be removed. Thus, the pyramid "depth" need not be fixed but can be allowed to vary dynamically from leaf node to leaf node, depending on th

22. As is probably obvious by now, such design tradeoffs are endemic to spatial database design.

23. The base pyramid nodes could actually point to an intermediate storage medium (for example, file) that actually holds the name lists.

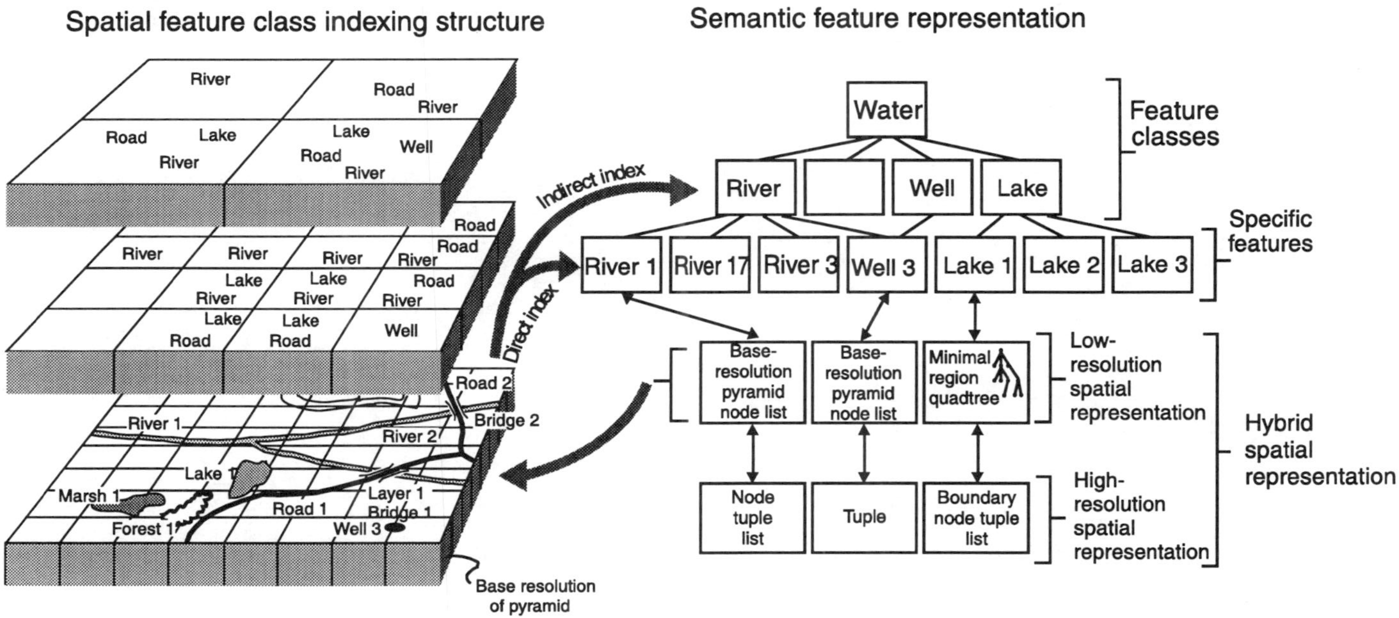

Figure 8.24 Integration of spatial and nonspatial representations.

distribution of spatial features. By proper choice of splitting and merging thresholds, database maintenance operations can be minimized.

Example. To demonstrate the advantage of pyramid-based search for classes of spatial features, consider the database query: Does a bridge exist within a particular spatial window? To simplify the analysis, a square spatial window is assumed. With a pyramid-based *object representation of two-dimensional space,* the above query can be answered by first interrogating the smallest pyramid cell that completely covers the query window. If the bridge feature bit associated with that node is 0, no bridge exists within the query window, and no further analysis is required. On the other hand, if the bridge feature bit is 1, further testing is required to answer the query. In general, the original rectangular window will be smaller than the pyramid node that completely covers the window. Because we need only identify the existence of at least one bridge within the window, the query does not require that either the location or the total number of bridges be determined. The worst case search requirements can be evaluated in terms of the block of the $n \times n$ pyramid base nodes that are within the original search window. If n is reasonably large, the top-down interrogation of this block of nodes requires a polynomial search-space size of order 4 where $i = \log_4 n^2 = 2 \log_4 n = \log_2 n$. Alternatively, using a single-resolution representation, exhaustive search of the same $n \times n$ cells could require an exponential search-space size of order $n^2 = 4^i$.

Inheritance. In order to *maintain* the *object representation of two-dimensional space,* the pyramid must be updated following additions, changes, or deletions of individual spatial features. Figures 8.25 and 8.26 depict the logical flow of a simple bottom-up inheritance mechanism for addition and removal operations, respectively. In general, additions are more straightforward than removal operations. With addition, the feature class already exists in the appropriate pyramid base node, no action is required. On the other hand, if the feature class does *not* exist, the feature bit "turned on." In the case of a removal operation, as shown in Figure 8.26, a determination must be made that there exist no other features of that class associated with the pyramid node. In an *indirect indexing scheme,* this can be a relatively expensive test. In a *direct indexing scheme,* since all specific features associated with a node are explicitly indexed, the number of features that must be tested tends to be considerably smaller.

After all base-level nodes have been updated, a recursive inheritance process must be invoked to update the higher level representations, as shown in Figures 8.25 and 8.26. Because both the pyramid and the low-resolution representation individual spatial features share an identical quadtree-based decomposition, spatially local changes to individual features require only relatively local changes to the pyramid representation. As discussed previously, implicit links between features (that is, spatial relationships) require no maintenance following database changes.

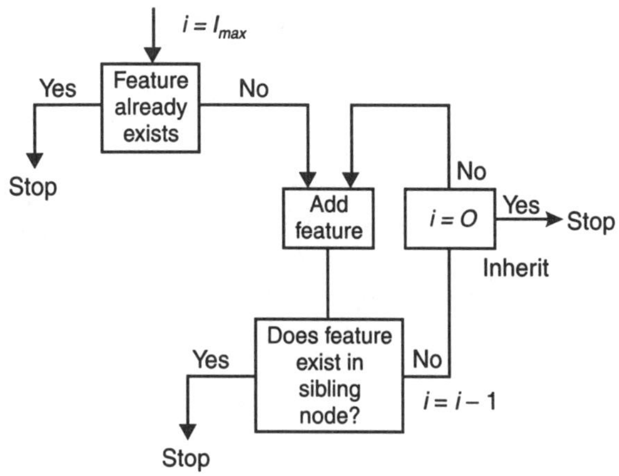

Figure 8.25 Pyramid inheritance for addition of a feature class where *i* is the pyramid depth.

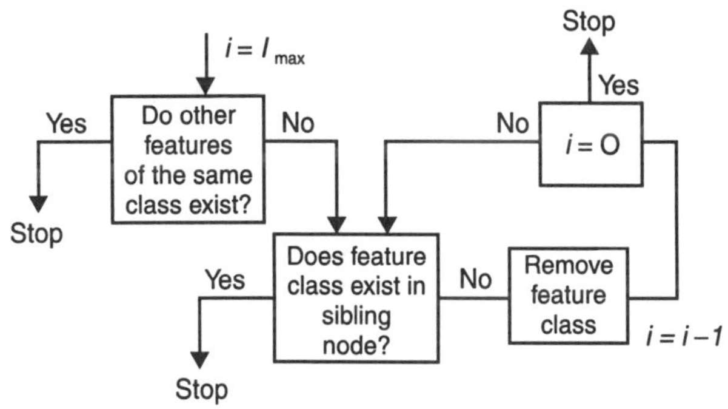

Figure 8.26 Pyramid inheritance for the removal of a feature class where *i* is the pyramid depth.

Consider the addition of the object bridge > class3 > bridge23 to an object representation of two-dimensional space. Because spatial search for the specific spatial feature bridge23 using a spatial feature class index would be far less efficient than *semantic object search* for the same object, inheriting specific features (bridge23) within the pyramid provides only limited benefits. On the other hand, the inheritance of the *bridge class hierarchy* facilitates highly efficient top-down spatial search for classes of bridges.

Rather than maintaining a single level-of-abstraction semantic description of spatial feature classes, the pyramid could provide a multiple level-of-abstraction description that depends on the resolution level. For instance, near the base of the pyramid, the full object class name (*bridge* > class3) may be relevant, while at some higher level of abstraction, only the super class *bridge* might be required. At an even higher level of abstraction, the bridge could be considered an element of a *transportation network*. In terms of storage efficiency, even though the semantic meaning of features at different levels in the pyramid changes, the information in each node remains efficiently bit-coded. While "high-resolution" semantic and spatial information tends to be rather *specific*, "lower resolution" semantic and spatial information tends to be more *general*. Thus, pyramid inheritance can effectively maintain the multiple level-of-abstraction character of both the spatial *and* the semantic aspects of information.

8.4.3 Integration of the Spatial and Nonspatial Representations

As we have seen, the *spatial object representation* consists of:

1. A high-resolution vector-based representation of individual point, line, and region features;
2. A hierarchical, true two-dimensional spatial index to individual vector representations;
3. A hierarchical, true two-dimensional spatial index to classes of spatial features.

Since the spatial indexing of features shares the identical regular region quadtree decomposition, the various elements of the spatial database can be readily integrated. The first two items comprise the hybrid spatial representation of individual spatial features; the third item provides the object representation of two-dimensional space.

In addition to a fully integrated spatial representation, seamless integration between the hybrid spatial representations and the semantic object representation is required. Because the semantic object representation points directly to the individual (named) hybrid spatial representations of points, lines, and regions, a simple one-to-one relationship exists between the two representations. The QUILT GIS [26] system and its successor, the spatial and nonspatial data (SAND) system [27] possess many of the characteristics of the integrated semantic/spatial object database kernel discussed here.

8.4.4 Database Kernel Summary

Figure 8.1 depicts the high-level organization of the integrated spatial and semantic object kernel. Table 8.5 illustrates the hybrid representation of individual spati features, consisting of a quadtree-based spatial index to one or more tables that define high-resolution vector-based representations of points, line segments, and region boundary segments. The hybrid representation supports straightforward search, association, and set operations among arbitrary combinations of point, line, and region features.

Point features are maintained as a single linear quadtree node that, in effect, specifies the general location of the point in terms of a quadtree-based bounding node plus an associated (x, y) offset that identifies the actual location of the point within the bounding node. Although the use of a hybrid representation for point objects is not actually required, logical consistency is preserved with the hybrid line and region representations.

Vector-based representations offer high-precision piecewise-linear approximations of *lineal features.* The hybrid lineal representation augments the traditional vector representation with a regular region quadtree-based spatial index. The low-resolution representation consists of a set of contiguous quadtree leaf nodes that cover the line segments. While a traditional bounding box approach would capture the entire feature within its minimum enclosing rectangle (which can be extremely large for extended lineal features), the hybrid spatial representation uses a *set* of logically disjoint bounding nodes that provide a coarse-resolution representation of the feature. The vertices of the line segments associated with each bounding node are stored as an ordered list of (x, y) offsets.

Region features possess both a boundary and an associated area. Thus, the region representation consists of both a high-resolution region boundary as well as a relatively coarse, multiple-resolution areal-based representation. The hybrid representation facilitates efficient search and set operations as well as high-precision analysis.

Because both the object representation of two-dimensional space and the representation of individual spatial features are metaphorically similar to the database form of the acetate-overlay paradigm, the database supports automated spatial reasoning that emulates manual analysis. The hierarchical character of the quadtree-based spatial representation of individual points, lines, and regions supports efficient, top-down focus of the search space. The high-resolution representation of points, lines, and regions supports memory-efficient, high-precision characterization and analysis.

Much like a book index that facilitates direct access to both general as well as to specific information, the two separate indexing schemes (pyramid-based *object representation of two-dimensional space* and the *quadtree-based representation of individual spatial features*) provide indexing to both general and specific spatial features, respectively. A book index, for example, may reference both general subject areas that might involve an entire chapter on digital filters, while a more specific reference might point to the material on a specific page that discusses the design of Butterworth digital filters. In much the same way, the pyramid facilitates access to classes of information that exist within regions of two-dimensional space, while the quadtree-based representation of individual spatial features index specific high-resolution spatial representations.

Section 8.4.2 discussed both *indirect* and *direct* pyramid-indexing schemes. An *indirect indexing* scheme supports access to all features of a particular class that are associated with the corresponding spatial region. A *direct indexing* scheme facilitates access to all *specific* spatial objects associated with the highest resolution layer in the pyramid. For all classes of spatial objects, the pyramid provides a *direct link* to either *semantic object classes* (for a pyramid that supports indirect indexing) or *specific semantic objects* (for a pyramid that supports direct indexing) as well as an *indirect link* through the specific semantic object to the individual hybrid spatial representations.

In summary, the spatial object representation consists of:

1. A *hierarchical regular representation of individual spatial features* that supports both low-resolution spatial reasoning and provides a spatial index to high-resolution spatial features;
2. A *high-resolution nonregular representation of spatial features* that supports high-resolution spatial reasoning;
3. An *object-oriented representation of two-dimensional space* that supports low-resolution spatial reasoning with respect to classes of spatial features as well as provides an index to individual spatial features.

As illustrated in Table 8.5, all low-resolution point, line, and region features share an identical quadtree-based spatial decomposition and all high-resolution point, line, and region features are stored in vector form. When such a spatial object representation is fully integrated with a semantic object representation, the resultant *semantic/spatial object database kernel* potentially supports efficient and highly intuitive search and manipulation of both spatial and nonspatial data.

8.5 HARDWARE ARCHITECTURE

For database management systems that rely on disk-resident data, data retrieval time depends on both search-space size and disk retrieval speed. To avoid competing for resources with application software, the entire database management system should ideally reside on dedicated hardware. Further performance improvements can be realized by distributing the *processing*, the *data*, and the *control* while minimizing the required communication among the processors. Because it minimizes processor interaction, communication requirements are minimized if all data sets and database software are maintained at the individual processing nodes. By distributing the data among a large number of processors, a significant portion of the database can reside within random access memory (RAM). The remaining data can be stored on disks associated with individual processor nodes. In addition to the speed advantage of RAM access relative to disk access, such distribution offers the potential for sophisticated concurrent search and manipulation.

Due to the recursive symmetry of the pyramid representation, storage, search, and manipulation of the *object representation of two-dimensional space* could be readily distributed to four processors operating on fully disjoint data sets. Such decomposition would permit concurrent search and manipulation operations. By maintaining a "control node" processor, search and manipulation requests that cross the data-set boundaries of the four "lower level" processors can be readily handled.

Additional concurrency and online RAM would be available with the addition of a 16-node processor network one level lower in this hierarchical process decomposition. The five processors at the first two process levels could maintain the higher level pyramid representation as well as support hierarchical control of the lowest level processors. This divide-and-conquer approach can be continued, where each increase in "depth" of the processor network produces an approximate four times increase in both processing concurrency (relative to that afforded by the previous layer) and overall on-line memory. If individual processors possessed 128 MB of RAM, a four-level, 85-node (1 + 4 + 16 + 64) processor network, for instance, would possess 10 GB of distributed random access storage plus powerful hierarchical processing potential. Because of the natural hierarchical and regular spatial symmetry of the quadtree representation, each processor node could effectively employ identical software.

The hierarchical structure of the processor network supports both centralized as well as highly distributed local control. Each process (except the root and leaf nodes) would have a single parent process and four offspring processes. Query and update requests directed at a particular processor are either acted on locally, sent to the parent processor, or distributed to one or more subordinate processors.

In addition to employing a pyramid-like processor network implementation for the spatial feature class indexing scheme, both the individual hybrid spatial object representations and the semantic object representations can benefit from additional distribution. Because the individual hybrid spatial representations are disjoint, the data sets could be distributed across a two-dimensional array of separate processor nodes.

Since a significant number of operations between processors may be required, a hierarchically organized processor network can offer significant advantages. If the upper level nodes of such a network served primarily a control role, a complete processor hierarchy would not be required. For instance, a truncated or partial processor hierarchy would permit the upper level nodes to control operations that extend beyond the boundary of the lowest level processor nodes. Because the topology of the semantic object database is not as well structured as the spatial object database, the use of a reconfigurable hierarchical processor network might best match the character of the semantic object database representation.

As with all high-concurrency architectures, the degree of concurrency achievable is highly dependent on the distribution of individual queries. Queries that address physically disjoint spatial data sets (resident on different processor nodes) can be performed concurrently, while low concurrency would be expected for spatially local queries. In addition, if too many levels of processor decomposition are used, a significant number of operations will occur at the boundaries of spatial data sets which require resolution operations across multiple processors. Efficient message-passing communication between processors is possible since all required data, reasoning knowledge, and local control knowledge are locally resident. By supporting distributed processing, distributed data (data handled by each processor is in local RAM or on a local disk), and distributed control, the recommended hardware architecture would exhibit the three critical elements that facilitate high-concurrency database search and manipulation.

For highly mobile forces or autonomous vehicle applications, a vehicle-centered, spatially organized database could be maintained in RAM by progressively deactivating the "trailing edge" processors and reassigning them to new spatial region of interest. In this way, a distributed spatial database would permit the dynamic migration of the overall digital database (residing on disk) to the multiprocessor network with minimal impact on database performance.

8.6 SUMMARY AND CONCLUSIONS

A database kernel that can support the development of both sophisticated and context-sensitive data fusion algorithms has been outlined. Since the representation, search, and manipulation of spatial data tends to eclipse other database requirements, the database design focused on the development of an effective and efficient object representation of both spatial feature classes as well as representations of specific point, line, and region features.

At the highest level of abstraction, the database kernel consists of the following two fully integrated object representations:

1. A conventional semantic object representation;
2. A true two-dimensional spatial object representation.

The spatial object representation consists of two distinct but fully integrated representations:

1. A hierarchical true two-dimensional object representation of relatively static two-dimensional spatial feature classes (the pyramid);
2. A hybrid spatial representation of individual point, line, and region features.

The hybrid spatial representation, in turn, consists of:

1. A hierarchical true two-dimensional low-resolution spatial representation;
2. A vector-based high-resolution spatial representation.

The hybrid spatial representation supports very natural two-stage spatial reasoning. In the first stage, the linear quadtree-based representation supports top-down low-resolution spatial reasoning with respect to specific spatial features. Stage 1 facilitates global problem solving and spatial focus of attention, offering potentially massive search-space reduction as compared to strictly nonhierarchical, non-true two-dimensional spatial representations. When higher resolution analysis is required, the second stage of analysis refines the products generated in stage 1 using high-resolution vector-based representations. The linear quadtree representation can be treated as a multiple-resolution spatial index to memory-efficient vector representations of individual points, lines, and regions.

As observed in Figure 8.10, the minimal region quadtree representation compares favorably with the pixel-boundary representation in terms of storage requirements and yet supports both hierarchical true two-dimensional search and areal-based set operations. Since both raster and pixel-boundary representations are grid-based, the high-resolution representation of points, lines, and regions can demand extremely small pixel sizes, which, in turn, increases memory requirements and decreases the efficiency of virtually all database operations.

Although vector (or equivalently, polygon) representations of points, piecewise-continuous lines, and piecewise-continuous region boundaries offer both high-accuracy and low-storage requirements, vector-based representations do not support spatial focus-of-attention, spatially anchored search, efficient set operations, or complex associations among combinations of spatial features. On the other hand, a hybrid representation, consisting of a vector representation and associated hierarchical, true two-dimensional spatial index, can capture the advantages of both data structures.

Even though both conventional raster and pixel-boundary representations are widely used spatial data representations, neither is employed in the semantic/spatial object database kernel. While a *raster* representation is a true two-dimensional spatial representation, raster representations do not support top-down, multiple-resolution reasoning and tend to be highly memory-inefficient when high-resolution representations are required. Since they store only the boundary cells of a region, *pixel-based boundary* representations are not true two-dimensional spatial representations. Although more memory-efficient than raster representations, such efficiency comes at the expense of both search and manipulation efficiency.

A conventional semantic object representation maintains a natural hierarchical relationship between atomic objects and progressively higher level abstractions. Taken together, the pyramid and the hybrid spatial representations of individual spatial features possess these same characteristics. Just as the top-down search through a semantic object hierarchy progressively focuses the search-space size by locating specific lower level semantic objects of interest, top-down search through the pyramid provides an identical capability for spatially organized knowledge. The full integration of both spatial and semantic object representations effectively provides both a semantic window into a spatial database as well as a spatial window into a semantic database. Important extensions to the database kernel, including query optimization and higher level user interfaces, are discussed in the next chapter.

References

[1] Samet, H., *Applications of Spatial Data Structures: Computer Graphics, Image Processing, and GIS*, Reading, MA: Addison-Wesley, 1990.

[2] Samet, H., *The Design and Analysis of Spatial Data Structures*, Reading MA: Addison-Wesley, 1990.

[3] Klinger, A., "Patterns and Search Statistics," *Optimizing Methods in Statistics*, J. S. Rustagi (ed.), New York: Academic Press, 1971, pp. 303–337.

[4] Gargantini, I., "Detection of Connectivity for Regions Represented by Linear Quadtrees," *Comput Math. Appl.*, Vol. 8, No. 4, 1982, pp. 319–327.

[5] Antony, R., "A Generalized Graph Representation of Expert System Knowledge Bases," The Harry Diamond Laboratories, Report DELHD-RT-RD-84-6, 1984.

[6] Antony, R., "Spatial Reasoning and Knowledge Representation," *Proc. Geographic Information Systems in the Government Workshop*, B. Opitz (ed.), Hampton, VA: A. Deepak Publishing, 1986.

[7] Hunter, G.M., and K. Steiglitz, "Operations on Images Using Quadtrees," *IEEE Trans. on Pattern Anal. and Machine Intelligence*, Vol. 1, No. 2, April 1979, pp. 145–153.

[8] Samet, H., "Neighbor Finding Techniques for Images Represented by Quadtrees," *Compute Graphics and Image Processing*, Vol. 18, 1982, pp. 37–57.

[9] Boggess, J.E., P.B. Nation, and M.E. Harmon, "Using Artificial Neural Networks for Data Compression of Color Information in Digitized Images," 1994 (personnel correspondence from the principal author).

[10] Dyer, C.R., "The Space Efficiency of Quadtrees," *Computer Graphics and Image Processing*, Vol. 19 No. 4, August 1982, pp. 335–348.

[11] Chubb, D.W.J., "A Parallel Implementation of Boolean Set Operations Using a Generalizable Vector Formatted Polygon Map Feature Representation," *Proc. 1994 Army Science Conf.*, Orlando, FL June 1994.

[12] Antony, R., "A Hybrid Spatial/Object-Oriented DBMS to Support Automated Spatial, Hierarchical and Temporal Reasoning," Chapter 3, *Advances in Spatial Reasoning*, S.-S. Chen (ed.), Vol. I Norwood, NJ: Ablex, 1991.

[13] Hunter, H., and K. Steiglitz, "Operations of Images Using Quadtrees," *IEEE Trans. on Pattern Ana and Machine Intelligence*, Vol. PAMI-l, 1979, pp. 145–153.

[14] Samet, H., "The Quadtree and Related Hierarchical Data Structures," *Computing Surveys*, Vol. 16 No. 2, 1984, pp. 187–260.

[15] Knuth, D.E., *The Art of Computer Programming, Vol. 3, Sorting and Searching*, Reading, MA: Addison Wesley, 1973.

[16] Finkel, R.A., and J.L. Bentley, "Quad Trees—A Data Structure for Retrieval on Composite Keys *Acta Informatica*, Vol. 4, 1974, pp. 1–9.

[17] Bentley, J.L., "Multidimensional Binary Search Trees Used for Associative Searching," *Comm ACM*, Vol. 18, No. 9, 1975, pp. 509–517.

[18] Nelson, R.C., and H. Samet, "A Population Analysis of Quadtrees with Variable Node Size," Report No. TR-1740, University of Maryland Computer Science Department, College Park, MD, December 1986.

[19] Warnock, J.E., "A Hidden Surface Algorithm for Computer Generated Half Tone Pictures," Computer Science Department TR 4-15, University of Utah, Salt Lake City, UT, June 1969.

20] Samet, H., and R.E. Webber, "On Encoding Boundaries with Quadtrees," *IEEE Trans. on Pattern Anal. and Machine Intelligence,* Vol. 6, No. 3, May 1984, pp. 365–369.

21] Tamminen, M., "The EXCELL Method for Efficient Geometric Access to Data," *Proceedings of the Nineteenth Design Automation Conference,* Las Vegas, June 1982, pp. 345–351.

22] Ballard, D., "Strip Trees: A Hierarchical Representation for Curves," *Comm. ACM,* Vol. 4, No. 5, May 1981, pp. 310–321.

23] Nelson, R.C., and H. Samet, "A Consistent Hierarchical Representation for Vector Data," *Computer Graphics,* Vol. 20, No. 4, August 1986, pp. 197–206.

24] Samet, H., and R.E. Webber, "Storing a Collection of Polygons Using Quadtrees," *ACM Trans. on Graphics,* Vol. 4, No. 3, July 1985, pp. 182–222.

25] Guttman, A., and M. Stonebraker, "R-Trees: A Dynamic Index Structure for Spatial Searching," Electronics Research Laboratory, College of Engineering, University of California, Berkeley, Memorandum No. UCB/ERL M83/64, October 14, 1983.

26] Shaffer, C.A., H. Samet, and R.C. Nelson, "QUILT: A Geographic Information System Based on Quadtrees," Center for Automation Research technical report CAR-TR-307, July 1987.

27] Aref, W.G., and H. Samet, "Extending a DBMS with Spatial Operations," *Advances in Spatial Databases—Second Symp.,* SSD'91, O. Gunther and H. J. Schek (eds.), 1991.

CHAPTER 9

Database Extensions

The last chapter outlined the organization of an integrated semantic and spatial object database kernel designed to support the development of both intuitive and efficient higher order canonical form fusion algorithms. This chapter presents a number of database kernel extensions intended to enhance both database effectiveness and efficiency. Section 9.1 discusses simple query language extensions for supporting spatially oriented queries. Section 9.2 outlines the underlying principles of a query-optimization strategy for mixed spatial and semantic constraints. Section 9.3 discusses the benefits of developing higher level generic as well as application-specific database shells as means to extend database functionality. Finally, Section 9.4 reviews contemporary database research directed at achieving higher level database functionality.

9.1 DATABASE QUERY LANGUAGE

Database query can be viewed somewhat equivalently as either:

1. *Conditional search;*
2. *Generalized templating;*
3. *Windowing;*
4. *Filtering.*

As discussed at the beginning of Part II, SQL is the current database query language standard. Various extensions to SQL have been developed to support spatial queries and abstract data types. A *verb-object-constraint set* grammar provides a simple, yet highly expressive, query form:

$$Find\ (n)\ f_1\ (objects)\ conditioned\ on\ \left\{\prod f_i\,(c_i)\right\} \quad 9.$$

where f_i represent transformations or modifiers and c_i represent constrair clauses for $i = 1,\ldots, I$. Equation (9.1) is merely a generalization of the SQL SELEC { } WHERE { } and will be referred as the *generic query form.*

Consider the following query:

Find all freshwater lakes of size 10 < *acres* < 50 *that are between* 50*km* and 100 *km City X* and *that are less than* 20 *km from an intersate highway* 9.

Query (9.2) has the following generic query form:

$$\begin{gathered} Find\ (all)\ f_1\ (lake > freshwater)\ (intersect) \\ f_2\ (city > City\ X > shape)\ (intersect) \\ f_3\ (road > interstate) \end{gathered} \quad 9$$

where f_1 is a semantic qualifier or filter on the attribute *surface_area* (10 < su face_area < 50) of the spatial object *lake freshwater* and f_3 is a spatial transformatic (20-km region growth) of spatial object *road > interstate.* Because object-oriented d tabases support object typing, ambiguous semantic references are often easily r solved. For example, although the object *lake > freshwater* might possess ma attributes, if *surface_area* is the only attribute that has the units "acres," the query c be unambiguously interpreted. Similarly, the modifier or spatial transformation (50-km to 100-km growth-ring) must be associated with the most appropriate *attr ute* of the object *city > City X.* In this case, the attribute *city > City X > shape* is an a propriate candidate, whereas the attribute *city > City X > population* would not be candidate.

As a second example, consider the query:

Find all wheeled_vehicle that are near (x_1, y_1, t_1) 9

In its normal semantic usage, *near* is a generalized, context-sensitive metric. will be discussed in more detail in Chapter 12, the metric *near* can be evaluated performing spatial search in the "vicinity" of point (x_1, y_1, t_1). The above query c be readily transformed into its generic form:

$$\textit{Find all moving\_target} > \textit{wheeled\_vehicle intersect } g_1^{\uparrow}(x_1, y_1, t_1) \qquad 9.5$$

where g_1 represents region growth with respect to a context-sensitive function "*spatially near*" point (x_1, y_1) and simultaneously "*temporally near*" time t_1. Using the semantic/spatial object database kernel, both spatial and nonspatial queries can be treated in a uniform manner.

9.2 QUERY OPTIMIZATION

Query optimization depends on a range of hardware, organizational, implementation, and representation issues. The focus of this section is on effective search and retrieval strategies that are supported by the integrated semantic/spatial object database kernel. In general, the "cost" of performing a query can be reduced by effectively matching the *internal* database search strategy and manipulation operations to the underlying data representations. For data that is accessible along appropriate query dimensions (either naturally or through artificial means such as indexing, use of bounding boxes, hashing tables), the search-space size is related to the size or extent of the query window. For data that is not effectively organized, the search-space size tends to be related to the size of the data set.

Because both the hybrid spatial data representation of individual features and the spatial index to classes of spatial features (the pyramid) rely on an identical true two-dimensional spatial decomposition, two fundamentally different approaches to spatial query evaluation are supported: *set operation–based* and *search-based.*

9.2.1 Set Operation–Based Approaches

In set operation–based query evaluation, clause conditions are treated as sets. Set operations associated with both the *object representation of two-dimensional space* and the *hybrid representation of individual spatial features* can be generated by performing simple Boolean operations with respect to the nodes of the quadtree-based spatial representations. In the case of the hybrid spatial representation, a second-stage of evaluation can be required to refine the resulting set-operation products.

Figure 9.1 shows an example of the intersection of two individual regions. Since the regions contain both interior nodes and boundary nodes, the first stage generates three classes of intersection products:

(1) (*region* 1 *interior nodes*) $\cap$ (*region* 2 *interior nodes*) (9.6)

(2) (*region i boundary nodes*) $\cap$ (*region j interior nodes*) $i = 1, 2, j = 1, 2$, and $i \neq j$

(3) (*region* 1 *boundary nodes*) $\cap$ (*region* 2 *boundary nodes*)

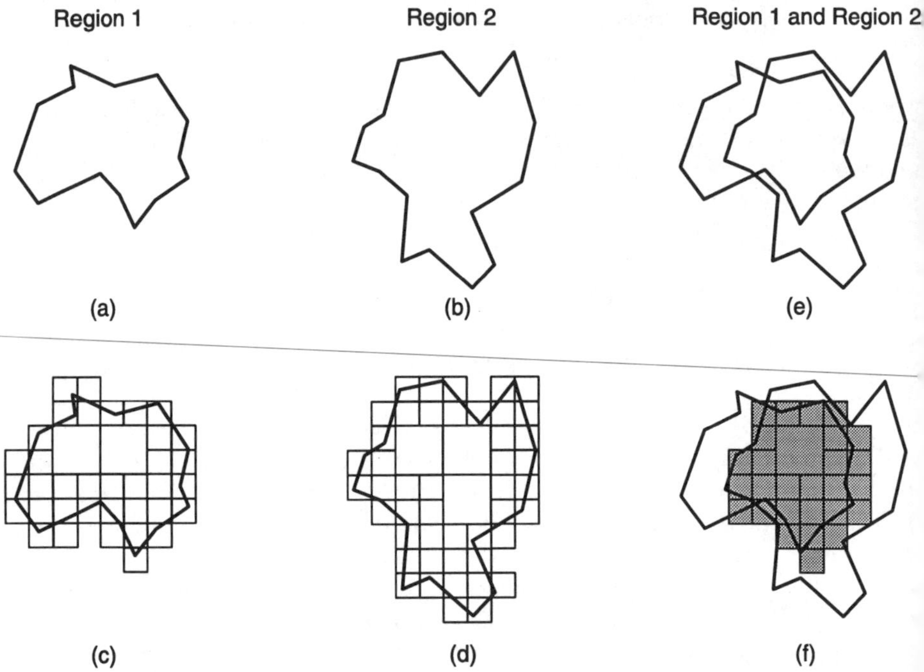

Figure 9.1 Region intersection using a quadtree-based spatial representation. Polygon representation of (a) region 1 and (b) region 2 are overlaid in (e). Minimal region quadtree-based repre sentations of (a) and (b) are shown in (c) and (d), respectively, and the resultant intersectio product is shown in (f).

Because the minimal region quadtree representations of the two regions ar fully registered, the result of the intersection or union of two *interior nodes* is an *int rior node*. The *intersection* of a *boundary node* and an *interior node* produces a *bounda node*; the *union* of a *boundary node* and an *interior node* produces an *interior node*.[1] F nally, the *intersection* or *union* of two *boundary nodes* yields a *boundary node*. The proc uct class for the intersection and union of all combinations of two-dimension spatial features are summarized in Table 9.1.

Figure 9.2 shows the three general classes of intersections that result from th first stage of the intersection operation for the regions depicted in Figure 9.1. I stage 2, all first-stage products that contain boundary nodes must be refined. Figur 9.3 illustrates the generation of a refined stage-2 product for the intersection of *boundary node* and an *interior node* and the intersection of two *boundary nodes*.

1. Recall that in the hybrid spatial representation, boundary nodes possess the identical line quadtree-node representation as interior nodes except that they point to a piecewise lineal rep sentation of the high-resolution region boundary contained within that node.

Table 9.1
Product of Boolean Intersections and Unions

Feature Classes	*Intersections (AND)*	*Unions (OR)*
Point × Null	Null	Point
Line × Null	Null	Line
Region × Null	Null	Region
Point × Point	Point or Null	Point(s)
Point × Line	Point or Null	Line or Point & Line
Point × Region	Point or Null	Region or Region & Point
Line × Line	Point, Line, or Null	Line(s)
Line × Region	Point, Line, or Null	Region or Region & Line
Region × Region	Point, Line, Region, or Null	Region(s)
Point × Line × Region	Point or Null	Region, Line, & Point
		Region & Line
		Region & Point
		Region

Two primary classes of set operation–based queries are considered: *sequential clause evaluation* and *optimal clause-order evaluation.* Figures 9.4(a–c) show three re-gions that will be used to illustrate the various set-operation evaluation strategies. Figures 9.4(d–g) demonstrate *sequential clause evaluation set operation–based query.* In his case, region 1 (which contains 14 nodes) is intersected (ANDed) with region 2 which contains 11 nodes) as shown in Figure 9.4(d). The resultant product ontains 9 nodes as shown in Figure 9.4(e). These 9 nodes are then ANDed with the nodes of region 3. Thus, the total number of nodes considered in step 1 is 5(14 + 11) and the total number of nodes in step 2 is 12 (9 + 3), for a total of 37 nodes.

Based on Table 9.1 and the above example, the following observations can be made.

1. Because *lines* are degenerate *regions* and *points* are degenerate *lines, regions* represent the *highest order* (two-dimensional) *spatial feature* and *points* represent the *lowest order spatial feature.*
2. An *intersection product* generally assumes the dimensionality of the lowest order feature class, while the *union product* assumes the dimensionality of the highest order spatial feature class.[2]

Limiting case exceptions to this rule can occur. For instance, if two regions intersect only along a portion of their boundaries, the product would be a line rather than a region.

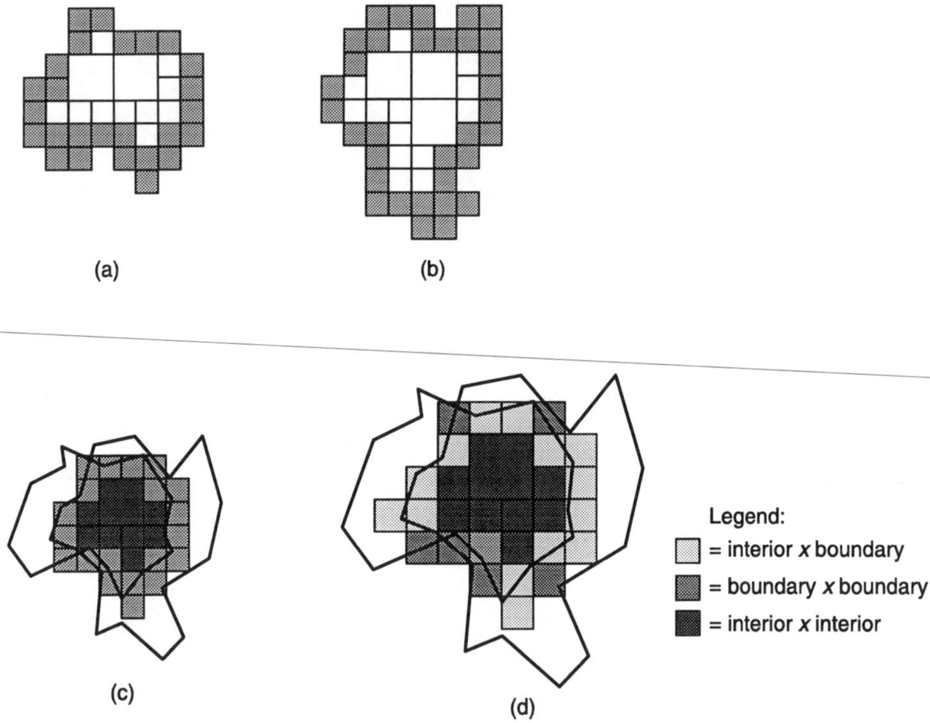

Figure 9.2 The shaded nodes of regions in (a) and (b) represent the boundary nodes from Figure 9.1(c, d), respectively. The intersection product is shown in (c). The AND operation is seen to produce three classes of intersection as shown in (d).

3. If a single null product occurs during the evaluation of concatenated intersections, the entire product is null.
4. Concatenated intersection operations generate products that monotonically *decrease* in size, while concatenated union operations generate products that monotonically *increase* in size.

The commutative, associative, and distributive laws of sets reveal that the *product* of concatenated set operations is independent of clause evaluation order. The *efficiency* of the evaluation process, on the other hand, can be significantly effected by clause evaluation order. Based on the above observations, the following three evaluation strategies can potentially improve query evaluation efficiency.

1. Intersection operations should be performed before union operations.

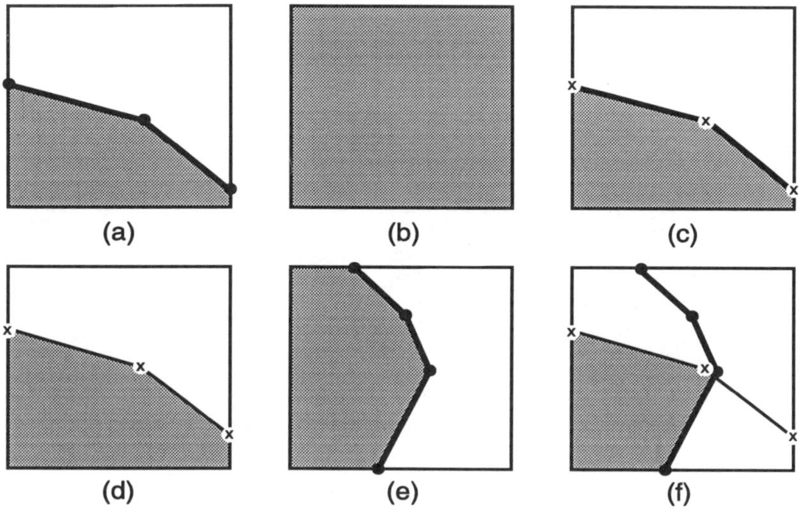

Figure 9.3 Two classes of boundary node intersections. The intersection of a boundary node (a) and an interior node (b) is shown in (c), and the intersection of boundary nodes (d) and (e) is shown in (f).

2. Since the lowest order feature controls the intersection product, the evaluation of concatenated intersections should begin by intersecting the two *lowest order features.*
3. Because the product of intersections between similar classes of sets (for example, two regions) is equal to or smaller than the size of the smaller set, concatenated intersections should begin with the two *smallest sized* sets.

For example, if a query requires the intersection of a region, two lines, and a oint, the intersection between the point (the lowest order feature) and the shortst line (the smallest set size of the next lower order feature) should be computed irst.

Thus, *optimal clause-order set operation–based query* evaluation requires that interections are performed before unions and that stronger constraints (representing hysically smaller sets and/or lower order sets) be applied before weaker contraints. *Worst case* set operation–based query evaluation, on the other hand, occurs hen unions are performed before intersections and when weaker constraints are pplied before stronger constraints. Since, in each case, weaker constraints were aplied before stronger constraints, the example shown in Figure 9.4 represents the orst case query-evaluation strategy.

Because the three regions shown in Figure 9.4(a–c) are each progressively naller than the previous region, the optimal clause order occurs when the clauses

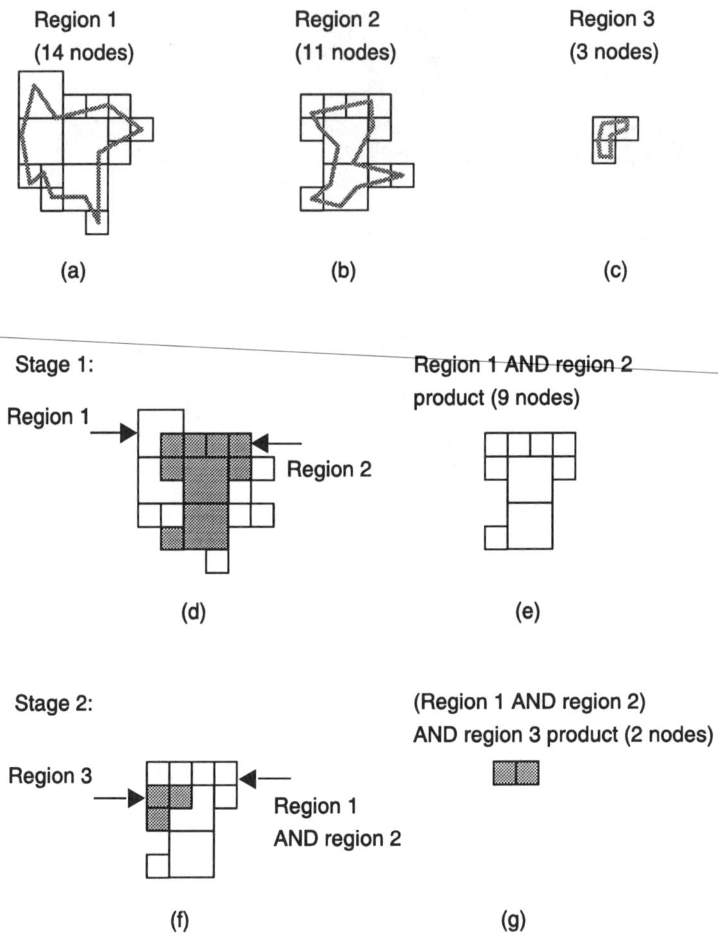

Figure 9.4 Sequential set operation–based intersection steps for the three regions depicted in (a–c).

are evaluated in reverse order as shown in Figure 9.5. In step 1, three region 3 nodes are ANDed with eleven region 3 nodes (14 nodes considered). Since the resultant product has two nodes, in step 2 these two nodes are ANDed with the fourteen nodes of region 1 (16 nodes considered). Thus, the total number of nodes considered is 30, rather than 37 nodes for worst case clause-evaluation order.

9.2.2 Constraint-Based Approaches

In constraint-based query evaluation, clauses are treated as database search conditions rather than as sets. The efficiency of constraint-based query is highly depend

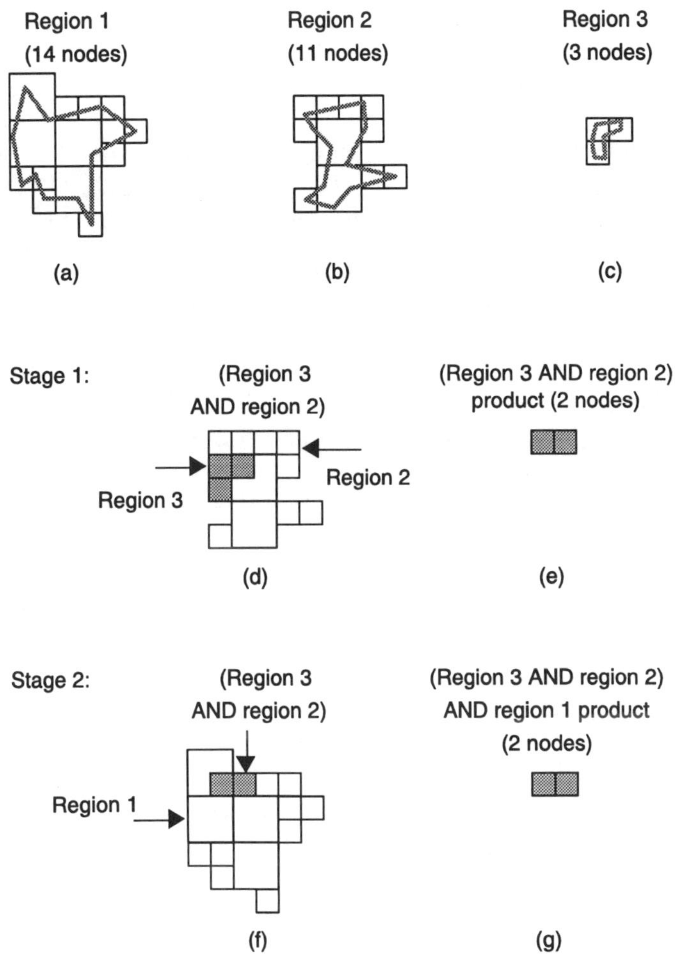

Figure 9.5 Optimal clause-order set operation–based intersection steps for the regions depicted in (a–c).

ent on the database organization. Because both spatial and nonspatial data are stored along natural search dimensions and the search dimensions are hierarchically organized, the integrated semantic/spatial object database kernel supports highly efficient constraint-based search.

Just as in set operation–based query, two classes of constraint-based search can be considered: *sequential constraint-based search* and *optimal clause-order constraint-based search.* Figure 9.6 demonstrates *sequential constraint-based search* for the evaluation of the same data set used in the set operation–based approach previously shown in Figure 9.4(a–c). Since it is impossible for a quadtree node that is not at least partially inside region 1 to be in the product of the intersection of (region 1 AND region 2), the only nodes of region 2 that can contribute to the product are those nodes that are at least partially inside region 1. Thus, region 1 can be treated as a spatial search window within which to search for region 2 nodes. The product of this operation becomes the spatial window within which to search region 3. Thus, rather than exhaustively ANDing all three regions, a search-based approach restricts the analysis to those nodes that can potentially be within the final query product at each stage of the evaluation process. Clearly, this search-based evaluation strategy requires a data representation that supports true two-dimensional spatially anchored search.

Query efficiency can be further enhanced if the clause-evaluation order is adjusted so as to minimize the total number of nodes that must be tested. Since, as discussed previously, region 3 < region 2 < region 1, Figure 9.7 demonstrates the advantage of searching region 2 within region 3 to produce the intermediate product that forms the spatial search window for interrogating region 1. In this case, only a total of five nodes need to be explored. Although all four query-evaluation strategies produce identical results, as shown in Table 9.2, each successive strategy is potentially more efficient than the previous strategy.

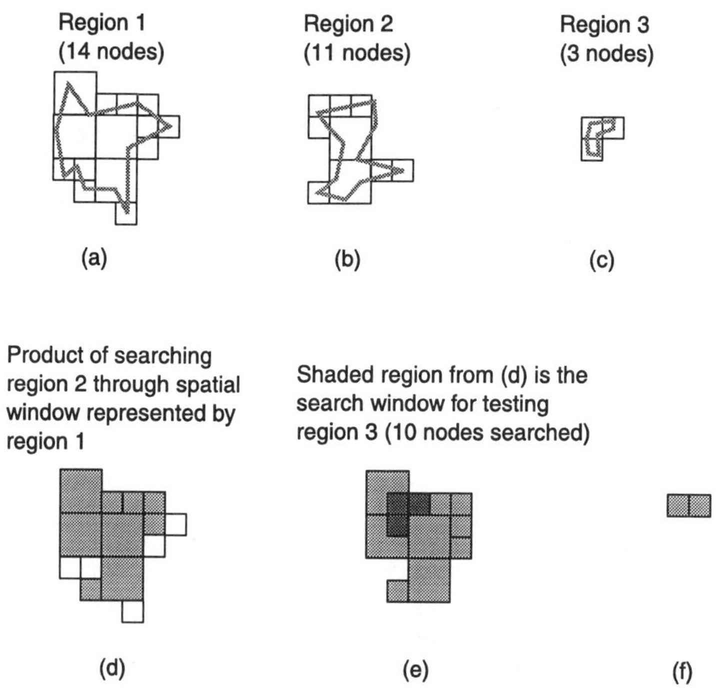

Figure 9.6 Sequential search-based intersection steps for regions depicted in (a–c).

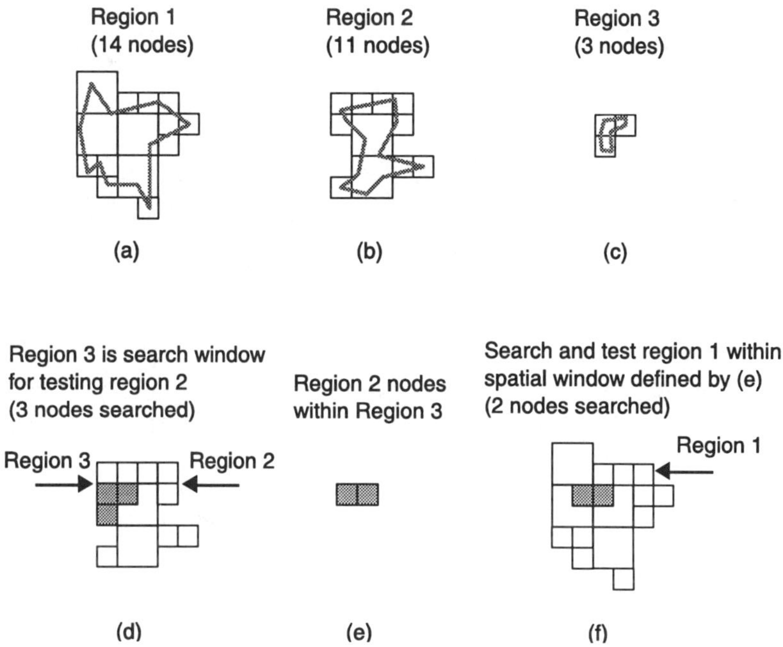

Figure 9.7 Optimal clause-order search-based intersection steps for regions depicted in (a–c).

Table 9.2
Comparison of Intersection Strategies Shown in Figures 9.3 to 9.6

	Stage 1 Operation	*Stage 1 Product (Nodes)*	*Stage 2*	*Total Nodes Considered*
Set operation (sequential)	14 region 1 nodes AND 11 region 2 nodes	9	9 stage 1 nodes AND 3 region 3 nodes	25 + 12 = 37
Set operation (optimal clause order)	3 region 3 nodes AND 11 region 2 nodes	2	2 stage 1 nodes AND 14 region 1 nodes	14 + 16 = 30
Search-based (sequential)	Search & test region 2 within 14 region 1 nodes	9	Search & test region 3 nodes within 9 stage 1 nodes	14 + 9 = 23
Search-based (optimal clause order)	Search & test region 2 within 3 region 3 nodes	2	Search & test region 1 within 2 stage 1 nodes	3 + 2 = 5

A second example is offered to illustrate an additional advantage of optimal clause-order, constraint-based search. Consider the following query that is based on the data set shown in Figure 9.8.

$$\begin{gathered}\textit{Find region X conditioned on}\\ \textit{east}\,(\textit{Marsh}\ 1)\ \ \textit{AND}\\ \textit{south}\,(\textit{River}\ 1)\ \ \textit{AND}\\ \textit{within}\ 1\,\textit{km of}\,(\textit{Bridge}\ 1)\end{gathered} \tag{9.7}$$

Since the first two constraints represent full half-plane sets (open sets), set operation–based approaches would typically approximate such clauses as large closed

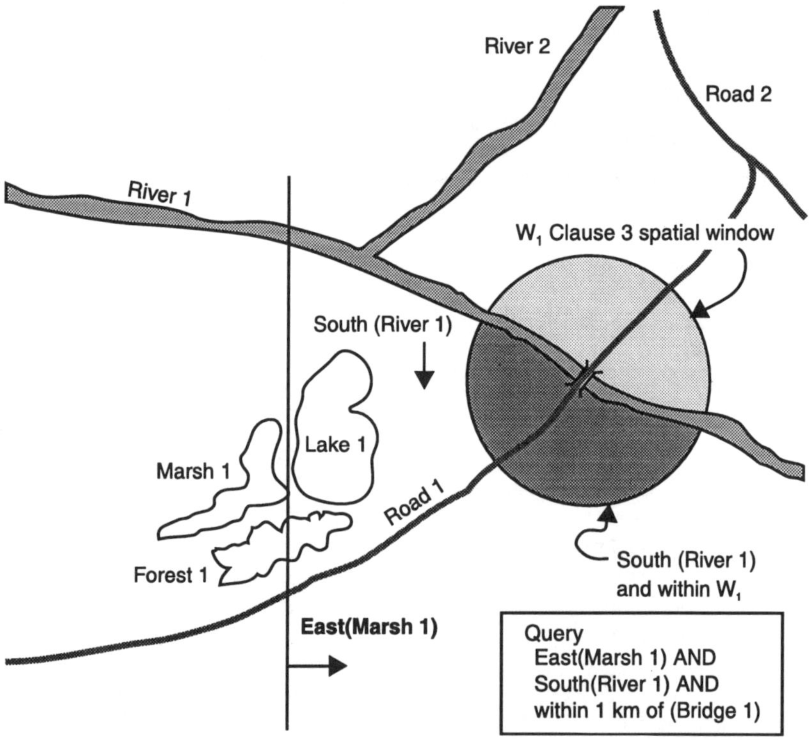

Figure 9.8 Solution space for query (9.7).

sets. Exhaustive AND operations between sets comprised of a large number of nodes tend to be highly inefficient.

On the other hand, optimal clause-order constraint-based search tends to be significantly more efficient. Because clause 3 has the least area, clause 3 is the selected spatial window within which to search the next strongest constraint. Because both constraints 1 and 2 are effectively the same "size," no order preference exists. Suppose River 1 is searched within the spatial window represented by clause 3. In effect, the query produces all linear quadtree nodes within W_1 that are "south" of the quadtree nodes of that portion of River 1 that is within W_1. A simple test of the root node of the spatial window W_1 verifies that W_1 satisfies clause 1. Since all regions that are {south(River 1) AND within 1 km of (Bridge 1)} satisfy the constraint east (Marsh 1), the cross-hatched region shown in Figure 9.8 is the final query product. Unlike the exhaustive testing of large open sets required by set operation–based query evaluation, optimal clause-order, constraint-based search permits highly focused evaluation operations.

Union operations that employ optimal clause-order, constraint-based search effectively employ an evaluation strategy that is the inverse of intersection. In this case, evaluation begins with the constraint that possesses the *largest* extent and proceeds to the constraint that possesses the least extent. The objective is to identify all parts of the next largest constraint that is not inside the first; the result of this analysis is then added to the first constraint. This testing process is recursively applied until all constraints have been evaluated. Because of the monotonically increasing character of the union operation, search-based union generation tends to be much less efficient than search-based intersection generation.

9.2.3 Generalization of Constraint-Based Query Evaluation

Because of the potential efficiency of *optimal clause-order constraint-based query*, this section investigates its applicability to mixed spatial and semantic queries. While set operation–based query evaluation involves the composition of sets, constraint-based query evaluation represents the composition of constraints. Although, as we have seen, the composition *product* is independent of the order of operations, the *efficiency* of the evaluation process can be significantly enhanced by the proper ordering of database search constraints.

In general, constraints can be either spatial or nonspatial. Spatial constraints are conditions on two-dimensional point, line, and region features (for example, east(line 20), within 1 km of (x_1, y_1)), while nonspatial constraints are conditions on nonspatial features (for example, number of companies in a division, equipment performance specifications, logistics requirements, tables of equipment). In addition, constraints can be either *open* or *closed.* Closed constraints represent bounded sets, while open constraints represent inherently unbounded sets.

The *extent* of a constraint is defined to be its intrinsic *size,* which for a semantic object is considered to be the number of objects within the class of interest. For instance, if the *target* data set contains 100 *fixed targets* and 100 *moving targets,* the *extent* of the target data set would be 200. The *extent* of a spatial constraint is defined as the number of quadtree nodes associated with its spatial representation at some specified resolution level.

If W is a semantic or spatial search criterion and Z is the constraint to be searched, the notation $W(Z)$ is used to capture the relationship between the two constraint sets. For a set of concatenated clauses C_i the optimal clause-evaluation order is defined as $\overline{C}_j$ where $i = 1, \ldots, I$ and $j = 1, \ldots, I$. Since $\overline{C}_1$ is the constraint with the least extent, in the first step of the query evaluation process, $\overline{C}_2$ is searched within the window defined by constraint $\overline{C}_1$. Assuming Boolean operations, the result of the query is

$$\begin{aligned} \overline{C}_1\,(\overline{C}_2) &= 1 \textit{ when } \overline{C}_2 \textit{ is within } \overline{C}_1 \\ &= 0 \textit{ otherwise} \end{aligned} \qquad 9.8$$

In other words, all linear quadtree nodes in $\overline{C}_1$ that are also in $\overline{C}_2$ are in the product of this operation. Since for intersections, $\overline{C}_1(\overline{C}_2) < \overline{C}_1 < \overline{C}_2$ (the product of the evaluation process is monotonically nonincreasing), the extent of $\overline{C}_1(\overline{C}_2)$ is always smaller than (or at most equal to) the extent of the next weaker constraint $\overline{C}_3$. Thus, after the first intersection, the optimal search window is always the product of the previous stage of the evaluation process.

Consider the intersection of the point, line, and region shown in Figure 9.9. If the extent of a spatial feature is measured at the apex of the object representation of two-dimensional space, for example, all three features would have an identical extent. However, if the extent is measured at the pyramid base-node level, the proper ordering of the constraints becomes readily apparent, as observed in Figure 9.10. The point is determined to have extent 1, the line feature has extent 14, and the region has extent 11. In this case, the optimal clause-evaluation order is 1, 3, 2. Because one of the clauses is a point, the product of concatenated intersections (by Table 9.1) is known a priori to be either a point or null. Thus, in this example, the optimal recursive evaluation strategy requires the following two steps.

1. Search the *region* within the single quadtree node that contains the point feature; since the region contains the node, continue; if it had not, return null.
2. Search the *road* within the single quadtree node that results from the first intersection product; since the intersection is null, the query returns a null product.

Optimal search-based query evaluation tends to be significantly more efficient than traditional set operation–based query evaluation methods that can require consideration of the full extent of each query clause.

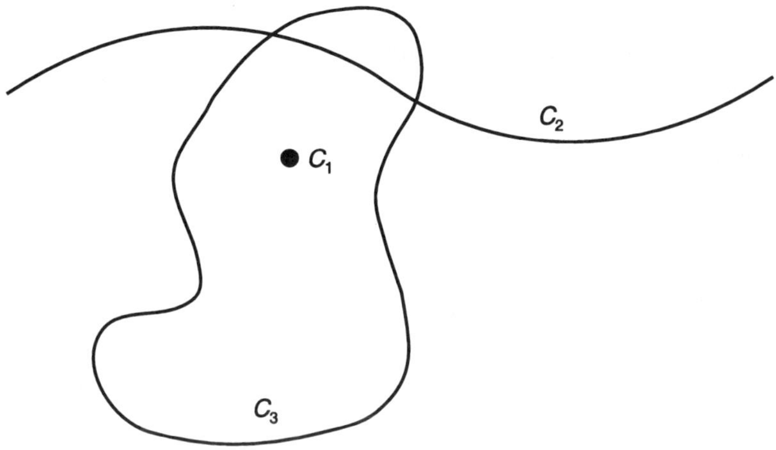

Figure 9.9 Intersection of a point, a line, and a region.

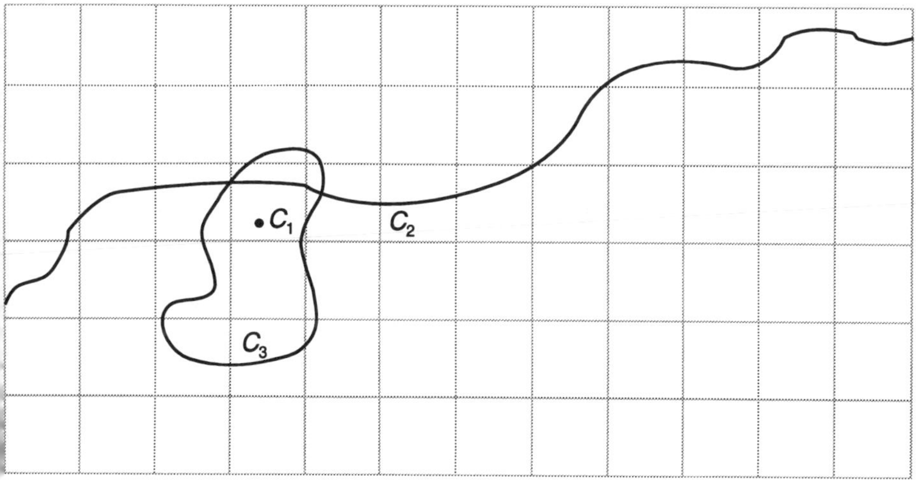

igure 9.10 Pyramid base-node depiction of the spatial features from Figure 9.9.

Consider the evaluation of a slightly more complex query:

Find all roads that are within 50 *miles of point X* and 10 *miles of region Y* 9.9

First, we define region *A* as a 50-mile-diameter circle about point *X* and region *B* as a 10-mile region growth of region *Y*. Since region *B* is smaller than region *A*, the optimal intersection of the two regions requires searching region *A* within the spatial window defined by region *B* (that is, *B*(*A*)). Figure 9.11 demonstrates the product of this operation for:

1. No overlap;
2. Partial overlap;
3. Complete overlap (region *B* completely contained within region *A*).

The product *B*(A) then becomes the search window for the road database. Because the database kernel supports true two-dimensional spatially anchored search, *only* that portion of the road database that falls within the spatial window defined by *B*(A) must be interrogated.

Next, consider the canonical form XV path-planning algorithm outlined in Chapter 3 that relied on mixed spatial and semantic search. In that example, top–down, true two-dimensional spatial search of the domain database was employed to determine if there existed an extended barrier to ground-mobility between the start node and the goal node. When a river obstacle was discovered, database search shifted from spatial search to semantic search in order to locate candidate bridges from the set of all bridges that cross the river.

Since the original query was spatially-sensitive ("Is there an extended barrier between two spatial locations?") and extended barriers are spatially extended classes of features (rivers, ravines, high mountains), the initial query is most efficiently handled via true two-dimensional spatial search of the object representation of two-dimensional space. On the other hand, although the bridges across the river could be located via spatial search along the river, semantic search of all bridges associated with *the* single river object of interest (River 1) tends to be considerably more efficient. Thus, for this example, mixed spatial/semantic search can be more efficient than either strict semantic-oriented or strict spatial-oriented search.

Next, consider the following query based on the data set shown in Figure 9.12:

Find all freshwater lakes of size 10 < *acres* < 50 *that are between* 50*km* and 100*km of City X* and *that are less than* 20*km an interstate highway* 9.10

As discussed in Section 9.1, the generic query form for the above query is

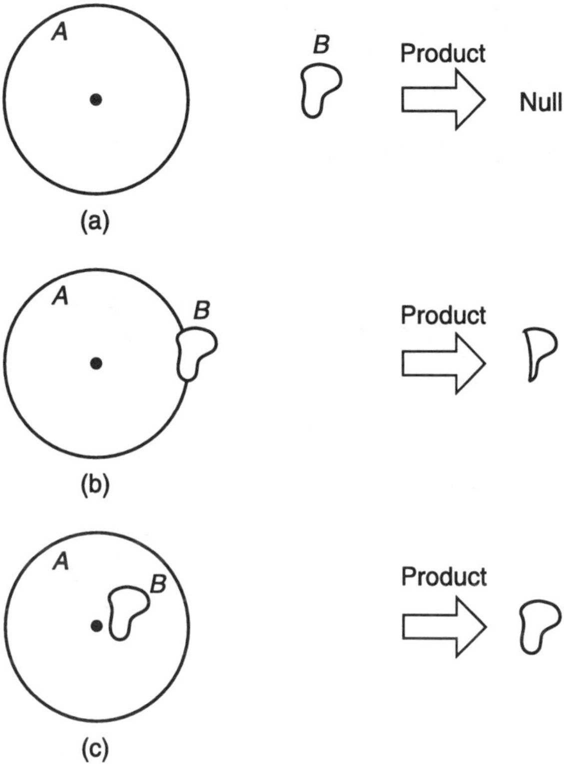

Figure 9.11 Depiction of query constraints for (a) nonintersection, (b) partial overlap, and (c) inclusion.

$$\textit{Generate all } f_1 \textit{ (region\_object > lake > freshwater)} \text{ (intersect)} \\ (f_2 \textit{ (region\_object > city > City X > shape)}) \text{ (intersect)} \\ (f_3 \textit{ (line\_object > road > interstate)}) \quad 9.11$$

In a *sequential set operation–based approach*, the clauses would be evaluated from the left to the right. In this case, since the first clause represents a semantic constraint, the entire lake database would be searched to find those lakes that satisfy the size constraint (by first ranking according to lake size and then testing against constraint f_1). The set of all candidate lakes would then be intersected with the set defined by clause 2. Finally, the product of the preceding operation would be intersected with the "region-grown" interstate highway network.

In contrast, *optimal clause-order constraint-based search* offers a much more efficient query-processing approach. The first clause of query (9.11) is treated as a se-

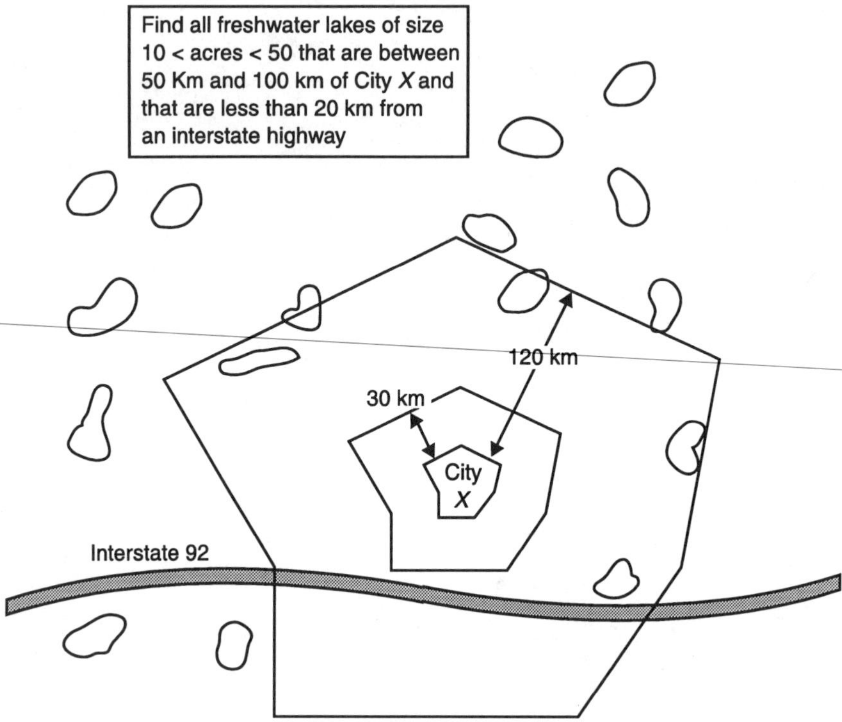

Figure 9.12 Database for query (9.10) depicting an interstate highway, city *X*, and a large number of lakes.

mantic object query under semantic constraints (f_1 is a constraint on the lake attribute *size*), while the next two clauses represent spatial constraints (f_2 and f_3) on spatial attributes of semantic objects.[3] The extent of clause 2 is related to the area of the annulus (50 km to 100 km) anchored about the region object and the extent of clause 3 is related to the area of the region-grown line object (0 km to 20 km).

A number of important considerations remain. First, we have not considered the cost of evaluating the hybrid representation of individual spatial features. In addition, because we have assumed that spatial extent is measured in terms of the number of pyramid base nodes, the extent of spatial constraints and semantic constraints are not directly comparable. In order to actually relate the "cost" of semantic queries to the "cost" of hybrid spatial representation queries, further analysis is re-

3. Although lakes possess spatial attributes (location, shapes) that are relevant to the overall query evaluation process, for the purposes of extent determination, these attributes are effectively secondary considerations. In this case accumulating the total extent of the union of all lakes would be highly inefficient.

quired. Although we will not attempt to resolve the issue here, the potential for developing effective evaluation criteria seems promising. In fact, relatively crude evaluation criteria may be sufficient since the purpose of constraint-set ordering is to support the development of an evaluation strategy. The overall efficiency of the algorithm is jeopardized if excessive efforts are expended attempting to optimize the evaluation strategy.

Suppose that clause 1 has semantic extent (measured at the pyramid base-node level) 10^5 (in other words, there are 10,000 lakes in the database) and that clause 2 has spatial extent 512, and clause 3 has spatial extent 1800. In this case, clause 1 possesses the largest extent. Because ponds and lakes can be quite common in many parts of the world and because interstate highways are highly extended lineal features, it is quite likely that a constraint associated with an individual city might possess the least extent.

Thus, the search window for the lake database is constructed by searching the interstate highway database within the spatial window defined by the 30 km to 120 km "annulus" about city *X* (the increase in the window size from 50 to 100 km to 30 to 120 km ensures that no region-grown interstate highway is missed by this window). For all sections of an interstate highway that intersect this search window, the actual spatial window depicting clause 3 is generated. Figure 9.13 illustrates this final spatial window overlaid on the natural and cultural feature data set.

Since robust and context-sensitive algorithms routinely demand the evaluation of constraints that can be represented as combinations of set operations to be applied to very large spatially organized data sets, improvements in database query efficiency can translate directly to improved algorithm performance. Although heuristic criteria must still be developed to permit direct comparison of the extent of spatial and semantic constraints, optimal clause-order, constraint-based query permits highly efficient mixed semantic and spatial database query.

9.2.4 Summary and Conclusions

For complex, concatenated queries, query optimization involves the following four-step process.

1. Transform the query to its generic query form.
2. Evaluate the *extent* of the constraint represented by each clause:
   a. Determine the number of semantic objects within the class;
   b. Determine the number of pyramid base nodes (or some other more appropriate spatial resolution level) associated with each spatial object constraint.
3. Use the clause with minimum *extent* as the search window (semantic or spatial) for the evaluation of the clause that possesses the next largest *extent.*

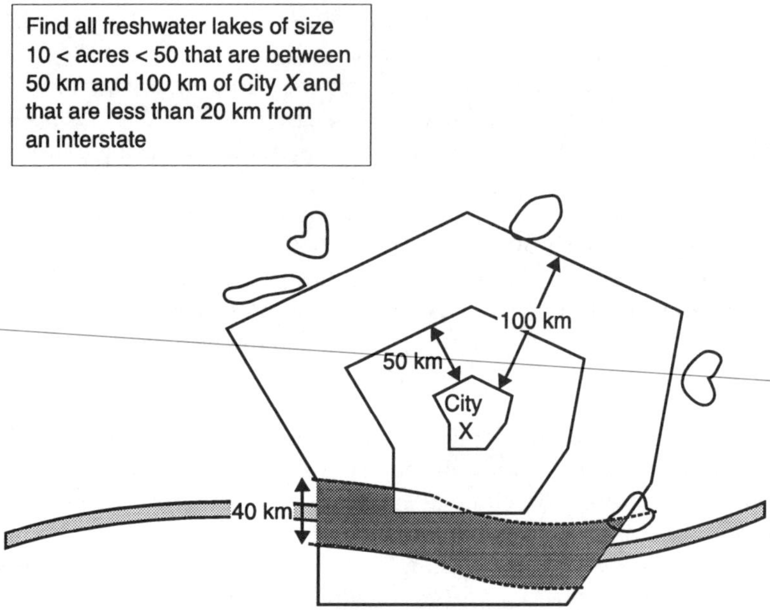

Figure 9.13 Spatial window for lake database query using optimal clause-order, constraint-based search.

4. Recursively evaluate the remaining clauses in order of increasing *extent* us- ing the search window developed from the product of the previous stage o the evaluation process.

Since optimal clause-order, constraint-based search relies heavily on true two dimensional spatial search, the evaluation strategy is well-supported by the seman tic/spatial object database kernel outlined in the last chapter.

The "best case" efficiency of the optimal query-evaluation strategy occurs whe the search-space size is equal to MIN(E_i) where E_i is the extent of each constraint, i 1, … , *I*.. The worst case query, on the other hand, occurs when the full extent of a query clauses must be considered. Thus, the upper bound on the relative perform ance efficiency f of the optimal strategy relative to the worst case strategy can be ex pressed as

$$f = \frac{\sum E_i}{MIN(E_i)} \qquad 9.1$$

For the example summarized in Table 9.2, $\sum E_i = 28$ *and* $MIN(E_i) = 3$; therefore, f 28/3 = 9.3. In this case, optimal clause-order, constraint-based search can be near an order of magnitude more efficient than exhaustive set operation–based query. As observed in Table 9.2, the total number of nodes that were required by the worst case evaluation strategy was 37 and the total number of nodes considered by the optimal clause-order constraint-based search approach was 5. Therefore, the relative efficiency of the optimal approach is 37/5 = 7.4, which is reasonably close to the performance upper bound (f = 9.3).

The relative improvement of optimal clause-order, constraint-based search relative to the worst case raster-based set-operation evaluation strategy can be computed by scaling (9.12) by d^2, where d is the ratio of the side length of the base pyramid node relative to the side length of a raster node. Therefore, the best case efficiency improvement f_r of optimal clause-order constraint-based search relative to exhaustive raster-based set-operation evaluation is

$$f_r = d^2 f = \frac{d^2 \sum E_i}{MIN(E_i)} \qquad 9.13$$

Thus, the efficiency improvement is proportional to both the size of the total constraint set ($\sum E_i$) and the search-space reduction potential of the pyramid representation relative to a full raster-based representation (d^2). For instance if d = 8, the "best case" improvement for the example summarized in Table 9.2 is f_r = (64) (9.3) = 595.2. Although (9.13) does not account for the analysis of the underlying vector-based representations, for large E_i and/or large d, optimal clause-order constraint-based search can potentially offer significant performance advantages.

9.3 HIGHER LEVEL QUERIES

The "building block" approach to high-level system development discussed at the beginning of Chapter 8 required:

1. The availability of the proper tools (object-oriented programming environments, higher level programming languages);
2. Sophisticated building materials (generic functional software, advanced database management systems).

Due to the lack of sophisticated building blocks, many data fusion applications have been developed essentially from the "bottom–up." As applications become increasingly more complex, the requirement for better tools, higher level languages, and reusable code becomes ever more crucial. Current high-level tools such as commercial object-oriented programming environments; computer-aided design (CAD) packages; computer-aided software environments (CASE) tools; Expert System shells; and statistical, mathematical, and signal-processing software packages provide important but limited high-level functional support to real-world data fusion applications. Object-oriented programming environments and Expert System shells, as mentioned earlier, generally provide only a basic *framework* for reasoning.

Software reuse requires not only good software engineering practices but an evaluation of what constitutes reusable functionality. Even when reusable functions are identified, complete dependence on software clearinghouses and nonenforceable "good coding practice" have obvious shortcomings. In spite of good coding practices, programmers often find they must significantly modify their own software for use in a similar application. Thus, more commercially supported application-oriented software is required. Since performance efficiency remains a critical issue, the required functionality must not compromise computational efficiency.

Because virtually all large-scale data fusion applications (as well as numerous other problem domains) employ a commercial database management system, building reusable functionality into future database systems potentially eliminates an algorithm developer's need to "reinvent" certain classes of generic functionality and would virtually guarantee software reuse. If reusable software modules were functionally organized into a hierarchy of shells built directly on top of the integrated semantic/spatial object database kernel, applications could effectively interact with the database at multiple levels of abstraction.

The addition of functionality to a basic database kernel merely represents the logical extension of the current trend toward enhanced database support. The database kernel would handle primitive operations, such as access control, data maintenance, data distribution, search, and query, while the higher level database shells would effectively reformulate high-level queries into lower level database operations. The higher level shells would become increasingly application-dependent with specialized interfaces, each optimized for specific classes of operations or operational domains (target tracking, classification, multisensor correlation, generalization/specialization, path planning). The user interface at each shell level could be tailored to the classes of generic reasoning performed at that level.

The availability of such higher level functionality could potentially both simplify and expedite application development, reduce the application-level communication requirements with the database, as well as ensure effective and efficient use of the underlying database kernel. An application developer would only need to create the functionality that was not supported by the current version of the database shell. By exploiting the unique characteristics of the integrated semantic and spatial object database kernel, such database shells could support both highly intuitive and highly efficient multiple level-of-abstraction algorithm development.

The advantage of such high-level interfaces can be illustrated using the Expert System paradigm. In this paradigm, rule conditions must be tested against the current problem state. For complex applications, the rule instantiation process can be far from trivial. Consider, for example, direct database support for the evaluation of the following high-level production rule:

IF [Target X is on a road and within 10 km of a rail line] THEN [·].

In this case, the location attribute of a target might be a represented as either a point feature or an uncertainty region; at a relatively low resolution, roads might be appropriately treated as a line object, while at a high resolution, multiple-lane highways might need to be treated as a region representation.

Rather than forcing an algorithm developer to interact with the database at the SQL-level, a high-level database shell could support the direct evaluation of such conditions. The existence of a database shell that could parse and evaluate complex queries would have widespread utility to users in many application domains. Just as a conventional database management system removes the user from the details of data storage and maintenance, the existence of sophisticated database shells would further elevate the level of interaction between the application and the data.

Much as in the database kernel design, the database shells require a top-down design approach to fully address relevant performance considerations and software standards. Each new shell release would support all previous functionality, while the enhancements would permit either higher level-of-abstraction database interaction or provide additional generic functionality.

Several high-level queries and their associated generic query forms, which illustrate the potential advantages of direct database support to complex, domain-sensitive queries, are offered below:

- *Generate a single path that supports high-speed car travel between region A and region B*; Generate (one) path conditioned on (from region *A* to region *B*; for car; for high-speed travel).
- *Generate the uncertainty function for a truck convoy that has traveled for* 5 *minutes from* (x_1, y_1) *along road* 81 *in the NW direction*; Generate location_uncertainty_function conditioned on (for convoy; from (x_1, y_1); for travel time = 5 minutes; for travel on road 81; for travel direction = NW).
- *Find the maximum and minimum straight line distances between region* 113 *and road* 81 *conditioned on west of point C*; Generate maximum distance and minimum distance conditioned on (from region 113; to road 81; west of point *C*).
- *Return the list of all terrain features intersected while traveling between* (x_1, y_1) *and* (x_2, y_2) *on a minimum terrain gradient path*; Generate all terrain_feature_classes conditioned on (from (x_1, y_1) to (x_2, y_2); for minimum terrain gradient path).
- *Create a map overlay named region* 1 *consisting of all regions that are either* (1) *forests and lakes or* (2) *forests and terrain gradient* < 10%; Generate region 1 conditioned on ([forests$\cap$lakes] union [forests$\cap h_1$(terrain_gradient)]).
- *Update the location of entity* 551 *to* (x_1, y_1, t_1); Generate entity_551 present_location conditioned on (present_location = "(x_1, y_1, t_1)").
- *Update the location of entity* 551 *to region formed as the intersection of region A and region B*; and Generate entity_551 present_location conditioned on (present_location = (region *A*$\cap$*region B*)).

In the next section, several classes of database extensions are discussed that are directed at creating more capable systems for both managing and accessing data and information.

9.4 CONTEMPORARY DATABASE ISSUES

Next-generation database management systems are expected to be considerably more powerful than the current-generation systems exemplified by the RDBMS. As the distinction between databases and knowledge bases continue to blur, future systems are expected to provide progressively higher level-of-abstraction products [1], including support for deductive functionality, multimedia, and object transformations.

Although the nomenclature of the various approaches to database system enhancement have not been widely standardized, Jeffery [2] proposed the following simple distinction between three classes of RDBMS extensions:

- RDBMS + object-oriented characteristics = extended DB system;
- RDBMS + logic programming = intelligent DB system;
- RDBMS + logic programming + object-oriented characteristic = expert DB system.

Research in both traditional and nontraditional database fields are likely to have a significant impact on future database management system development. Several of these research areas are discussed in this section from the perspective of the last several chapters.

9.4.1 Information Retrieval versus Data Retrieval

Vast quantities of information are generated daily in science, industry, and medicine, to name just a few. This information represents a wide variety of formats (text, graphics, imagery). Although much of the information is already stored in digital form, an even larger percentage will be in the future. The information may reside at dozens or even hundreds of separate locations, under completely different data management systems. For the information to be readily available to those who require it, a much more powerful means of information retrieval is required than currently exist (for example, data mediators, intelligent database networks). Since the required information is potentially both distributed and heterogeneous (noncommensurate), data fusion technology can potentially play an important role in both information combination and analysis.

The traditional approach to text retrieval has been *keyword search*, which is the string version of the rigid pattern-match paradigm discussed in Part I. Keywords can be combined using Boolean operators, and keyword synonyms can be employed to help generalize the template-match process. However, more powerful methods are required to support retrieval based on the content and meaning. The exploration hese more capable methods has generally fallen under the heading of *natural lan-uage understanding.*

The metaphorical similarity between data fusion and speech recognition was utlined in Chapter 1. Tactical data fusion was seen to involve the progressive gen-ralization of hierarchical sets of information for the purpose of developing a multi-le level-of-abstraction situation awareness. Since speech recognition involves the rogressive abstraction of a single phrase, sentence, paragraph, or document, natu-al language understanding can be viewed as the generalization of speech recogni-on. Both tactical data fusion and natural language understanding seek to develop robust, context-sensitive, and multiple level-of-abstraction interpretation through e effective combination of relevant information.

For example, consider the high-level database query:

Which Pacific Rim countries have the highest standard of living?

ven though the answer to this query is not explicitly stored in the database, rele-ant pieces of information that support the evaluation process may be present. For xample, the per capita income, the total gross national product, total exports, aver-ge life expectancy, and average income levels may be available. In addition, infor-ation may exist that permits an assessment of the level of development of a ountry's infrastructure, including roads, hospitals, and schools.

Databases and knowledge bases typically store information that supports multi-le applications. Specialized interpretations of data and information can involve omplex reasoning processes that are analogous to the puzzle-solving process de-ribed in Chapter 2. The root of many failures in both manual and automated data sion and natural language understanding has been highly local, context-insensi-ve interpretations of information.

Rather than posing strictly SQL-level queries, intelligent and expert database stems seek to provide a higher level-of-abstraction user interaction with the under-ing database. Thus, data retrieval involves a form of inference as opposed to atches on rigid templates or constraints. Deductive databases rely on deductive gic. Database systems that employ rule-based query mechanisms are often referred as expert databases. Rather than strictly data-handling and query-processing les, an active database system performs some level of autonomous analysis. Al-ough the classification of the various approaches are not always crisp, the goal of of these approaches is to extend the traditional notion of the database system om that of a data manager to that of an information service.

4.2 Multiple Media Databases

general, information takes many forms: text, lists, tables, graphs, signals, static im-es, video, sound, and three-dimensional objects. Multimedia databases seek to ercome the representation limitations of conventional relational database man-ement systems. This book has stressed the inherent advantages of natural data presentations both in terms of their support to application development as well as to data access and manipulation efficiency. Full integration among the multiple data types has been shown to be a key factor in facilitating application development and enhancing application efficiency.

9.4.3 Database Mining/Automated Pattern and Trend Discovery

Database mining involves the exploration of data patterns and associations. The database exploration process is enhanced by data representations and manipulation tools that support roam, zoom, and perspective change along conceptually relevant search dimensions. For instance, by viewing family income as a histogram indexed by residence location, the most exclusive residential subdivisions in a large city can be determined. By viewing enemy communication nodes by location, percent time on-the-air, and radio type, critical communication nodes can often be determined.

Although certain associations might be anticipated, in other cases, the results can be more serendipitous. For instance, the comparison of photographs of bystanders at the scene of suspicious fires might identify candidate arsonists as those individuals that are present at multiple fires. Similarly, a significant and unexpected increase in vehicular activity, RF emissions, or telephone conversations within a relatively small geographic region can provide doctrinally significant political, strategic, or tactical indicators. Because of its reliance on highly natural spatial and semantic data representations that preserve implicit associations among data, an integrated semantic/spatial object database kernel facilitates the development of highly intuitive database views. Aided by such representations, various classes of automated techniques can be applied to learn rules, patterns, and constraints from the data.

9.4.4 Database Support for Reconciliation of Contradictory Data

Due to the inherent uncertainty, ambiguity, and dynamic nature of both the data and the decision process, many complex information-processing applications require multiple-hypothesis reasoning. Effective data storage and search dimensions permit associations to be efficiently formed and easily broken. As will be discussed in detail in Chapter 12, the semantic/spatial object database kernel supports efficient evaluation of complex spatial, temporal, and semantic relationships between data and hypotheses as well as among competing hypotheses.

9.4.5 Distributed Database Issues

Interest in distributed databases tends to arise from two different requirements. First, data needed by a given application is often physically resident at multiple, spatially separated locations. Second, due to performance considerations, high-concurrency database implementations can be required to meet operational requirements. The database outlined in Chapter 8 lends itself to highly distributed implementations that support both applications.

The development of distributed database systems involves a wide range of is-ues. For instance, when a node in a distributed database is shutdown or its commu-ication link is disrupted, the currency of that segment of the system is ompromised. Once the node comes back "on-line," the node's database must be econstituted. Because the semantic/spatial object database kernel employs highly atural data representations, full reconstitution of both the spatial and semantic ata tends to require only highly localized updates.

In distributed processing applications, communication requirements among rocessors must be kept to a minimum. Since distributed database maintenance nd image transmission over band-limited channels are analogous problems, dis-ibuted database maintenance can benefit from a consideration of some of the un-erlying principles of image compression for both single-frame and motion-video nages. In the case of single-frame images, various image compression techniques e used to remove redundancies so that the transformed image achieves an opti-al, application-dependent tradeoff between bandwidth compression and image egradation. In fact, some bandwidth reduction can be achieved at no loss of image iality. For example, rather than transmit a string of *n* identically valued pixels, the istence of a *run* of length *n* is transmitted. Since the smaller the number of quanti-tion levels, the longer the expected run lengths, the largest compression ratios ould be expected for binary images.

For more general applications, other compression techniques tend to be more owerful. The cosine transform, the wavelet transform, and fractal-based compres-on schemes are three popular techniques that have been used to achieve band-dth reduction. In these approaches, the coefficients of the transformed image are ansmitted and the image is reconstructed at the remote location by applying an in-rse transform. In general, the higher the compression ratio, the poorer the qual-of the reconstructed image.

Analogous to "change-detection" encoding of adjacent pixels in a single-frame age, motion video can be compressed by transmitting only those pixel values that ange from frame-to-frame. Since many pixel values will not change every frame, ce an initial two-dimensional image has been communicated to the end-user, nsiderably less bandwidth is generally required to transmit subsequent frames. leo-teleconferencing systems, for example, employ both "within-frame" and ame-to-frame" compression. Even higher bandwidth compression is possible if only relative change of a pixel value (rather than its actual value) is transmitted. Thus, h an *m*-bit pixel representation (for example, 8-bit gray-scale), the *relative* change in a el's value on the average may require less than *m* bits to encode.

Variations on this theme of "change-detection" are used in other application mains as well. For example, Doppler frequency offsets are extracted in MTI radar ng a form of pulse-to-pulse subtraction. In order to control false alarms due to wly varying changes in the background noise or clutter level that would not be re-ved by pulse-to-pulse subtraction (spatially and temporally local estimate), con-nt false alarm rate (CFAR) processing is employed to adjust a system's sensitivity as a function of temporally and/or spatially varying noise and clutter environmen (spatially and temporally global estimate).

Much like motion video, where the background may be relatively static, th long term domain knowledge base in data fusion applications tends to be relativel static (geographic and cultural feature databases). Over this background, conside able changes can occur representing both short-term and medium-term knowledg Thus, the dynamic element of a database or knowledge base is analogous to movin objects in video.

In general, change-encoding operates at the basic data-element level. Becaus a moving object in imagery is not syntactically separate from its background, *chan detection* is generally associated with "pixel-level" changes over time.[4] While the bas data element of raster images is a pixel, the data elements in a database system ha a richer structure. Since database objects are syntactically differentiated from th balance of the database, more efficient change updating is possible. For instance, a target moves from point *A* to point *B*, the logical object that resided at locatic (x_A, y_A) can be removed and placed at location (x_B, y_B). Thus, only the minimal s of information that is required to keep all database nodes current must be transm ted. In this case, the location attribute of a particular icon that had been associat with location *A* must now be associated with location *B*. Because the structure data elements is known, database change updating tends to be both more efficie and more effective than picture-element compression. Just as in simple run-leng encoding, such change-updating schemes reduce the information bandwidth wi no loss in information content.

Just as in imagery compression, where the most efficient pixel-level codi scheme is sought, in updating two-dimensional spatial databases, the most efficie data-element (logical object) coding is sought. Since, as discussed in Chapter 8, ve tor-based point, line, and region-boundary representations tend to be the m memory-efficient spatial representations, vector representations are natural can dates for transmitting spatial feature change updates. Because the spatial indexi provided by both the pyramid and low-resolution spatial representations of indiv ual spatial feature can be generated locally at each node, such information need n be transmitted.

Upon reception of a spatial object update, the quadtree-based indexing str ture can be either created or updated as appropriate and any required changes the semantic object representation and the pyramid representation can be ma For example, if a new region feature must be added to a remote database, t boundary list of the region (relative to an appropriate absolute reference point

4. One school of image-compression research pursues the development of a knowledge-based sys that would generate a higher level-of-abstraction description of the scene (for example, image derstanding) prior to encoding. If achieved, very high compression ratios would be possible si the transmission of symbolic information may replace much of the high-resolution, high frame- information required by more conventional approaches.

ransmitted. At the receiving database, the full hybrid spatial representation is cre-ted and appropriately linked into the pyramid and the semantic object database.

.5 SUMMARY

'his chapter has discussed a number of extensions to the database kernel outlined ı Chapter 8. A framework was developed to support query optimization for mixed patial and semantic clauses. For concatenated intersection operations, optimal lause-order, constraint-based search was shown to minimize the required search-pace size by effectively eliminating the interrogation of any spatial node or seman-c object that does not satisfy the strongest constraint imposed by the entire onstraint set.

The development of a hierarchy of database shells built on top of the database ernel was recommended as a means of evolving higher level tools to support future pplication development. Much of the generic (spatial, temporal, and hierarchical) ıd application-specific reasoning discussed in Part III could be incorporated ithin such database shells.

In the earlier chapters, data fusion was seen to be a critical component of situ-ion development, speech recognition, and image understanding. In this chapter, atural language understanding and database retrieval were added to the rolls. An nportant theme of this book is that data fusion plays a critical role in numerous as-ects of machine-based reasoning.

References

] Brodie, M.L., and J. Mylopoulos, "Artificial Intelligence and Databases: Dawn, Midday or Sunset?" *Canad. Inform. Processing,* July/August 1992, pp. 25–27.

Jeffery, K., "The Expert Database Jungle," *Expert Database Systems,* K. Jeffery (ed.), New York: Academic Press, 1992.

CHAPTER 10

Database Issues Summary

art I of this book emphasized the importance of developing *effective* algorithms that re *practical* to implement. In general, data fusion algorithms interpret and abstract oth *sensor-derived* information (short-term knowledge) and the current *situation de-ription* (medium-term knowledge), both of which are highly context-sensitive. Be-ause they support robust performance over a wide range of *explicit* inputs as well as ipport implicit *sensitivity* to both static and dynamic domain context, model-based gorithms that are sensitive to general context knowledge tend to be more *effective* ian rigid, context-free or highly context-specific algorithms. In addition to support-ig robust and context-sensitive reasoning, model-based algorithms tend to be ex-nsible to new domains or to changes in the characteristics of an existing task.

To be *practical*, an algorithm must achieve adequate computational perform-ice to meet the operational requirements of the objective application. For many oplications, practical automation demands both high-performance hardware and igh-performance software.[1] Since context-sensitive, model-based algorithms can re-uire access to and manipulation of extensive domain knowledge bases, they tend to ace significant demands on the supporting database management system. As a direct sult, the development of *practical* data fusion applications requires both inherently ef-ient algorithms as well as efficient database support to those algorithms. Tradition-ly, databases have served as repositories for information that is searched for, thdrawn from the database, and manipulated by a set of algorithms. As we have seen, e development of highly sophisticated data fusion applications can require access to assive domain knowledge bases. Since the search and withdrawal of significant por-ons of such data sets would be highly inefficient, application efficiency can be en-inced by performing much of the analysis process within the database system itself. ie last several chapters have explored numerous issues associated with efficient data-

Hardware considerations, for the most part, are outside the scope of this book.

base support to such "embedded" applications. In the last chapter, a highly efficien semantic/spatial database kernel, enhanced with a set of higher level functiona shells, was recommended as the core element of future data fusion applicatior development.

10.1 DATABASE REPRESENTATION

Since complex problems typically possess a multiple level-of-abstraction character algorithm efficiency can often be improved by employing top-down, global prob lem-solving strategies. A top-down, constraint-satisfaction-based doctrinal templat ing paradigm, for example, tends to be both more effective and more efficient thai myopic, single level-of-abstraction approaches that depend on highly combinatori composition operations and back-end optimization.

Support for multiple level-of-abstraction reasoning requires multiple level-o abstraction data representations. Since spatial data (maps, images, situation dis plays) and semantic data (facts, rules, tables of equipment, system parameters) ar the two primary data classes in most data fusion applications, the use of hierarchica representations of both spatial (multiple resolution) and semantic information (hi erarchical object-oriented representations) tends to enhance algorithm efficiency The integrated semantic/spatial object database kernel outlined in Chapter 8 prc vides effective support to both such representations.

Because scenario-driven "proof of concept" demonstrations typically utiliz very limited non-sensor-derived information, database performance may not surfac as a significant issue. Even when laboratory demonstrations do employ non-senso derived context knowledge, typically only relatively small, *context-specific* knowledg bases are utilized. Thus, while prototype algorithms might appear to meet real-tim operational requirements, full-scale applications may not if they require access t and manipulation of significantly larger domain databases.

Thus, the development of efficient data fusion systems and subsystems re quires both inherently efficient algorithms as well as efficient algorithm/dat base interaction. Efficient database support, in turn, demands highly focuse search and manipulation of both spatially organized and nonspatially organize data. Databases that do not support highly focused search and manipulation ca significantly impede the development of sophisticated fusion algorithms. An ir herently efficient target-tracking algorithm that is sensitive to the distance be tween individual radar detections and the closest road, for example, may not b practical to implement if the context-instantiation process requires the exhau tive search and analysis of a large road network. Because both the databas search-space size and the computational requirements can scale exponential with problem size, neither faster hardware nor small-scale parallelism can con pensate for performance problems that are due primarily to inefficient databas search and evaluation operations.

In Chapter 7, database principles associated with both algorithm support and algorithm performance were discussed. Algorithm-related requirements include:

1. Support for the development of *intuitive algorithms*;
2. Support for the development of *robust* and *context-sensitive algorithms*;
3. Support for the development of *computationally efficient algorithms*;
4. Provision of *high-accuracy data representations.*

Database performance requirements include:

1. *Memory efficiency*;
2. *Overhead efficiency*;
3. *Search efficiency*;
4. *Association efficiency*;
5. *Complex query efficiency*;
6. *Implementation efficiency.*

Because there exist significant dependencies among these ten requirements, a practical database design requires effective compromises across these requirements.

Three fundamentally different database organizations were considered. In a *query-product* organization, the preprocessed response to all queries is explicitly maintained. In a *single representation, on-line query* organization, all underlying information is maintained in a single data structure and queries are evaluated on-line. In a *multiple-representation, on-line query* organization, the underlying information is maintained in multiple data-type dependent structures and queries are evaluated on-line. Only the latter organization was seen to provide the necessary degrees of freedom to achieve an effective compromise across all the algorithm and database requirements.

Since support to robust and context-sensitive algorithms can require the maintenance and manipulation of massive spatially organized data sets, the database design presented in Chapter 7 was driven predominately by spatial-reasoning requirements. Table 7.2 summarizes the relationships between the database design objectives and spatial data representation characteristics. *Intuitive algorithm development, algorithm performance efficiency*, and *database search efficiency, indexing efficiency, maintenance efficiency, association efficiency, complex query efficiency*, and *implementation efficiency* all benefit from the use of hierarchical, true two-dimensional spatial representations. On the other hand, *representation accuracy* and *memory efficiency* are seen to benefit from the use of nonhierarchical, nontrue two-dimensional spatial representations. Therefore the development of a hybrid spatial representation that incorporates both a hierarchical, true two-dimensional spatial representation *and* a nonhierarchical, nontrue two-dimensional spatial representation can provide a simple, yet effective, compromise among these competing requirements.

A significant limitation of conventional database systems is their inability to support highly context-sensitive algorithms that require the search and manipulation of massive long-term declarative domain knowledge bases. The *object representation of two-dimensional space* seeks to overcome this limitation by supporting efficient top-down true two-dimensional spatial search of classes of domain features. Figure 10.1 shows a simplified block diagram of the semantic/spatial object database kernel. At the highest level of abstraction, the database consists of a hierarchical *object representation of two-dimensional space* and a hierarchical representation of *semantic objects.* The *semantic object* component possesses all the usual object characteristics, such as hierarchical organization, encapsulation of methods, and message passing–based communication between objects.

Just as the semantic object *water* maintains the common characteristics of *rivers, lakes,* and *oceans*, a spatial object summarizes the general characteristics of its offspring by cataloging the class of all objects associated with its spatial domain. As a result, an *object representation of two-dimensional space* provides a search-efficient data structure that effectively summarizes the *spatial organization* of semantic objects and their attributes. Just as top-down search through the semantic object hierarchy helps focus the search space with respect to lower level-of-abstraction objects, so does top-down search through the *object representation of two-dimensional space* focus the search space with respect to spatially organized objects.

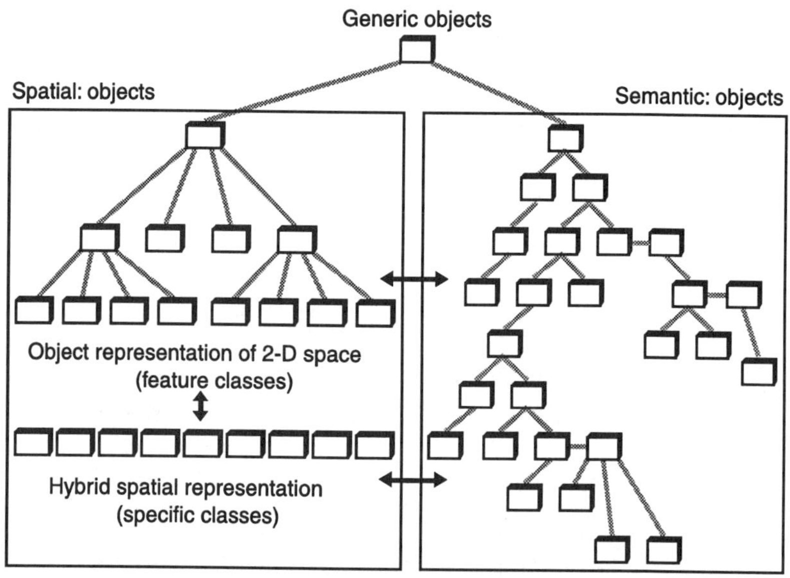

Figure 10.1 High-level view of the semantic/spatial object database kernel.

At the bottom of both the semantic and spatial object hierarchies are individ-ıal atomic objects. In general, spatial objects can be treated as spatial attributes of emantic objects. Thus, a particular lake (lake 1) is an atomic semantic object, while he two-dimensional spatial representation of that lake is an atomic spatial object. igure 10.2 depicts the relationships between the spatial and semantic repre-entations of these two objects.[2] In this case, the semantic object *lake* 1 contains two patial attributes (*location* and *shape*) that point to the hybrid spatial representation f lake 1. The *location* of lake 1 is coarsely specified by its root-node location with re-pect to the global and local origin offsets. The root node serves the role of a tradi-ional bounding box but possesses additional important properties due to the yramid's hierarchical grid-based character.

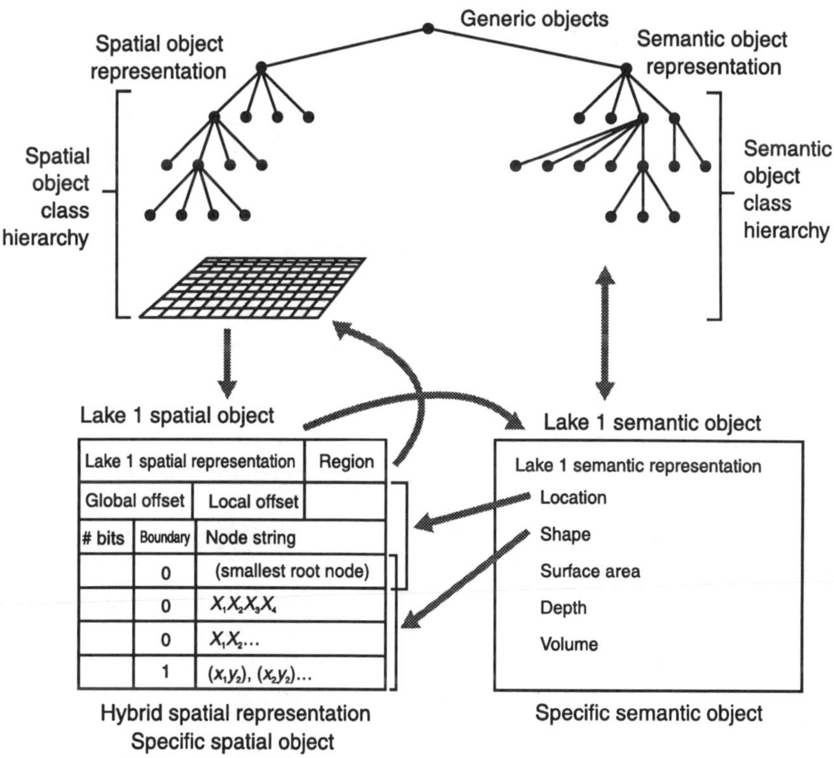

ure 10.2 Summary view of the relationship between the spatial object and semantic object representations.

Although a number of design details have been presented, our objective has been to motivate the general principles rather than to provide a comprehensive treatment of a detailed database design.

The node-string list effectively provides a multiple-resolution representation o the *shape* of the lake. When the relevant offsets are applied to the root node, bot the *shape and* the *location* (within either the local or global grid system) are uniquel specified. When a higher resolution representation is required, the vector repre sentation associated with the lake boundary nodes is utilized.

Although indexing structures, by their very nature, represent redundant info mation (the index is derived from the information being indexed), they can signif cantly enhance information access. Library indices and keyword retrieval searc systems avoid random or sequential search through a potentially vast number c books, magazines, and conference proceedings. Similarly, the *object-oriented repr sentation of two-dimensional space* provides a fully registered, multiple-resolution s mantic description of two-dimensional space that serves as a hierarchical spati index to *classes* of spatial features. With no spatial-indexing capability, the determ nation of "the closest river to a particular query point" might require exhaustive tes ing of much of the river database.

The *object representation of two-dimensional space* serves not only as a hierarchic *spatial index* to higher resolution representations of individual features but as a mu tiple-resolution *spatial representation* of classes of spatial features. Because all pyram cells that contain lakes have a particular node descriptor bit set to 1, the pyramid e fectively provides an efficiently searched, multiple-resolution spatial data structu that represents the *union* of all individual lake representations. The hybrid repr sentation of individual spatial features is utilized when either (1) the resolution the pyramid representation of *classes* of spatial features is inadequate, or (2) *speci* spatial objects (and not just object *classes*) are of interest.

Because of the potential size of the pyramid data structure (of order n^2, whe n is the number of cells along one border of the base resolution grid), pyram maintenance requirements can significantly affect the overall efficiency of the da base kernel. When changes occur to the underlying data set, all pyramid offspri nodes and superordinate nodes on the path to the root node must be updated. E cause the changes to a leaf node remain within a well-defined branch of the da structure, the pyramid can be considered a relatively dynamic data structure. Ho ever, maintenance requirements can be avoided altogether by indexing only re tively *static* long term knowledge. Because the size of the dynamic short-term a medium-term knowledge bases are typically many orders of magnitude smaller th the size of the long-term declarative knowledge base, the benefit of indexing *namic* spatial data tends to be substantially less than the advantage of indexing *st* spatially organized knowledge.

Since the hybrid representations of *individual* spatial features and the pyran representation of *classes* of spatial features are based on the identical quadtree-bas spatial decomposition, the two representations are intimately related. The r node of the individual spatial representations, for example, point either directly indirectly to a node in the pyramid. If the root-node size of a specific spatial feat is equal to or larger than the size of the base nodes of the pyramid, the hybrid s al representation points directly into the pyramid. If the root node is smaller than pyramid base node, the root node of an individual spatial object points indirectly) a single pyramid base-node.[3] By treating an individual spatial representation as a patial window into the pyramid data structure, spatially oriented associations, such s intersections, proximity, and inclusion, are readily evaluated.

While an index file in a RDBMS supports efficient one-dimensional indexing f table-based data, the pyramid serves as an efficient hierarchical two-dimensional idex into *classes* of point, line, and region features (with *indirect* pyramid indexing) *individual* spatial features (with *direct* pyramid indexing).[4] The quadtree-based omponent of the hybrid spatial representation provides an efficient hierarchical o-dimensional index to high-resolution, vector-based representations of individ-l spatial features as well as an index into the pyramid. Because all cells within a ngle layer of the pyramid are inherently disjoint, pyramid-based data structures nd themselves to distributed implementations. Since the representations of indi-dual spatial features are self-contained, the hybrid spatial representations can be stributed as well.

Human problem-solving approaches are typically much more computationally ficient than traditional algorithmic approaches. Thus, it may be possible to en-nce algorithm efficiency by emulating at least certain aspects of human problem lving. For example, given the data representation shown in Figure 10.3, a human n immediately identify all roads that are *within region* 1 or *north of region* 1 without tensive analysis. On the other hand, working from a vector-based representation the same data set, a human would have great difficulty performing the same task. us, algorithm efficiency can be highly dependent on the form of the data to be alyzed. Because the semantic/spatial object database kernel provides natural rep-entations of both spatial and semantic data, the database supports the develop-nt of algorithms that emulate the highly *intuitive* semantic and spatial reasoning rformed by humans. Because it facilitates highly *efficient* data access and manipu-ion of both semantic and spatial information, the database facilitates the develop-nt of both highly intuitive and highly efficient data fusion algorithms.

Three spatial representation classes have dominated much of the work in spa-reasoning: *raster, vector,* and *pixel-boundary* representations. Raster data structures a single-resolution, true two-dimensional, areal-based representation. Vector a structures provide a memory-efficient, high-accuracy, nontrue two-dimensional resentation of point and lineal (such as lines and region boundaries) features. el boundary-only data structures store points as individual pixels and lines and ions as the ordered list of pixels that cover the lines and region boundaries, pectively.

Progressively higher level parents of a linear quadtree node can be generated by recursively truncat-ing the node-string notation of the original node.

Because of the efficiency of tree search, most RDBMS systems actually employ *B*-trees as their inter-nal, underlying data structure.

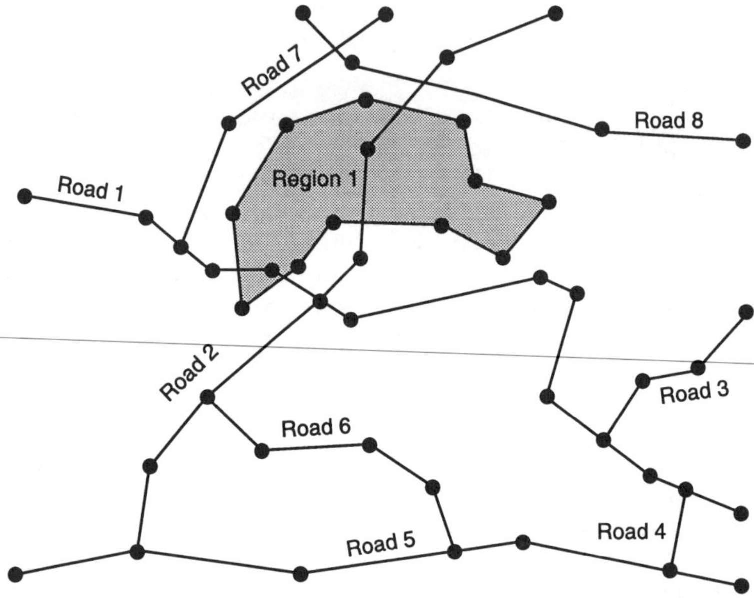

Figure 10.3 Partial road network database depicting spatial search window described by region 1.

The hybrid spatial data structure recommended in Chapter 8, on the oth hand, employs a *minimal region quadtree* representation that supports efficient, to down variable-resolution true two-dimensional spatial search and manipulation a an associated vector-based representation; the latter representation supports hi accuracy analysis. Although raster representations support true two-dimensio spatial search and manipulation, they are single-resolution spatial representatio that do not support either:

1. Top-down spatial indexing and search;
2. Top-down set operation and manipulation.

Both strict vector and pixel-boundary representations possess significant li tations as well, including:

1. Both are nontrue two-dimensional spatial representations;
2. Neither representation preserves spatial relationships among features;
3. Neither representation is areal-based.

As a result, vector and pixel-boundary representations do not support eit true two-dimensional spatial search or highly efficient set-operation computati Because both raster and pixel-boundary representations are fundamentally g based, both are relatively inefficient for storing high-resolution representations of point, line, and region features.[5] Although pixel-based boundary representations are more memory efficient than their raster-based counterpart, memory efficiency is achieved at the expense of both search and manipulation efficiency.

In summary, the key properties of the semantic/spatial object database kernel are as follows:

1. The low-resolution representations of specific spatial features are stored along natural two-dimensional query dimensions that supports:
   a. Multiple-resolution, true two-dimensional spatial search;
   b. Data association based on implicit spatial relationships.
2. The preservation of natural hierarchical relationships among both spatially organized and nonspatially organized objects permits efficient bottom-up/top-down, mixed spatial/semantic object reasoning.
3. Since both the pyramid and the hybrid spatial representation share the identical quadtree-based spatial decomposition, the spatial indexing of both individual spatial features and classes of spatial features are fully integrated.[6]
4. The semantic object representation, the spatial object representation of two-dimensional space, and the hybrid spatial representations are fully integrated.
5. The database kernel effectively provides a spatial window into the semantic database as well as a semantic window into the spatial database.

This seamless integration of diverse data representations is a significant departure from the common viewpoint that future database systems must merely *support* multiple data types.

The semantic/spatial object database kernel accommodates top-down, bottom-up, as well as single level-of-abstraction reasoning with respect to both semantic objects and two-dimensional space. Because both the semantic and spatial object representations are fully integrated, the database permits the development of sophisticated and highly intuitive algorithms that rely on both spatial and non-spatial data. Table 10.1 summarizes the general characteristics of both the *object representation of two-dimensional space* and the *hybrid representation of individual spatial objects.* The combination of the two representations embraces the full spectrum of spatial representation attributes.

---

5. In this sense, the pixel-boundary data structure is actually a variant of a raster representation. However, pixel-boundary representations can also be portrayed as a uniformly sampled variant of a vector-based representation. Thus, the pixel-boundary representation shares characteristics of both vector and raster representations.

6. The low-resolution representation can be treated as a natural extension of the pyramid for the indexing of *specific*, rather than *classes* of, spatial objects.

Table 10.1
Summary of the Characteristics of the Proposed Database Kernel Demonstrating the Diversity of Spatial Representation Attributes Supported

Database Element	*Spatial Representation Attributes*											
	Regular	*Nonregular*	*Hierarchical*	*Nonhierarchical*	*Areal*	*Nonareal*	*2-D*	*Non-2-D*	*Explicit*	*Implicit*	*Specific*	*Composite*
Object representation of 2-D space (pyramid)	•		•		•		•		•			•
Hybrid representation of specific object												
•Low-resolution index	•		•		•		•		•		•	
•High-resolution spatial representation		•		•		•		•		•	•	
Composite representation: pyramid & hybrid spatial representation	•	•	•	•	•	•	•	•	•	•	•	•

Table 10.2 provides a coarse comparison between the traditional spatial data representations and the semantic/spatial object database kernel with respect to six of the ten database design criteria. In this table, a data structure is considered "good" if it fully supports a database requirement. At the other extreme, it is considered "poor" if it provides very little direct support. For instance, although it is both memory-efficient and of potentially high precision, a purely vector-based data structure does not support efficient spatial search. By contrast, raster-based representations support spatial search but are neither memory-efficient nor inherently of high precision. When the classification is neither clearly either "good" or "poor," the data structure support is considered to be "moderate." For instance, in terms of memory efficiency considerations for a given resolution level, pixel-boundary representations tend to be better than raster representations but worse than vector-based data structures. The table indicates that the hybrid spatial representation provides an effective compromise across a wide range of database performance requirements.

Table 10.2
Spatial Data Structure Comparisons

	Representations			
Design Criteria	*Vector-Based**	*Raster-Based*	*Pixel Boundary*	*Semantic/Spatial Object Kernel*
Memory efficiency				
Data storage efficiency	Good	Poor	Moderate	Moderate
Overhead efficiency	Good	Good	Good	Moderate
Spatial search efficiency				
Specific features	Moderate	Good	Moderate	Good
Feature classes	Poor	Good	Poor	Good
Association efficiency				
Specific spatial features	Poor	Good	Moderate	Good
Spatial feature classes	Poor	Good	Poor	Good
Complex Boolean query efficiency				
Specific features	Moderate	Good	Moderate	Good

Table 10.2 (continued)

	Representations			
Design Criteria	*Vector-Based**	*Raster-Based*	*Pixel Boundary*	*Semantic/Spatial Object Kernel*
Complex Boolean query efficiency (continued)				
Feature classes	Poor	Moderate	Poor	Good
Accuracy	Good	Poor	Good	Good

* Includes associated bounding boxes.

10.2 QUERY OPTIMIZATION

Query optimization seeks to enhance query response time. In general, query clauses include both spatial and semantic constraints that are linked by operators. Spatial constraints modify two-dimensional point, line, and region features (east(line 20), within 1 km of (x_1, y_1)), while nonspatial constraints define or modify nonspatial features (number of companies in a division, equipment performance specifications). For data that is stored along the appropriate search dimensions (either naturally or through artificial means), the search-space size tends to be related to the "size" or the "extent" of the query constraint; otherwise, search costs tend to be related to the data-set size. Since the quadtree-based spatial representation effectively represents a hierarchical spatial index, no separate indexing structure is required.

In addition to search-dimension issues, query efficiency is enhanced if database manipulation operations, such as set operations and region growing, are matched to the internal data representations. The top-down intersection of two multiply-connected regions, for instance, tends to be considerably more efficient when data are maintained in fully registered minimal region quadtree representations than as boundary-only data structures.

Although the *product* of a query involving concatenated clauses is independent of either the clause order or the evaluation methodology, the order of clause processing as well as the evaluation methodology can significantly affect the *efficiency* of the evaluation process. Given the region quadtree-based data structures employed by both the pyramid and the hybrid spatial data representation, two fundamentally different approaches to spatial query evaluation are supported: *set operation–based* and *search-based.* Set operation–based query evaluation can be treated as composition of sets, while search-based query evaluation can be treated as composition of constraints. In the former approach, the query is evaluated by exhaustively performing Boolean operations on sets that define the individual query clauses. In the latter approach, the query is evaluated by treating clauses as database search constraints. Search-based evaluation techniques tend to be considerably more efficient than set operation–based approaches. For both set operation–based and search-based approaches, query efficiency can be enhanced by the proper ordering of the constraint clauses. Because the database kernel supports true constraint-based search of both semantic and spatial objects, the database supports the development of efficient query optimization strategies for mixed semantic and spatial queries.

10.3 SUMMARY AND CONCLUSIONS

Part I demonstrated that automating many data fusion tasks virtually demands the use of the higher order canonical problem-solving form approaches. Part II argued that there currently exist significant shortfalls in the existing information-processing infrastructure to support these more sophisticated applications. In Chapter 6, current database systems were shown to provide only limited support to the development of more robust, context-sensitive, and efficient algorithms. Chapter 7 offered a top-down view of database issues that are relevant to the development of both effective and efficient fusion algorithms. A database kernel was introduced in Chapter 8 that overcomes many of the limitations of relational database management and GIS systems. Chapter 9 described several database kernel extensions, including query optimization for mixed semantic and spatial queries, and higher level database interfaces. This chapter has highlighted many of the important principles discussed in Part II.

In Part III, the algorithm development principles from Part I and the database principles from Part II are brought together to illustrate that, given the recommended infrastructure enhancements, it should be both feasible and practical to develop robust, context-sensitive, and efficient data fusion algorithms.

PART III

Putting It All Together

Although the previous two parts have focused almost exclusively on military applications, as suggested several times, there exists a much broader range of data fusion applications. The fusion of multispectral imagery and environmental monitoring data support land use, air quality, and earth resource studies. The fusion of multiple two-dimensional X-ray scans in *computer aided tomography* (CAT) produces three-dimensional X-ray images of dense body tissues. Similarly, *magnetic resonance imaging* (MRI) constructs a three-dimensional image of predominately soft tissues. The fusion of CAT and MRI images generates composite images that depict both hard and soft tissue. The monitoring of complex automated manufacturing operations, such as aircraft engine assembly, requires the fusion of information from numerically controlled machine tools, automated inspection devices, and quality assurance tests on individual subassemblies.

While the above examples are simple extensions of the tactical data fusion process to other application domains, numerous less obvious applications exist as well. Consider, for example, the production of "on-demand" electronic news products. Because most newspapers and magazines are directed at a rather broad audience, an individual reader is likely to have little or no interest in many of the articles, moderate interest in some, and considerable interest in only a few. The articles of most interest to the reader are likely to possess inadequate detail. Although specialty news formats exist, the material is still prepared for a relatively broad cross section of readers.

In the near future, however, highly personalized electronic news services will generate products based on an analysis of a client's education and background as well as subject-matter interests. The generation of such products represents a form of multiple-source data fusion. In this case, the major information sources are press releases (*medium-term knowledge*) and relatively static background information (*long-term declarative knowledge*). An automated electronic news service would filter the news source, correlate potentially relevant information from separate sources, template the material based on the user profile, and abstract the information to achieve the level of detail and focus desired by the customer. For example, a client may be generally interested in all science-related news but have a particular interest in a comprehensive treatment of major breakthroughs in astronomy, with a focus on radio astronomy. In addition, the same client may be interested in all available experimental findings in superconductivity research with a focus on potential commercial applications.

As opposed to using simple keyword search, the system could employ intelligent agents that attempt to "understand and interpret" the information being scanned. Because of the hierarchical nature of interest profiles, the highest level template constraints effectively represent a coarse "filter" on the information sources, while the progressively lower level constraints fine-tune the products that pass the coarser filters. The information products can be further refined by consideration of important contextual constraints, such as the previous information products that have been provided to the client, the relative importance of the information, as well as through direct feedback from the client.

The fused information product would provide a concise, coherent, comprehensive, and highly focused presentation of current news releases (medium-term knowledge) interpreted within context (with respect to both the medium-term knowledge and the static long-term declarative knowledge bases) that is consistent with the client's interest profile (level-4 system-level constraints and objectives). In addition to highly customized products, since the traditional preparation, printing, and distribution time have been effectively eliminated, very timely product delivery is possible.

Thus, a wide range of real-world problems can be viewed as the fusion of similar and dissimilar information. The final four chapters in this book attempt to demonstrate both the generality and the utility of the concepts that have been developed in Parts I and II. Chapter 11 reviews many of the important algorithm and database principles that were developed in the first two parts. Chapter 12 discusses spatial, temporal, and hierarchical reasoning from the perspective of their underlying role in data fusion algorithm development. Chapter 13 presents detailed examples of several sophisticated data fusion algorithms to demonstrate that it is possible to develop both effective and practical algorithms for complex, real-world problems. Finally, Chapter 14 summarizes some of the key concepts and issues that have been presented, suggests an approach for evolving more integrated approaches to machine reasoning, and offers a number of recommendations regarding the need for continued development of an enabling environment for more effective information analysis.

CHAPTER 11

Summary of Algorithm and Database Principles

1.1 THE FUNCTIONAL FUSION MODEL

'he tactical situation development process currently relies heavily on the manual nalysis of sensor-derived information. Because of the time-critical nature of many ntelligence products, the limited number of available problem-domain analysts, nd the ever-increasing volume of sensor data, there exists both extensive requirenents as well as considerable opportunities for increased data fusion automation. In he short term, the development of visualization tools and decision support systems an be expected to enhance analyst productivity. In the long term, however, many outine, tedious, and time-critical functions require increased automation. This ook has discussed a wide range of algorithm and database issues associated with uch automation.

In this section, several different views of the functional fusion model depicted n Figure 1.1 are offered. Figure 11.1, for example, emphasizes both the sequential rocess flow and multiple level-of-abstraction character of the first three fusion levs and the process control-like role of the fourth fusion level. Figure 11.2 depicts a artial taxonomy of the external information sources that provide input to the tactial data fusion process. Figure 11.3 offers a partial taxonomy of the first three levels f the fusion process.

In the object-centered view offered by Figure 11.3, level 1 is associated with the nalysis of *individual objects*, level 2 with *organizations of objects*, and level 3 with the *igher level analysis of those organizations*. In level 1, data from available information ources are associated with *individual objects* (a particular tank, command post, misle launcher, or radar system). The *attributes* of the object (size, velocity, type, class) re progressively refined as additional information becomes available. The *current ehavior* of the individual objects (current trajectory, transmission frequency, operat-

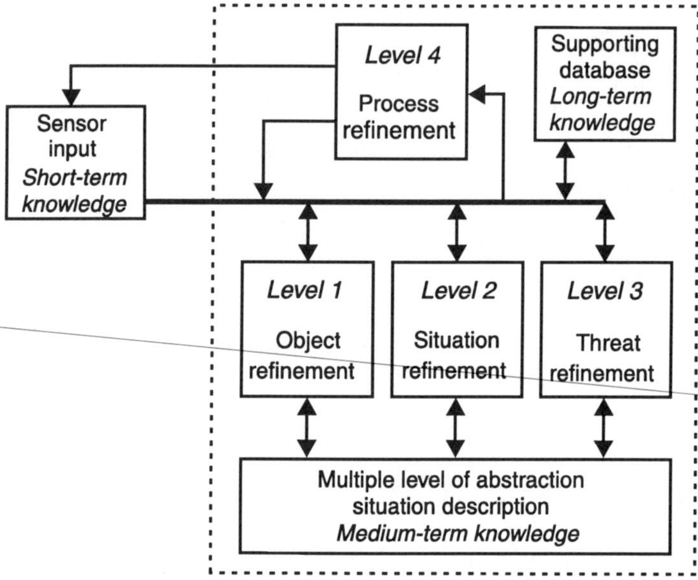

Figure 11.1 High-level view of the situation development process depicting the four levels of fusion an the major information sources.

ing mode, and flight path characteristics) are analyzed to attempt to understan behavior.

In level 2, *individual objects* are associated with *organizations of objects* (comp nies, platoons, brigades, divisions). The *attributes* of these composite objects are pr gressively refined as information becomes available. The *current behavior* of th composite objects are evaluated (a bridge crossing, a reconnaissance overflight, un reorganization), and their *projected behavior* is hypothesized (create a diversion, infi trate a city, capture a power plant). The disposition of the level-2 objects, couple with both the current and projected behavior of those objects are progressively a stracted in level 3. Thus, the first three fusion levels provide a progressive abstra tion of lower order information.[1]

Figure 11.4 illustrates the multiple level-of-abstraction situation refineme process resulting from the detection of emissions from several radios. In level 1, th specific locations of individual *emitters* are identified and progressively refined by tr angulation or other means. At the next higher level-of-abstraction (level 2), *rad networks* are identified based on traffic analysis and emitter-specific characteristic

1. This viewpoint does not preclude the direct utilization of high level-of-abstraction information fro an external source. For instance, the visual sighting of a river-crossing event could directly feed level-2 analysis function with only indirect effects on level-1 processes.

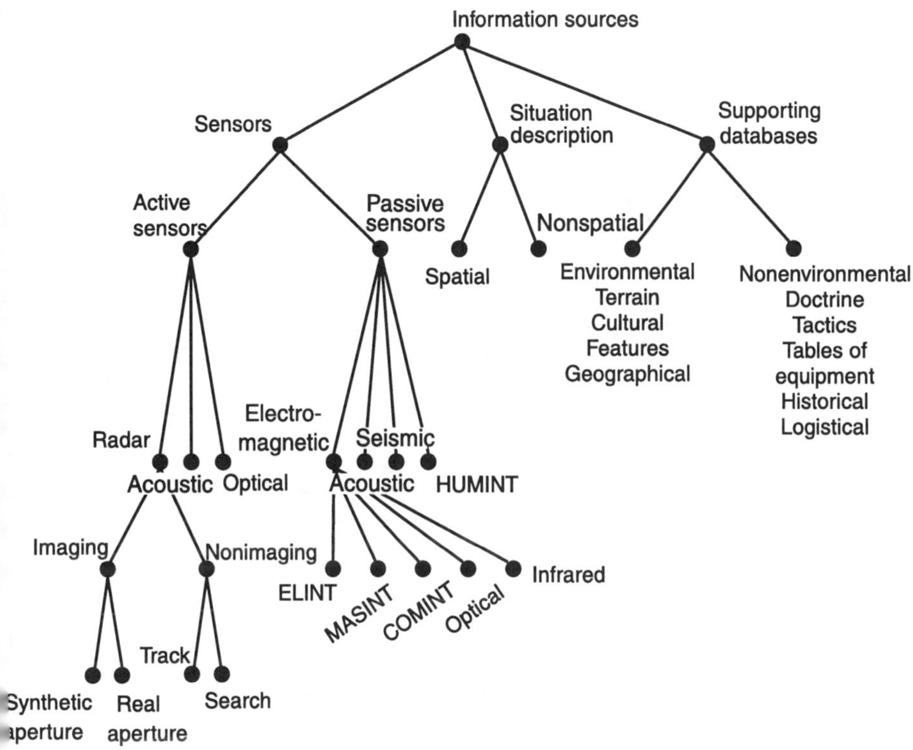

igure 11.2 Partial taxonomy of the three main information sources that support situation development.

'inally, the units that would most likely be associated with such radio networks are lentified based on supporting sensor-derived information, historical data, and docrinal knowledge. The analysis of this higher level situation product supports threat etermination (level 3) and command decision making.

Figure 11.5 provides an abstract view of the fusion process that emphasizes the ritical roles of non-sensor-derived knowledge sources. Figure 11.6 associates a numer of broad classes of knowledge with the various fusion levels. As observed in this gure, knowledge of individual sensor characteristics and capabilities tend to be ıost relevant at levels 1 and 4, whereas knowledge of enemy tactics and doctrine nd to be most relevant to the higher level processes. Some knowledge classes, such geographic and cultural feature information, on the other hand, tend to be idely applicable across all four fusion levels.[2] Figure 11.7 illustrates that both spa-

This observation highlights the critical role of non-sensor-derived domain knowledge in supporting sophisticated data fusion applications. The relevance of the database issues presented in Part II becomes clear when the potential size of such data sets is considered.

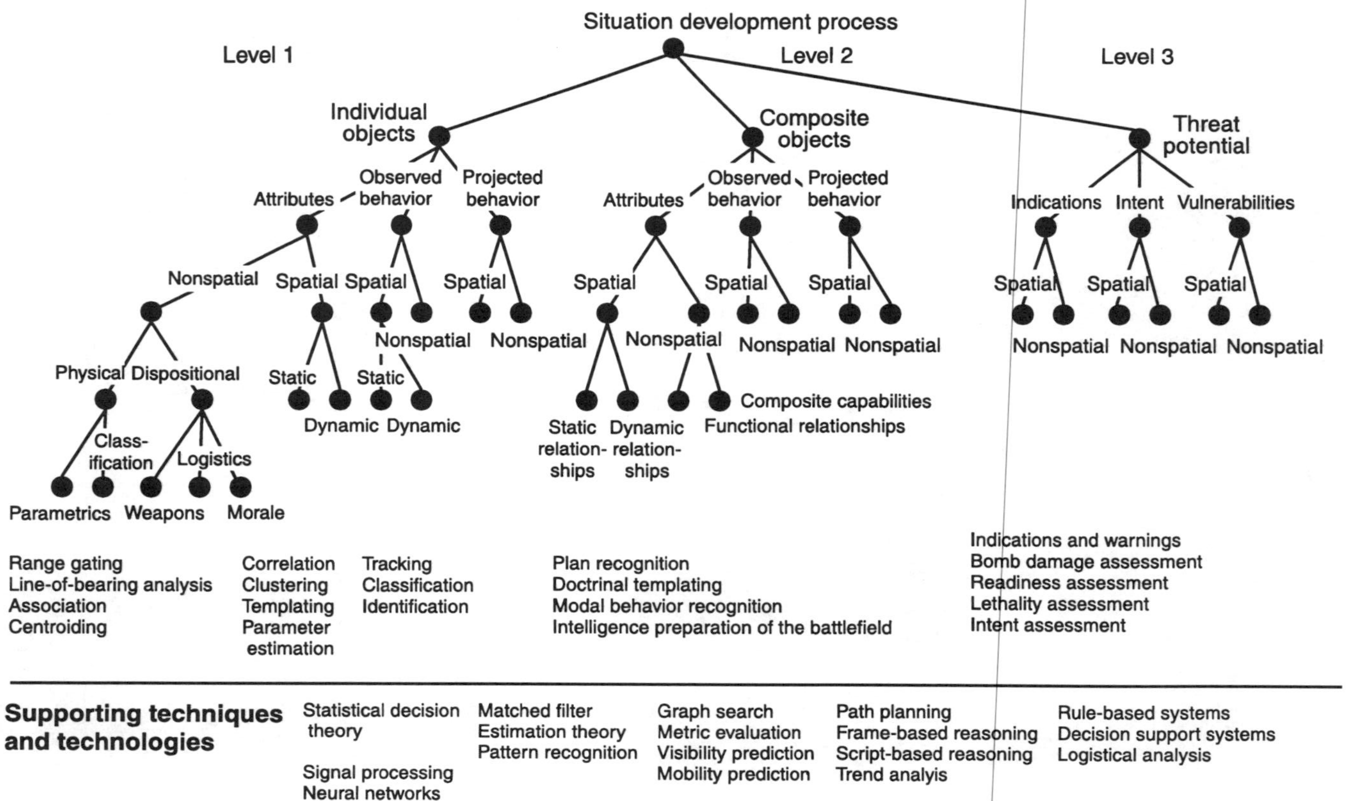

Figure 11.3 Partial taxonomy of an object-centered view of the situation development process demonstrating both the hierarchical and sequential character. Supporting problem-solving approaches are shown across the bottom.

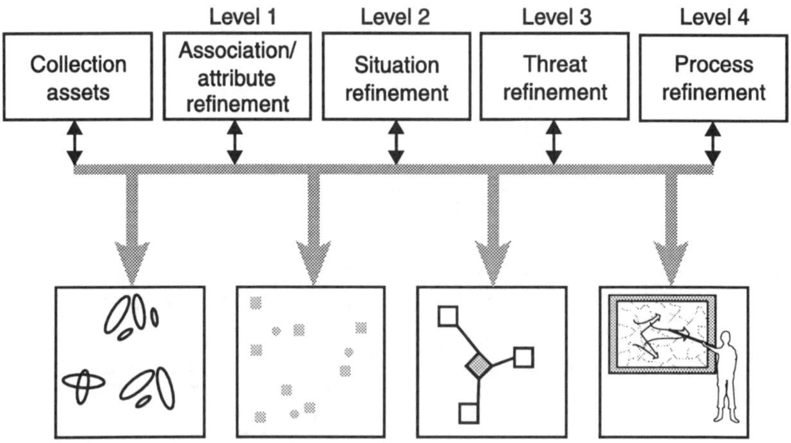

Figure 11.4 Example depicting the multiple level-of-abstraction character of the fusion process and its products.

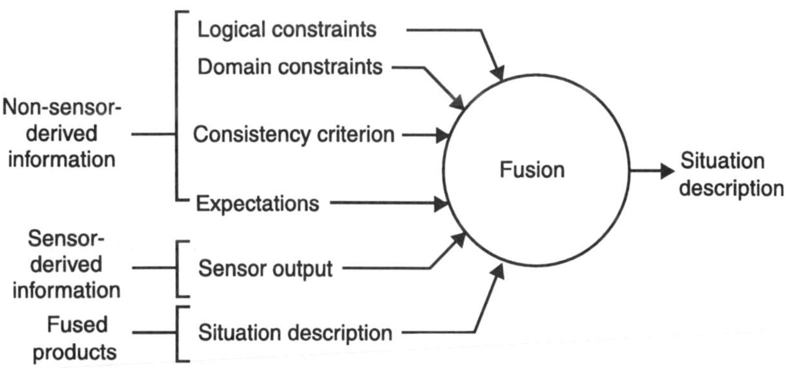

Figure 11.5 Abstract view of the fusion process that emphasizes the role of non-sensor-derived information.

tial and nonspatial information are normally associated with the three primary classes of knowledge.

In general, data fusion can be performed in a highly distributed data environment (information maintained at collection sites, intermediate processing sites, and fusion centers). Thus, information relevant to a particular fusion task might be resident within multiple distributed databases. Because of bandwidth limitations, providing all possible information to all possible processing sites might be infeasible. Therefore, there must exist a means of both acquiring and sharing information.

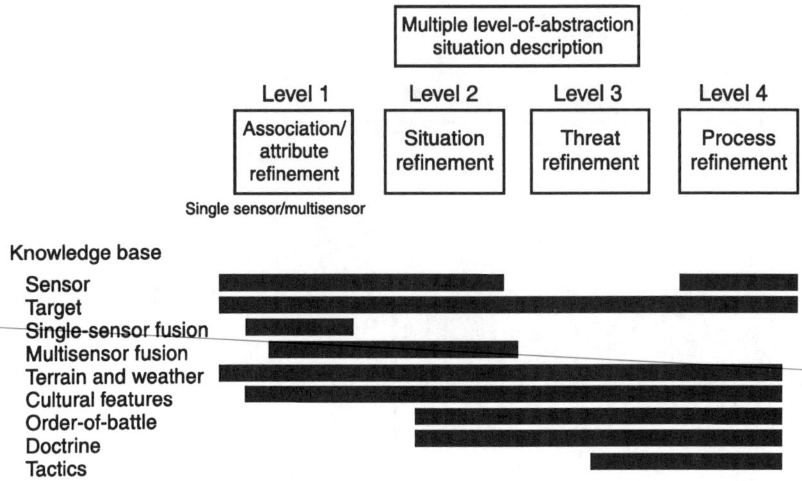

Figure 11.6 Classes of knowledge that can enhance the interpretation of sensor-derived knowledge.

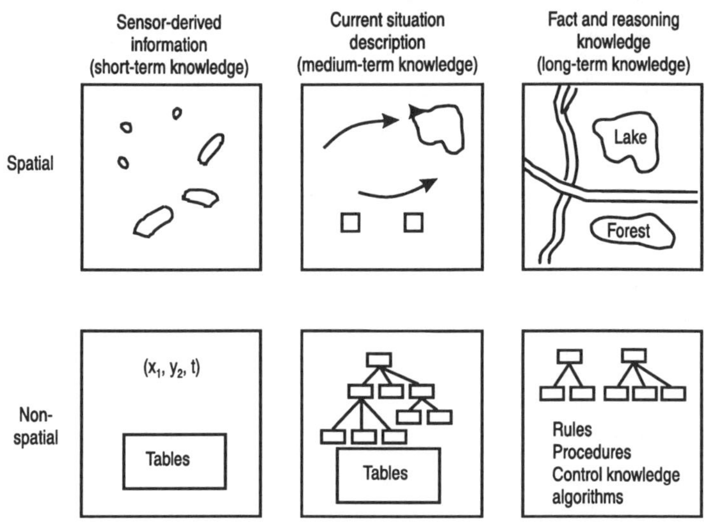

Figure 11.7 Illustration of various classes of spatial and nonspatial short-, medium-, and long-term knowledge.

Since, at a high level of abstraction, queries can be separated into either object-centered (provide all available information about a specific entity or class of entities) or spatial-centered (provide all information associated with a specific location or that is within a specific region), the organization of the spatial/semantic database naturally supports efficient data access (*data pull*) as well as data sharing (*data push*). In *data pull*, a simple and effective indexing scheme is required to broadcast information requirements. In *data push*, a similar indexing scheme is required to permit the recognition of standing requests for types or classes of information.

Since the hierarchical organization of the fusion model depicted in Figure 11.3 is naturally matched to the hierarchical data model for the semantic/spatial database kernel depicted in Figure 10.1, a highly effective and efficient taxonomy-based data access scheme can be developed that is motivated by the object-oriented view of the data fusion process and supported by the integrated semantic/spatial data model.

11.2 INFORMATION COMBINATION: COMMENSURATE DATA SETS

Although the fusion of similar classes of information might seem to be more straightforward than the fusion of dissimilar information sources, this is not always the case. Three separate examples are offered to illustrate the varying degrees of difficulty of fusing information from multiple sensors. First, we consider an example of the relative simplicity of fusing electronic intelligence (ELINT) and real-time synthetic aperture radar (SAR) imagery. Although dissimilar sources, the fusion process is relatively straightforward because both sensors provide relatively wide-area surveillance and highly complementary information. For example, once the information from the two sensors are fully registered, a line-of-bearing obtained by the ELINT sensor can directly be overlaid on the SAR image. If the line-of-bearing intersects only a single object of interest in the image, the radar system class as well as its precise location may be identified. This information, in turn, might permit other objects near the radar to be identified, such as the missile launchers that are normally associated with track-while-search radars.

The fusion of information from two or more identical sensors can represent a significant challenge. For example, consider the fusion of data sets obtained from two identical but spatially separated forward-looking infrared (FLIR) radars. Although FLIR has relatively high-azimuth and elevation resolution, it possesses poor-range resolution. Because the range and view angles to individual targets are different for the two sensors, the two data sets may be difficult to combine.

Finally, consider the fusion of light-amplified and FLIR images for the purpose of enhanced night vision. Light amplification devices effectively amplify the intensity of an optical image. Although such devices permit users to "see" in very low light conditions, they provide limited definition of personnel and equipment in deep shadows or in extremely low ambient light conditions (heavy cloud cover with no moon, inside buildings). FLIR devices, on the other hand, sense the thermal radiation of objects rather than their reflected light. Thus, thermal imagery readily detects humans, vehicles, and operating equipment that are at a higher temperature than the ambient background. While light-amplification techniques provide an adequate depiction of the general character of a particular scene, as well as support the identification of objects that are in motion, infrared sensors can detect stationary humans and operating equipment that may not be readily apparent in a light-intensified optical image.[3]

Since light-amplification techniques depend on the presence of ambient light sources, while infrared imaging relies strictly on an object's thermal properties, "shadows" in optical and infrared images are caused by different phenomenon. Thus, the effective combination of the output from these two sensors could potentially produce images that appear to be illuminated by both point (moon, street lights) and diffuse sources (internal thermal properties, star light). At least conceptually, these two sources could be fused using a pixel-by-pixel combination of the two images. However, since the resolution, sensitivity, and field-of-view of the images are likely to be very different, image registration can be a significant problem. In addition, the fact that the sensors measure entirely different phenomenon further complicates the fusion process.

In general, the development of fully commensurate data sets requires, at a minimum, three-dimensional spatial registration. If the data varies over time (moving targets), temporal registration may be required as well. For sensors that inherently provide true three-dimensional measurements with similar resolutions, registration among sensor measurements can be accomplished by applying appropriate scaling, rotation, and translation operations. Thus, the return from spatially separated radar systems that provide high-azimuth, elevation, and range resolution (surveillance, tracking, synthetic aperture radar) can be straightforwardly registered if the location of the individual sensors is accurately known (by relying on global positioning satellite (GPS) receivers, for instance).

When sensors resolve targets in only two spatial dimensions, effective sensor registration requires that the third dimension be supplied via physical constraints or direct estimation. Although monocular satellite imagery directly resolves only two spatial dimensions, the third dimension (range) can often be estimated. For example, the range to the top of a particular building in an image can be determined as the range to the earth's surface minus the building height, which is proportional to the length of the shadow cast by the building. In the case of video imagery, the range to a target can often be estimated based on the relative size of known objects in the field of view (vehicles, street signs, road width).

For sensors that inherently supply only a single spatial dimension, the other two dimensions must be supplied by other means. Although a direction-finding sensor provides only time and azimuth resolution, the range and elevation of emitters can often be estimated. In the case of netted direction-finding sensors, the emitter

3. Because of the characteristics of human perception, objects are generally more easily detected if there is relative motion between the object and the background.

range is typically determined via triangulation or time-difference-of-arrival techniques. For ground-based emitters, the elevation can be determined from the terrain elevation database once its azimuth and range are known.

11.3 NON-DoD APPLICATIONS

As discussed previously, numerous applications have analysis requirements that are imilar to tactical data fusion problems. Consider, for instance, the development of an urban traffic control system designed to maximize the safe flow of traffic within a network of roads while providing optimal routing recommendations for police and emergency vehicles. Just as in the tactical data fusion problem, the relevant *situation description* is developed by fusing sensor information from multiple sources (elevated and land-based imaging systems, human observers, radar, ultrasonic, magnetic, pneumatic). Traffic control measures would be provided by lane control signs, variable speed limits, traffic lights, message boards, as well as direct communiation with future on-board vehicle control systems.

In such an application, the sensor data represent highly dynamic *short-term nowledge*; accident locations, traffic-flow bottlenecks, and roadway construction sites epresent *medium-term knowledge*; the road networks and the location of key static doain features, such as parking garages and residential communities, and business enters represent *long-term declarative knowledge*. Just as in the tactical situation awareess problem, the traffic situation has both a stochastic component (the spontaneus decision of an individual to drive from his/her home to a shopping mall) as well s a deterministic component (traffic flow increases out of factories at the end of all ıifts). The deterministic element can be dependent on the time of day, the day of ıe week, and even the season (most factories operate limited weekend shifts, shoping malls are crowded during the Christmas season). Since many phenomena ossess *both* inherently deterministic and inherently nondeterministic components, either strictly causal-based nor strictly statistical-based algorithms might be lequate.

Although originally motivated by the tactical data fusion problem, the funconal-level fusion model offered in Chapter 2 can be seen to be equally appropriate a traffic-control application. The detection, tracking, and classification of trafficow patterns (the identification of high vehicle-density areas) as well as potential lynamic) bottlenecks or obstructions to the free flow of traffic (a natural gas line ak) represent level-1 processes. The abstraction of local traffic-flow patterns into a ore global picture of the traffic conditions represents a level-2 process. In level 3, e overall "threat" to the free-flow of traffic imposed by the entire ensemble of sigl failures, water main breaks, accident cleanups, fire alarms, road repair, and adrse weather conditions would be developed. Analogous to the tactical fusion plication, the high-level analysis of the current situation depends on an assessent of the capability of repair, cleanup, and control measures.

Just as in the tactical application, level 4 supports top-down control of levels to 3 for the purpose of enhancing the quality and the timeliness of the fusion prod ucts. Achieving data fusion system objectives within resource constraints can de mand significant tradeoffs. Just as mission-specific requirements and limite resources are inherent elements of the tactical situation awareness problem, th highway-control system is similarly constrained. Rather than collecting larg amounts of irrelevant and redundant information that ultimately burdens the anal sis process, sensors should ideally collect only highly relevant information. Both th tactical situation development and the traffic-control system possess a number c significant complications, including:

1. Information requirements are dynamic, time-sensitive, and often conflicting;
2. Sensor resources are limited, both in coverage and in capability;
3. Only certain sensors may be under direct control of the data fusion system
4. The capability of level-1 to -3 fusion algorithms is inherently limited;
5. The computational resources of any fusion system are limited.

Clearly, the objective and scope of a fusion system must be commensurate wit its resources. While a large-scale tactical data fusion system, such as the Army Al Source Analysis System (ASAS), for instance, has extensive resources and a broa multiple-echelon support role, a lower echelon tactical system would have mo constrained resources as well as more highly focused objectives. Similarly, in th smart highway application, traffic flow must be optimized given limited sensors an control measures, as well as highly dynamic operational requirements, such as r routing traffic around accidents or distributing traffic among parallel arteries minimize road-construction delays. A traffic control system designed for a large me ropolitan city would require considerably more resources than one designed for moderate-sized city.

11.4 CONVENTIONAL APPROACHES

Tactical data fusion involves a wide range of specific functional tasks, such as targ detection, track association, attribute refinement, classification, doctrinal templ ing, traffic analysis, determination of enemy intent, and evaluation of an adversar possible courses of action. Over the years, a wide range of technical problem-solvi paradigms have been exploited in an attempt to automate these functions. As e denced by the following statements, an equally wide spectrum of viewpoints exi with respect to the applicability and relative merits of the various technologies:[4]

4. The technical literature is replete with such highly polarized viewpoints. Chapter 4 offered some sight into the credibility of these particular assertions.

1. "The movement of forces would be difficult to simulate in a neural network; although abstract processes, such as situation assessment and planning, are good applications" [1].
2. "While neural networks may provide an effective solution for tasks involving the interpretation of low-level data, they appear not to be appropriate for traditional symbolic domains" [2].

Traditionally, data fusion automation is approached using either (1) deterministic algorithms, (2) statistical-based algorithms, or (3) artificial intelligence techniques. Although all three of these technical approaches have been successful when applied to relatively restricted problem domains (car repair based on decision-tree analysis, optimal signal detection based on statistical decision theory, and medical diagnosis based on Expert System approaches, respectively), none have been fully successful when applied to dynamic, multiple level-of-abstraction problems that demand context-sensitive interpretation of fragmentary, uncertain, inaccurate, and partially erroneous data.

Target recognition based on the bottom-up transformation of raw sensor data often fails due to the inadequate information content of the sensor-derived information. Typical statistical-based approaches do not provide adequate performance for abstract decision-making tasks that are highly sensitive to the underlying domain context. Although multiple cooperating Expert Systems have been recommended for solving complex problems, the relatively static nature of the production-rule paradigm does not match the dynamic, process-oriented character of many data fusion tasks.

Neural networks are often recommended for problems that do not lend themselves to either conventional algorithmic or Expert System–based solutions. On the surface, neural nets offer at least two benefits over alternative approaches: massive parallelism, and no requirement for knowledge elicitation and algorithm development (neural networks are trained by examples). Although analog implementations do promise high-speed throughput, the ability of such networks to "learn" to solve problems of general complexity has not been fully realized. Adequate training samples must be available, and considerable training time can be required. For many real-world problems, a comprehensive set of training samples might be difficult or even impossible to generate. Rather than expecting networks to learn completely from examples, "important properties must be built-in or 'hard wired,' perhaps to be tuned by experience, but not learned in any statistically meaningful way" [3].

Although multiple-stage neural networks are theoretically capable of generating high-dimensionality decision surfaces, the network will fail to achieve acceptable performance if the effective dimensionality of input data is too low to permit separation of input classes. Because the neural networks provide relatively rigid transformations of the input data, such networks are not well suited to analyze phenomenon that exhibit complex dynamic behavior. Thus, undifferentiated neural systems tend to be best suited for applications that are relatively context-insensi tive, static, and that lack complex, systematic dependencies among critical problen attributes.[5]

Since observed activity in a tactical environment appears to be nonpredictable statistical-based reasoning approaches often appear to be quite natural. However there often exists more order than can be accounted for by strictly statistical-basec models. Numerous examples have demonstrated that the weather, doctrine, tactics terrain, cultural features, and logistical considerations tend to constrain the behav ior in reasonably deterministic ways. Thus, fusion algorithms can require both de terministic and nondeterministic elements.

As a simple example, consider the road-following vehicle example presentec in Chapter 1. In this example, a strictly statistical-based algorithm predicted that a wheeled vehicle that was, in fact, following a hard-surfaced road, would be detecte next in a large lake on the far side of a mountain. Since the purely statistical-base tracker was insensitive to the domain within which the target is embedded, the algc rithm produced erroneous conclusions. However, because the behavior of a vehicl is not truly predictable (the vehicle might accelerate, decelerate, or stop altogethe between observations), a strictly deterministic algorithm might not be fully ade quate either. However, by building a statistically based model on top of a set of car didate target behavioral models (road-following, minimum terrain gradien goal-seeking) that are sensitive to domain context, a potentially much more intu tive and effective algorithm results. In this case, the candidate behavior mode could be viewed as deterministic problem-solving constraints on a nondeterministi reasoning approach. An additional shortcoming of purely statistical-based ap proaches is that they often rely on simplifying assumptions (statistical ind pendence, complete knowledge of all a priori probabilities, and condition probabilities).

In general, data fusion demands the use of more domain knowledge than utilized by deterministic and statistical approaches and more procedural-oriente reasoning than is offered by conventional AI-based approaches. Consider, for in stance, the development of a doctrinal template that could be used to identify a div sion-level force. If two particular elements within a division are known to requi line-of-sight (LOS) communications, and no subordinate units can be separated a maneuver barrier, the doctrinal templating process must be sensitive to the d main features (rivers, marshes, forests, elevation, weather) that affect both LOS ar maneuver. Thus, data fusion effectively demands high-level symbolic reasoning th is supported by a procedural-oriented context-instantiation process. In convention automated approaches, the target database is typically correlated with a "comple set" of rigid templates that generally ignores any such contextual information.

---

5. Limited reasoning in context is possible if the relevant contextual information is explicitly suppli as input during both the training *and* the operation of the network.

11.5 GENERAL DATA FUSION PRINCIPLES

n the next three sections, the key principles that have been developed in the last en chapters are identified and organized. In this section, three categories of *general lata fusion principles* are presented as they relate to (1) information requirements, 2) problem-solving principles, and (3) fusion system products. Specific *algorithm deelopment principles* are presented in Section 11.6, and specific *database principles* are presented in Section 11.7. As has been observed numerous times, these principles pply quite generally to many forms of machine-based reasoning.

1.5.1 Information Requirement Principles

n this section, information requirements associated with the data fusion process are eviewed. Information sources have been loosely categorized as either sensor-deived or non-sensor-derived. Figure 11.5 presents an abstract view of data fusion as ne composition among sensor-derived and non-sensor-derived information.

ensor-derived information can be inherently fragmentary, multiple level of abstraction, and reundant. The puzzle-solving metaphor presented in Chapter 1 illustrated the inhernt hierarchical nature of information. The existence of several puzzle pieces nformation, measurements) showing tree bark and other pieces containing fragents of branches and leaves, coupled with a conceptual model of tree (a tree consts of a bark-covered trunk, branches, and leaves), supports the identification of ees in the scene. The recognition of a forest requires the existence of a large numer of trees of certain stature (a first-year Christmas tree farm would not qualify) nd spatial distribution. Once the high-level determination is made that a forest exts, there might be little value of collecting and analyzing additional information out individual trees or their subparts. Thus, unless a more detailed analysis of the rest is required (mix of tree species, forest density, tree-height distributions), the llection and analysis of additional information would be highly redundant.

Similarly, in a tactical situation awareness application, a moving target indicar (MTI) radar might report the presence of several moving vehicles of unknown entity, a human observer might report the visual sighting of a small group of arored personnel carriers, an ELINT sensor might establish the existence of a speic radar system typically associated with a particular unit class, while an imagery nsor might supply evidence of a division-level force. Although the various pieces information represent different viewpoints, perspectives, and levels of abstracn, each sensor could be observing elements of the same organization. Once an equate understanding of a particular element of the situation description is develed, further collection and analysis provides an ever-diminishing "return on investent." The objective of both the puzzle-solving task and the tactical situation areness task is to construct a coherent multiple level-of-abstraction picture from information fragments by effectively controlling both the information collectioı and analysis process.

Adequate information is required to support effective decision making. Attempts to autc mate data fusion tasks that humans perform with apparent ease have often been unsuc cessful, in part, because the algorithms tend to be *information-limited.* Sinc *information-limited* problem solving generates *underconstrained* solutions, the uncer tainty (or entropy) of the resulting situation description tends to be high. Numeı ous examples have been offered to demonstrate that, regardless of the problem-solvin paradigm utilized, reliable decision making requires adequate information. In genera the fusion process relies on three general classes of information:

1. Short-term knowledge (sensor measurements);
2. Medium-term knowledge (current situation description);
3. Long-term declarative and procedural knowledge.

Although no general metric exists for computing the minimum informatio requirements for a given problem, complex decisions tend to require more (c higher dimensional) information than simpler problems. In general, algorithm pe formance tends to improve as the dimensionality of information relevant to a dec sion-making task approaches the dimensionality of that decision process. The tot information is often maximized using a suite of sensors that measure different face or features of the domain. For example, since a Doppler radar provides an estima of the radial velocity of a moving target, an independent estimate of the target range rate by some other sensor would provide little "value added." The fusion pro ess is not enhanced because the new measurement does not significantly increa the information available to the decision process. On the other hand, a sensor th measures the electromagnetic emissions from a target under track would provide i dependent information that might significantly enhance the situation awarene process. Thus, in order to increase the effective dimensionality of the informatic available for decision making, the additional information must be both relevant ar at least partially uncorrelated with the existing information.[6]

The road-following vehicle-tracking problem that has been discussed sever times illustrates the advantage of increasing the total dimensionality of the inform tion that is available to decision making.[7] Because a purely statistical-based decisic process is underconstrained, a tracking algorithm that depends only on sensor-c rived inputs might be unable to track ground-based vehicles along a highly serpe

6. Redundant information can be confirmational and might be necessary for certain problems wh highly dynamic state changes are expected.

7. This example has been referred to numerous times for two separate reasons. First, the exampl easy to both explain and understand. Second, the example serves as a common thread for tying gether many of the basic principles of both algorithm development and database support.

ine road. However, with the addition of relevant non-sensor-derived domain knowl-
dge (roads, terrain, weather), additional constraints become available.

As a second example, recall the multiple target-tracking example offered in
hapter 6. Based on the knowledge of the local terrain and cultural features, a num-
er of relatively high-confidence decisions were made that include the following:

1. Based on their speed and apparent association with a hard-surfaced road, the tracks from three of the targets were identified as members of a convoy of wheeled vehicles.
2. Based on their overland travel speeds and the mobility resistance in the vicinity of the tracks, two targets were identified as tracked vehicles.
3. Because the speed of one of the targets was too high to be ground-based and too low to be a fixed-wing aircraft, the target was identified as a helicopter.

Although the sensor-derived information in this problem provided only direct evi-
ence of the existence of a target and an estimate of its radial velocity, the use of sup-
orting non-sensor-derived domain knowledge allowed the observed target behavior to
e interpreted with respect to sophisticated context- and class-specific target models.
remote sensing applications, domain context is generally established using a priori data-
ses. Although sensory fusion in biological systems is a natural metaphor for multi-
e-sensor fusion, biological and artificial fusion systems establish domain context
very different means. In general, human problem solvers establish problem con-
xt based predominately on sensory-derived information. For instance, in the road-
ossing example of Section 1.6, the visual sense provides the information required
identify both the presence and location of moving vehicles as well as establish
uch of the relevant domain context (height of the curb, width of the road, posi-
n of the crosswalk, day or night, weather conditions, road surface type).

In the case of remotely sensed data, the domain conditions are generally not
rectly provided by the sensor data.[8] Thus, much of the domain context must be es-
blished by relying on a priori knowledge bases. In tracking a road-following vehi-
, a human analyst would typically rely on a visual interpretation of a road map of
e region. In a similar fashion, a context-sensitive tracking algorithm could investi-
te the existence and character of a road network in the vicinity of a target based
the search of the available database. In an off-road vehicle tracking application,
local mobility characteristics would be evaluated by accessing the previously es-
lished terrain, vegetation, and soil-type databases in the vicinity of the current
get (based on the location provided by the sensing system) and the current
ather conditions.

As noted earlier, terrain/elevation data, cultural features, and other domain characterizations are often originally developed using sensors that collect relatively static long-term knowledge. Real-time data collection of short-term knowledge generally involves a different class of sensors.

To achieve maximal benefit from available information, information fragments must b synergistically combined. Although employing adequate information is vital to success ful decision making, adequate information alone does not guarantee that it is effec tively utilized. As illustrated in Part I, the higher order canonical form algorithm tend to support more synergistic fusion of information than the lower order canon cal forms. In the bridge-building example in Section 1.3, a rigid canonical form classification paradigm would not be capable of fully exploiting the diverse an highly complementary character of the sensor observables nor the knowledge of th local terrain and road network, the possible missions/functions of the variou pieces of equipment, and the organizational characteristics of an engineer batta ion. In other words, while adequate information might be *necessary* to solve a specifi problem, in general, it is *not sufficient.* In this particular example, a highly intuitiv higher order canonical form algorithm was outlined that produced a sophisticate context-sensitive interpretation of the available sensor-derived information.

A second example is offered to emphasize the importance of effective utiliz tion of all relevant information. Suppose that an *ELINT* collection system establish a line of bearing (LOB) to a particular radar system, a *communications intercept* pr vides the identification of a unit class that could possess such a radar, and an airborr synthetic aperture radar locates several targets along the identified LOBs. The fusion these three pieces of evidence is complicated by (1) the *disparity of the features,* (2) *reg tration errors* among the sensors, and (3) *resolution differences* of the systems.

Because it might not be possible to map these three classes of features into a ful registered, common feature set, typical transformation-based decision functions (e.g linear or nonlinear discriminant functions) tend to be suboptimal. Human analys however, successfully fuse such diverse information by employing sophisticated reaso ing "models" that are sensitive to supporting contextual knowledge.

Suppose, for instance, that based on terrain and foliage considerations, on specific segments of the LOB could support an ELINT detection (due to terrai masking and foliage-penetration considerations). By intersecting each of the targ uncertainty regions with the LOB fragments, many of the candidate targets suppli by the synthetic aperture radar would be eliminated. By consideration of the logis cal constraints (accessibility, slope, elevation) and doctrinal knowledge (cover a concealment requirements, radio communication requirements with higher co mand elements), a further refinement of the radar detections might be possib Thus, in spite of possible registration errors and resolution differences, the thr fundamentally different types of information can be effectively combined.

11.5.2 Problem-Solving Principles

Human analysts fuse disparate information using approaches that tend to be hig intuitive, robust, context-sensitive, temporally sensitive, and efficient. Algorithm velopers must seek to achieve these same characteristics.

Effective data fusion automation requires a holistic view of evidence combination. Since the analysis of individual sensor reports can lead to ambiguous, inconsistent, and highly local interpretations of observables, evidence combination can benefit from a global, multiple level-of-abstraction perspective. A famous poem by Godfrey Saxe [4] provides an excellent metaphor that emphasizes the need for a holistic approach to information fusion. The poem describes an attempt by six blind men to gain a first-hand understanding of an elephant. The first man happens to approach the elephant from the side and surmises that an elephant must be like a wall. The second man touches the tusk and imagines that an elephant must be like a spear. The third man touches the trunk and decides an elephant must be like a snake. The fourth man feels the knee and determines that an elephant must be like a tree. The fifth man chances to touch an ear and decides that an elephant is like a fan. The sixth man grabs the tail and decides that an elephant must be like a rope. While the conclusions drawn by each of the men are entirely reasonable given the highly local nature of their observations, a coherent picture of the elephant fails to emerge. A considerably more useful interpretation results if the local observations are combined for the purpose of generating a more global perspective.

The puzzle-solving metaphor presented in Chapter 1 illustrated that complex dependencies can exist among information fragments and the completed situation description. When viewed from a highly local perspective, it may not be possible to determine whether an individual puzzle piece is even relevant to a particular task (a given piece may actually belong to an entirely different puzzle). Even if the information *is known* to be relevant, a highly local interpretation may not resolve whether a blue piece represents sky, water, or some other feature. Only when viewed holistically can the information content of a single piece be assessed and its value properly judged.

Problem task requirements should drive problem-solving paradigm selection. Attempts to develop fusion algorithms have traditionally relied on lower order canonical form approaches. Because many real world fusion tasks demand both robust and context-sensitive performance, such algorithms tend to be inadequate. As we have seen, problem requirements must dictate the selection of a problem solving paradigm. The canonical problem-solving form taxonomy discussed in Chapter 2 provides a categorization of the spectrum of possible algorithmic approaches. Table 11.1 offers a simple evaluation criterion for matching problem requirements to canonical problem-solving forms.

Powerful fusion algorithms can often be constructed by mimicking the character of human problem-solving approaches. The development of artificially intelligent machines has been an intriguing possibility since the dawn of the computer age. Much of the work in AI has focused on the goal of achieving general machine intelligence. Expert systems, strict logic-based constructs, exhaustive search of state-space graphs, and artificial neural networks have been extensively studied and applied to a wide range of appli-

Table 11.1
Summary Characteristics of the 16 Canonical Problem-Solving Forms

	Algorithm Characteristics							
	I & II	*III & IV*	*V & VI*	*VII & VIII*	*IX & X*	*XI & XII*	*XIII & XIV*	*XV & XVI*
Robustness	No	No	No	No	Yes	Yes	Yes	Yes
Context sensitivity								
Declarative	Explicit	Explicit	Explicit	Explicit	Implicit	Implicit	Implicit	Implicit
Control	Explicit	Explicit	Implicit	Implicit	Explicit	Explicit	Implicit	Implicit
Control	Rigid/ local	Rigid/ hier- archical	Flexible/ local	Flexible/ hier- archical	Rigid/ local	Rigid/ hier- archical	Flexible/ local	Flexible/ hier- archical
Dynamic behavior	No	No	Limited	Limited	Limited	Limited	Yes	Yes

cations. Unfortunately, each problem-solving technique tends to mimic only a certain aspect of the human-reasoning process.

The development of reasoning paradigms that approach the capability of humans has been handicapped by a variety of technical problems, including hardware limitations (speed, memory size), the limited nature of our understanding of biological intelligence, a reliance on complex graph-based or nonintuitive machine-oriented data structures, and the lack of a robust and highly intuitive certainty calculus.

While most attempts to automate data fusion tasks have employed either highly combinatoric numeric-based algorithms or relatively simplistic symbolic-based algorithms, humans typically fuse information based on robust and context-sensitive reasoning models that tend to be both efficient and general enough to accommodate domain variants and extensions to novel situations. Highly robust, context-sensitive, and efficient algorithms can often be developed by simply mimicking the *character* of manual problem-solving approaches. Numerous examples were offered in Part I to illustrate the potential of such approaches; the next two chapters offer additional examples.

Situation development entails a compromise between the "cost" of information production and the reliability and quality of the resultant products. Whether the process is performed manually or is automated, both the nondeterministic nature of many real-world phenomenon and the largely expectation-based character of the analysis process prevents the elimination of all uncertainty in the situation description. Although the acquisition and analysis of additional sensor and non-sensor-derived information tends to increase the reliability, accuracy, and quality of the fusion product, the cost of such improvement can increase much faster than the value added by the analysis. As there is no formal method for characterizing the intrinsic *value* of specific information or the *information requirements* of a particular decision process, an ill-defined tradeoff exists between information analysis cost and fusion product quality and reliability.

11.5.3 Fusion Product Principles

This section summarizes a number of principles associated with fusion product production.

Both the input sensor data and the fusion system products tend to be context-sensitive. Biological systems develop situation awareness by analyzing sensory data and the evolving situation characterization within context. For instance, a child will almost instinctively run after a ball thrown to him as long as it remains on the playing field. However, if the ball rolls into a busy intersection, the impulse to retrieve the ball is tempered by an assessment of the potential danger of following the ball into the intersection. In both instances, the child wants the ball, but the two contexts elicit dramatically different behaviors. Both the analysis process and the resultant fusion product can be highly sensitive to domain context.

An adequate interpretation of sensor data can require deep knowledge of the domain. As discussed in Chapter 2, data fusion is often treated as a filtering or parameter-estimation problem. However, for many tasks, sensor data must be interpreted using extensive knowledge of the problem domain. As mentioned in the discussion of the puzzle-solving metaphor presented in Chapter 1, the identification and placement of an individual puzzle piece typically depends on information that is *not* explicitly contained within the piece. For example, the proper interpretation of a brown piece that depicts a tree bark texture requires the existence of a reasonably comprehensive model of a tree.

Situation awareness products are inherently multiple level-of-abstraction. In the tactical data fusion domain, the situation awareness product consists of the location and disposition of enemy forces on the battlefield as well as a characterization of current and potential enemy actions. Since a division consists of multiple battalions, a battalion consists of multiple brigades, and a brigade consists of multiple companies, a division-level force can be treated as a collection of multiple level-of-abstraction domain objects. In addition, since a division-level attack plan consists of multiple subplans that are executed by subordinate elements, battlefield events typically consist of coordinated multiple level-of-abstraction actions. Thus, both the entities and their observed behavior typically exhibit a multiple level-of-abstraction character.

11.6 SPECIFIC ALGORITHM PRINCIPLES

The fusion process model proposed in Chapter 2 led to the problem-solving class taxonomy summarized in Table 11.1. Fusion algorithms based on a given canonical form were shown to exhibit similar characteristics with respect to *robustness, context sensitivity,* and *efficiency* . For instance, since both Kalman filters and artificial neural systems represent canonical form I approaches, both exhibit similar properties. For complex applications, neither approach supports the development of robust, context-sensitive algorithms and neither possesses a sophisticated control element that supports computationally efficient global problem solving. Although the two algorithms arise from completely different technical disciplines, both share the fundamental characteristics of all canonical form I class algorithms.

The algorithm principles associated with (1) robustness, (2) context sensitivity, (3) efficiency, (4) dynamic problem solving, and (5) extensibility are now summarized.

11.6.1 Robustness

Because they use only *specific* forms of procedural knowledge, canonical forms I to VIII represent fundamentally nonrobust problem-solving forms. The remaining eight forms (canonical forms IX to XVI) use more *general* forms of procedural knowledge that supports the development of more robust algorithms. The first eight canonical forms are fundamentally non-model-based, while the last eight forms support model-based reasoning. For fusion tasks that do not exhibit significant dependencies among problem-solving attributes, non-model-based algorithms are sufficient. However, for problems that exhibit critical dependencies among either the sensor-derived or the non-sensor-derived problem attributes, more general reasoning knowledge is required to effectively capture those dependencies.

11.6.2 Context Sensitivity

Regardless of the particular technology employed, conventional data fusion algorithms are typically insensitive to the underlying context or, at best, sensitive to only a specific set of context states. The former approach can lead to underconstrained algorithms, while the latter can lead to the development of "nominal" solutions. Since the behavior of military organizations tends to be sensitive to the *current, relevant* context conditions (the current weather conditions, the specific limitations on line-of-sight communications based on the present locations of all cooperating units), nominal solutions tend to be inadequate. Because different contexts can evoke drastically different behaviors, algorithms that employ *general context-sensitive* knowledge can be required.

Since context sensitivity inherently requires conditional dependence between the algorithm and implicit domain knowledge, the eight lower order canonical forms, which do not model dependencies among attributes, support rather limited context sensitivity. The higher order forms, on the other hand, can support context-sensitive reasoning. In particular, canonical forms I to IV are fundamentally non-context-sensitive, forms V to VIII support context-sensitive control knowledge, forms IX to XII support context-sensitive declarative knowledge, while forms XIII to XVI support fully context-sensitive algorithms.

Although trained human analysts can recognize military organizations from aerial imagery, a neural net trained on a nominal set of unit patterns might fail to recognize a given unit when the individual entities that make up the unit are affected by real terrain, foliage, weather conditions, the presence of nearby waterways, and attrition. While a human analyst readily accommodates dependencies among both sensor-derived and non-sensor-derived features, the lower order canonical problem-solving approaches do not.

1.6.3 Efficiency

As we have seen, many data fusion tasks that humans perform relatively effortlessly have not been successfully automated. Even for those tasks that have yielded to automation, highly combinatoric algorithms are often required to achieve satisfactory results. Due to the time-sensitive nature of many data fusion products and the limited computational resources available in the field, algorithm efficiency is a critical aspect of application development.

For relatively complex problems, algorithms that rely on specific procedural knowledge tend to be less efficient than algorithms that employ general procedural knowledge. Algorithms that use rigid control tend to produce less efficient solutions than those that employ more flexible control. Finally, algorithms that employ multiple level-of-abstraction control tend to be more efficient than those that employ single level-of-abstraction control. As a result, lower order canonical form algorithms rely on relatively myopic, local control mechanisms, the highest-order canonical form problem-solving approaches can support true global problem solving.

1.6.4 Dynamic Problem Solving

Although the lowest-order canonical forms support implicit temporal reasoning (signal processing, range-gating, statistically based target tracking), the higher order forms support various degrees of explicit temporal reasoning. Canonical forms I to V are appropriate for implicit temporal reasoning processes, canonical forms V to XII support limited explicit temporal reasoning, while canonical forms XIII to XVI support fully dynamic reasoning.

11.6.5 Algorithm Extensibility

Since complex, real-world application domains often evolve over time (new tactics are developed, new sensors are added, stealth technologies are fielded), data fusion algorithms must be extensible to changes in the problem domain. Because model-based algorithms accommodate complex dependencies between problem-solving attributes, the higher order canonical form problem-solving approaches tend to be much more extensible than the lower order form approaches.

11.7 DATABASE PRINCIPLES

In this section, database principles are organized with respect to representation and reasoning.

11.7.1 Representation Principles

An integrated semantic and spatial object representation can support the development of both effective and efficient algorithms. The development of robust algorithms can require the use of sophisticated, context-sensitive reasoning models. Numerous examples have demonstrated that algorithm efficiency can be highly sensitive to data representation. As a result, problem-solving efficiency inherently involves both algorithm-related and database-related issues. Part I discussed the development of both sophisticated and efficient algorithms, while Part II outlined database organizational and representational issues associated with efficient data storage and retrieval.

A hybrid database kernel that combines an object-oriented representation of semantic objects with an object-oriented representation of two-dimensional space has been shown to provide efficient support to many classes of data fusion algorithms. The semantic object representation inherently supports hierarchical semantic reasoning, while the object representation of two-dimensional space inherently supports efficient top-down spatial reasoning. Because the spatial and semantic representations are fully integrated, the database permits efficient evaluation of queries involving mixed spatial and semantic attributes.

Effective database search requires support for efficient problem-dependent search dimensions. The assignment of a target detection to a particular track file, for example, could require consideration of a large number of candidate track files.[9] Rather than exhaustively testing each track file, database search dimensions

9. Candidate tracks are those tracks that are "close" to the new report based on metrics such as spatial/temporal proximity, speed, velocity, cross section, or target class.

should permit direct access to those tracks that are "close" to a new target detection. For targets that are "close" with respect to Euclidean distance, highly efficient data access is possible if the spatial representation preserves the inherent two-dimensional spatial organization of the data. Since single-dimensional vector-based representations do not preserve such two-dimensional spatial relationships, data that is "near" in two-dimensional space might not be "near" in the database regardless of the indexing scheme. A *true two-dimensional* spatial data representation, on the other hand, naturally preserves spatial relationships among two-dimensional spatial features.[10]

In general, the use of *natural data representations* support inherently powerful and efficient database search dimensions and reduce the requirement for "artificial indices." The relational model, for instance, provides a very natural representation for both representing and organizing table-based data. Likewise, a true two-dimensional spatial representation is a natural representation for two-dimensional points, lines, and regions. An object-oriented representation is a natural representation of semantic objects, and a true three-dimensional representation would be a natural representation of solid objects.

On the other hand, *B*-tree and vector-based representations would *not* be considered to be natural representations for two-dimensional spatially organized information. Although a *B*-tree offers a highly efficient search structure for one-dimensional data, the representation does not capture the inherent dependency between the *x* and *y* coordinates of two-dimensional spatial data. Likewise, vector representations of point, line, and region boundaries do not preserve true two-dimensional spatial search dimensions or the inherent spatial relationships within the data set.

Because data fusion inherently requires the association between dynamic (moving vehicles, time-varying events) and static domain elements (buildings, missile silos, geographic, and cultural features), the efficiency of the fusion process is enhanced if both the dynamic situation description and static long-term factual and reasoning knowledge are maintained in fully compatible forms.

A fully integrated semantic and spatial object database representation supports query optimization for mixed spatial and semantic constraints. Sophisticated, higher order canonical form algorithms can require database search and manipulation of complex, mixed spatial and nonspatial data sets. The canonical form XV path-planning example discussed in Chapter 3, for instance, relied on mixed spatial and semantic search. Spatial search was employed first to identify whether a river existed between the initial location and the goal state. Next, semantic search was employed to determine the locations of all bridges associated with this ground-maneuver barrier. With fully in-

10. As discussed in Chapter 8, a true two-dimensional spatial representation stores points as points, lines as lines, and regions as the boundary plus the enclosed area as well as inherently preserves all spatial relationships among the data sets.

tegrated spatial and nonspatial data representations, a query-evaluation plan can be constructed based on the search-space "size" of each constraint or condition. Thus, query optimization for mixed spatial and nonspatial queries involves the determination of the proper sequencing of database operations so as to minimize the overall query-evaluation "cost."

The preservation of natural storage dimensions supports the development of efficient search, overhead, and association operations. Since natural representations maintain both highly intuitive and efficient data-search dimensions that preserve inherent spatial relationships among the data, highly efficient search and association operations are possible. Since such representations do not reindexing following additions and deletions, overhead efficiency is enhanced.

Set operations, inclusion, and proximity relations benefit from the use of a hierarchical, areal-based, true two-dimensional spatial representation. Because set operations among two-dimensional regions involves areal-based relationships among the regions, the use of explicit areal-based spatial representations tend to enhance computational efficiency. In addition to being areal-based, the hybrid spatial representation is both hierarchical and true two-dimensional. Because of these properties, the database supports set-operation computation based on highly efficient *optimal clause-order constraint-based search.* Rather than computing intersections and unions using exhaustive Boolean AND and OR operations, respectively, optimal clause-order constraint-based search can be implemented using highly efficient multiple-resolution, true two-dimensional spatial search.

Separation of data sets with respect to the four classes of context sensitivity can enhance database maintainability and extensibility. As discussed in Chapter 2, knowledge may or may not be sensitive to domain context. Two different classes of context-sensitive knowledge were defined. *Context-specific* knowledge is knowledge that applies under specific conditions, while *context-general* knowledge is conditioned on or dependent on current context conditions. While the former knowledge class is often not readily extensible to new problem domains or to changes in an existing domain, *context-general knowledge* is naturally extensible. Just as *general declarative knowledge* supports the development of model-based algorithms that are sensitive to explicit, sensor-derived information, *context-general knowledge* supports the development of algorithms that are sensitive to implicit, non-sensor-derived information.

For knowledge that is insensitive to context (either *context-free* or *generic knowledge*), database maintenance operations tend to be limited. On the other hand, context-sensitive knowledge may require extensive database maintenance operations can be required. However, since *context-general knowledge* applies across a wide range of context conditions, such knowledge may require little or no modification as the context changes. Therefore, only *context-specific knowledge,* which is valid only under very restricted conditions, possesses extensive maintenance requirements.[11] Thus, the logical separation of *context-specific knowledge* from the three other classes of knowledge can potentially facilitate database maintenance operations.

11.7.2 Reasoning Principles

Storing data along natural storage dimensions supports the development of highly intuitive algorithms. A human's ability to perform complex reasoning often relies on powerful, yet highly intuitive, reasoning models based on natural representations of knowledge. Intuitive machine-reasoning paradigms can benefit from the use of natural data representations as well. As previously discussed, a hierarchical, true two-dimensional spatial representation supports the development of algorithms that mimic the efficient, top-down spatial reasoning style that is characteristic of human problem-solvers. Although the spatial data sets shown in Figures 6.8 and 6.9 are technically equivalent, the latter representation supports the development of much more intuitive, as well as more efficient, spatial-reasoning algorithms.

Context-sensitive reasoning can require maintenance and access to large, relatively static domain knowledge bases. Because context-sensitive algorithms rely on non-sensor-derived domain knowledge, such algorithms effectively operate in a higher dimensional information space than comparable noncontext-sensitive algorithms. As demonstrated by numerous examples presented in both Part I and Part II, context-sensitive algorithms can demand the maintenance and search of extensive spatially organized as well as nonspatially organized data sets. Because the required spatially organized data sets can be orders-of-magnitude larger than the nonspatially organized data, this book has focused on search and manipulation efficiency of spatially organized data.

Both algorithm and database inefficiencies can cause algorithm-performance bottlenecks. The Euclidean traveling salesman problem was used to illustrate the implications of large-scale combinatorics. Dependence on highly combinatoric problem-solving approaches generally leads to inefficient algorithms. However, even when computationally efficient algorithms are employed, inefficient algorithm performance can occur if the supporting database system is not well matched to the algorithm requirements. Thus, algorithm performance depends on both the inherent computational efficiency of the algorithm and efficiency of the required database support.

A conventional DBMS offers limited support to complex data fusion applications. Chapter 2 demonstrated that data fusion can be viewed as the composition among predomi-

---

11. Because it applies only under highly restricted conditions, context-specific knowledge can be likened to rigid templates.

nately database-resident knowledge. As such, data management issues represent a critical element in both manual and automated data fusion systems. Traditionally, data fusion analysts have maintained much of their medium-term and long-term databases in various analog formats (3-in × 5-in index cards, on wall-sized maps, or in manila file folders). As the information-handling requirements increased, database management systems replaced many manually maintained databases. Although the human analyst's *data management* role has been reduced by such a conversion, *data access* bottlenecks can persist at the man-machine interface. Commercial database management systems have attempted to reduce this bottleneck by simplifying and expediting human/computer interaction with specialized views and graphical user interfaces.

In spite of such enhancements, however, the volume of information to be analyzed is often so large that it can overwhelm virtually any purely manual analysis process. Thus, many data fusion tasks require at least limited automation. While conventional relational database management systems were seen to adequately support the lower order canonical form algorithms, they tend to be insufficient for the higher order canonical form approaches that can demand retrieval and manipulation of massive spatially organized domain knowledge bases.

In general, the more sophisticated the algorithm, the more database efficiency tends to affect the efficiency of the overall fusion process. Although not all fusion tasks are as time-critical as many tactical situation awareness tasks, database issues remain an important element of all applications that involve either large-scale or sophisticated spatial, temporal, and/or hierarchical reasoning.

High-accuracy spatial reasoning requires high-accuracy spatial representations. Nonregular spatial decompositions support efficient representation of high-accuracy spatial data. Vector-based spatial representations, for example, are a memory-efficient, high accuracy representation of real-valued (noninteger) points, lines, and piecewise-continuous region boundaries. Regular spatial decompositions, on the other hand, can have considerably larger memory requirements to achieve equivalent accuracy.

The recommended hybrid spatial representation provides an effective compromise between the advantages of regular spatial decompositions (it preserves spatial relationships and distance metrics and supports two-dimensional set operations) and the advantages of nonregular spatial decompositions (high-accuracy, memory efficiency). The low-resolution representation effectively serves both as a regular multiple-resolution spatial representation as well as an efficient hierarchical, areal-based spatial index to vector-represented high-resolution points, lines, and regions.

Natural data representations support algorithm extensibility. The ease of maintaining sophisticated data fusion algorithms can depend critically on the character and capability of the supporting database management system. Because natural data representations facilitate the logical separation of context-sensitive and noncontext-sensitive knowledge, as well as the maintenance of both intuitive and modular knowledge forms, natural data representations enhance database extensibility by ensuring that small changes in the problem space do not require extensive changes to the knowledge base.

11.8 SUMMARY

The first part of this chapter provided several alternative views of the functional fusion model originally presented in Chapter 1, offered insight into sensor data registration, discussed nontactical data fusion applications, and reviewed the limitations of conventional problem-solving paradigms. The latter sections catalogued many of the data fusion, algorithm, and database principles that were identified in Parts I and II.

In the next chapter, spatial, temporal, and hierarchical reasoning are discussed in some detail. The infrastructure required to support the development of more sophisticated fusion algorithms relies on the automation of these three reasoning classes. Chapter 13 presents detailed algorithms for several common data fusion tasks that are based on the principles reviewed in this chapter.

References

[1] Kramer et al., "Cellular Automata Program," Alliant Techsystems, Inc., Final Technical Report No. RL-TR-92-229, New York: Rome Laboratory, Air Force Systems Command, Griffiss AFB, September 1992.

[2] Barnden, J., and J. Pollack, "Introduction," *Advances in Connectionism and Neural Computation Theory*, Vol. 1, J. Barnden and J. Pollack (eds.), Norwood, NJ: Ablex, 1989.

[3] Geman, S., E. Bienenstock, and R. Doursat, "Neural Networks and the Bias/Variance Dilemma," *Neural Computation*, Vol. 4, 1992, pp. 1–58.

[4] Saxe, J.G., "The Blind Man and the Elephant," *The Poetical Works of John Godfrey Saxe*, Boston, MA: Houghton, Mifflin and Company, 1882.

CHAPTER 12

Hierarchical, Spatial, and Temporal Reasoning

As discussed throughout this book, data fusion applications rely heavily on spatial, temporal, and hierarchical reasoning. This chapter extends the previous treatment of these reasoning classes, as well as highlights the support provided by the spatial/semantic object database kernel outlined in Chapter 8.

12.1 HIERARCHICAL REASONING

In its most general sense, a *hierarchy* implies a ranking or ordering process. Both the natural world and human society exhibit a strong hierarchical character. Because of the evolutionary development of biological species, both plant and animal life inherently possess a taxonomical organization. Since individual organisms contain multiple organs, each organ consists of numerous cells, and each cell contains numerous functional elements, individual biological species possess a hierarchical internal organization as well. Large organizations of humans, whether clubs, associations, businesses, or governments, typically have a president, one or more vice presidents, multiple department heads, and so forth. Entities at each successively lower level of these organizations possess more focused responsibility and a correspondingly narrower span of control. The distribution systems of package delivery companies, airlines, and freight haulers, are typically hierarchically organized. Military units, tactics, and doctrine typically have a strong hierarchical organization. A high-level tactical behavior, such as "defend" or "attack," normally consists of multiple subordinate tactics.

Hierarchical reasoning is so natural, in fact, that it often underlies other forms of reasoning. Speech recognition, for instance, involves a hierarchy of analysis problems that spans the spectrum from acoustical, phonetic, phonemic, syllabic, lexical, syntactic, and semantic to pragmatic. The automated analysis of handwritten text is often approached using low-level pattern matching, character-level analysis (parsing for insertion, omission, replacement, and reversal errors), and higher level syntactic (grammar) and semantic analysis (meaning). A carotid pulse waveform consists of systolic and diastolic segments; each of these segments, in turn, can be decomposed into a hierarchy of subpatterns that can be used to identify classes electrocardiogram waveforms [1]. Picture processing is often treated in a hierarchical fashion, extending from segmentation to geometric models and finally to various relational structures [2].

Human perception involves a hierarchical reasoning process that relies on both bottom-up and top-down reasoning. For example, a forest is recognized as a composite object that consists of numerous trees; individual trees contain roots, a trunk, and numerous branches; and branches, in turn, contain numerous leaves. Thus, the discovery of a fragment of *tree bark* may provide partial evidence for the existence of a *forest.* Similarly, the observation of a forest establishes the expectation for lower level-of-abstraction entities, such as trees, as well as leaves and branches.

The resolution, accuracy, and perspective of sensor-derived information tends to vary from sensor to sensor. In a tactical application, for instance, high-resolution sensors might be capable of discriminating individual moving or stationary vehicles, thus providing intelligence with respect to a low-level force element (an individual tank), while other lower resolution sensors might provide evidence for the existence of only higher level entities (motorized infantry). Imagery and forward-observer reports are capable of identifying even higher level entities (division-level force).

Hierarchical data structures provide a natural means of maintaining hierarchically organized data. A *tree* representation is the simplest hierarchical data structure. In a tree, each entity has at most a single parent and an arbitrary number of offspring. Because of its well-defined structure, tree representations support efficient search and retrieval operations. As discussed in Chapter 8, a *binary tree* is a highly efficient data structure for one-dimensional data retrieval; the *regular region quadtree* is an efficient data structure for retrieval of areal-based two-dimensional data. In general, tree-based data structures are very natural data representations for information that can be recursively decomposed into mutually exclusive subsets. In the remainder of this book, all references to hierarchically organized information will assume tree-based representations.

Situation awareness development often demands both bottom-up and top-down reasoning. In terms of bottom-up reasoning, for example, the existence of a group of closely spaced tanks apparently operating "in concert" may indicate the existence of a tank company. Multiple, closely spaced tank companies that appear to be operating in a coordinated fashion with their associated support elements suggest the presence of a tank battalion.

Hierarchical representations can support both the conceptualization and the development of highly efficient multiple level-of-abstraction or multiple-resolution algorithms. For example, top-down search of the object representation of two-dimensional space supports efficient spatial search for specific classes of domain features. Many forms of diagnostic reasoning depend on an underlying tree-structured representation. Because object-oriented representations can effectively capture the hierarchical structure that underlies complex problems, such representations support:

1. Single level-of-abstraction reasoning;
2. Higher level-of-abstraction reasoning (*generalization*);
3. Lower level-of-abstraction reasoning (*specialization*).

Figure 11.3 illustrates the hierarchical character of the functional fusion model presented in Chapter 1. At the highest level of abstraction, the entire fusion process is represented by a single node. The four offspring nodes represent the four fusion levels. Each of these nodes, in turn, can be recursively decomposed into numerous functions and subfunctions. In addition to the "vertical" (within-level) hierarchical decomposition, since information is progressively refined and abstracted within the first three fusion process levels, the fusion model possesses a "horizontal" (sequential) multiple level-of-abstraction character as well.

Optimal solutions for finite-state problems are often developed by exhaustively generating all admissible solutions, followed by the selection of the "best" solution with respect to a selected evaluation metric. For instance, the canonical form I path-planning algorithm and the optimal solution to the Euclidean traveling salesman problem (TSP) both rely on exhaustive generation-and-test. Because the state-space size can be extremely large for such problems, such combinatorial optimization-based approaches may be impractical for large-scale problems.

To overcome the combinatories often associated with determining optimal solutions, approximate methods and heuristic evaluation criteria are often employed that seek more efficient algorithms that generate "near-optimal" solutions. The A* algorithm [2], for example, employs a heuristic evaluation function for controlling the combinatorics associated with large finite-state problems. Although the A* algorithm and its variants can reduce the computational complexity of many tasks, the control strategy employed by such algorithms is generally highly local. The implementation of true global control of the problem-solving process, on the other hand, can often permit the generation of efficient solutions to highly complex, real-world problems. Both the canonical form XV path-planning algorithm and the canonical form XVI doctrinal templating algorithm outlined in Chapter 3 employed efficient global-reasoning strategies. For many real-world problems that possess extremely large state spaces, the latter approach generally leads to much more practical algorithms than either an exhaustive generate-and-evaluate paradigm or approximation methods that rely on rather local evaluation strategies.

The Euclidean traveling salesman problem is a simple example of a large-state space problem for which there exists no efficient local strategy that guarantees a globally optimal solution. By relaxing the requirement for global optimality, numerous reasonably efficient suboptimal strategies have been developed. Consider a simple top-down problem-solving strategy for the TSP. Suppose the city location database exhibits the hierarchical clustering depicted in Figure 12.1.[1] (The separation between the clusters has been emphasized for illustrative purposes.) Because of the existence of "nested clusters," an efficient hierarchical path-planning algorithm can be developed that treats the overall problem as a set of relatively independent subproblems.

In general, the optimal clustering of a set of N objects into K-member groups based on spatial proximity can require as many as $(N!/(K!\ (N-K)!))$ locational comparisons. For example, for $N = 100$ and $K = 5$, as many as ten million possible tests $(100!/(5!95!))$ could be required. However, for data that is maintained in a true two-dimensional spatial representation, spatial clustering can be performed using efficient top-down, spatially anchored search.

Suppose the data set shown in Figure 12.1 is viewed at the coarse-resolution level shown in Figure 12.2. All grid cells that possess at least one data sample are

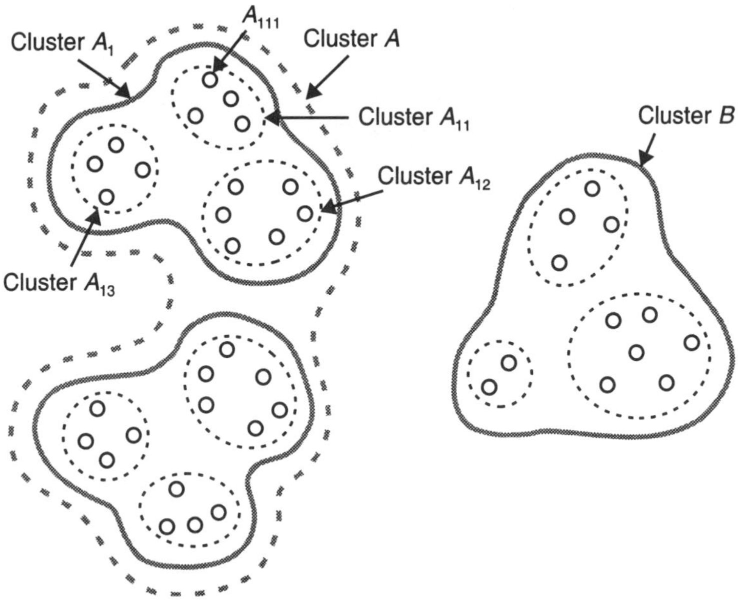

Figure 12.1 City location database for the hierarchical TSP algorithm. The small circles represent city locations.

1. A hierarchical city-location distribution is often a realistic assumption since a constellation of smaller cities around a larger city is a very common characteristic of sparsely populated regions.

then viewed at the next higher resolution. This recursive search and refinement process continues until the spatial resolution reaches twice the dimension of the largest expected cluster size (or some other heuristic stopping criterion). At this point, because all lowest level clusters will be within a single (highest resolution) quadtree node of each other, the lowest level clusters can be efficiently isolated using standard nearest-neighbor algorithms.

Next, the individual clusters can be aggregated by viewing the centroid of each cluster at progressively coarser resolution levels and reapplying the cluster-finding algorithm. Because it explores only those regions of space that possess mass, a hierarchical cluster-finding algorithm that employs multiple-resolution true two-dimensional search tends to be highly efficient. Individual cities are assigned to clusters based on local, multiple-resolution spatial search. Because the clustering algorithm effectively performs top-down problem decomposition, disjoint subregions will be identified as long as the individual clusters are reasonably distinct.

Next, conventional exhaustive path-length optimization is used to discover the minimal length path within each of the lowest order clusters. Since the number of

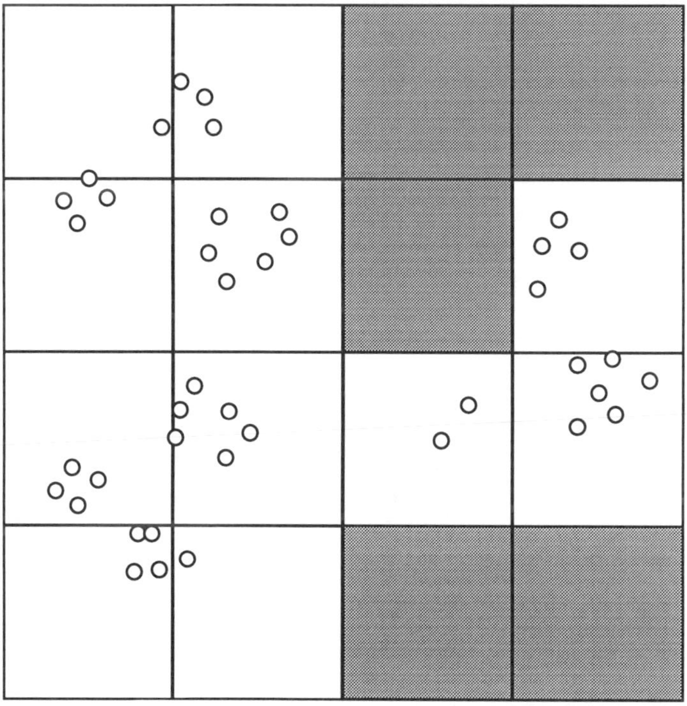

igure 12.2 Top-down, quadtree-based cluster-finding algorithm for the example depicted in Figure 12.1.

cities within a cluster is potentially relatively small, only low-order combinatorics are generated. Next, the optimal path between the centroids of the lowest order clusters are computed. This process is repeated recursively until the minimal-length path between the highest order clusters has been generated. The approximate minimum-distance path for the entire city data set is the "best" path between clusters combined with the "best" path within each cluster.

The efficiency of a hierarchical path development algorithm depends heavily on the character of the data set. If a city data set is uniformly distributed, for instance, the hierarchical clustering algorithm would generate only a single cluster and the algorithm just proposed would offer no computational advantage over the conventional exhaustive compute-and-test approach.[2] However, when an exploitable organization does exist, the hierarchical path development algorithm can offer substantial computational advantages. As mentioned previously, hierarchical reasoning is often an underlying element of other classes of reasoning. In this case, the path development algorithm represents a form of hierarchical spatial reasoning.

12.2 SPATIAL REASONING

Spatial reasoning plays a critical role in numerous data fusion applications. Although we have already encountered numerous examples of spatial reasoning, this section extends the earlier discussions and emphasizes the important role that the semantic/spatial database kernel can play in automating spatial reasoning. Seven distinct combinations of spatial features are illustrated in Table 12.1. In addition, the table provides an illustration of a set-operation, path-development, and metric-evaluation query for each spatial feature combination. For instance, a query such as

Find all *vehicles* at *road intersections*

is a "point-point" class query since both *vehicles* and *road intersections* would normally be represented as point features. Because "at" can be interpreted as *intersection* (or *disjunction*), the query involves a set operation between the vehicle and road intersection data set. Similarly, the query

Find the distance between *lake* 1 and *forest* 3

is a "region-region" class query since both lakes and forests are region features. "Distance between" is a metric-based relation.

As discussed in Chapter 8, the spatial/semantic object database was designed to support two-stage spatial reasoning. For tasks that do not demand high-resolution solutions, only the higher level layers of the pyramid representation or the quadtree-based representation of individual point, line, or region features might be

2. In fact, the algorithm would be less efficient due to the overhead required to identify that no hierarchical clusters existed.

Table 12.1
Examples of Natural Language Query Representing Set Operations, Path Development, and Metric Computation for Seven Combinations of Spatial Feature Classes

Class	*Category*	*Feature*	*Relation*	*Feature*
Point-point	Set	Vehicle	At	Road intersection
	Path	Vehicle	Path to	Tower
	Metric	Vehicle	Distance	City
Point-line	Set	Vehicle	On	Bridge
	Path	Vehicle	Path on	Road
	Metric	Convoy	Near	Railway station
Point-region	Set	Vehicle	In	Forest
	Path	Vehicle	Path to	St. Louis
	Metric	Tower	Near	Town
Line-line	Set	Convoy	On	Road
	Path	Convoy	Path along	Treeline
	Metric	Convoy	Near	Shoreline
Line-region	Set	Convoy	North of	Staging area
	Path	Convoy	Path to	Rail yard
	Metric	Convoy	Distance to	Town
Region-region	Set	Island	In	Lake
	Path	Lake	To	Air field
	Metric	Forest	Near	Town
Point-line-region	Set	Vehicle on road in town		
	Path	Convoy path to bridge in staging area		
	Metric	Vehicle distance from road in forest		

required. However, when refinement below the maximum-resolution quadtree representation is required, the high-resolution point, line, and region representation must be utilized. Thus, the first stage of spatial reasoning capitalizes on the inherent advantages of a regular, multiple resolution, true two-dimensional spatial representation. The second stage refines the product of the first stage using the memory-efficient, high-resolution spatial representation.

Six classes of spatial reasoning are discussed below.

12.2.1 Set Operations

Because set operations underlie many aspects of spatial reasoning, set-operation efficiency was a critical element of the database design discussed in Chapter 7. In general, sets can contain a countable or uncountable number of elements. Classical Boolean sets possess two states; data is either inside the set (the set membership is 1) or outside the set (the set membership is 0). Fuzzy sets [3], on the other hand, are multivalued sets where data possesses a degree of set membership in an interval (typically [0, 1] or [–1, 1]). To contrast them with "fuzzy" sets, Boolean sets are often referred to as "crisp" sets. Figure 12.3 demonstrates that multivalued membership functions are a generalization of Boolean set functions.

Since crisp-set operations among nonspatially organized information tend to be relatively straightforward and crisp-set operations among spatially organized information was treated extensively in Part II, the focus of this section is on fuzzy-set operations among spatial and nonspatial information. In general, Boolean sets can represent *absolute constraints* or *absolute modifiers,* such as *to the west of, within* 1 *km, nearest, before, more, to the left of, higher,* or *closer.*[3] The absolute constraint *within* 1 *km of* point (x, y), for instance, is defined by the set of all points inside a 1-km circle drawn about the point. All points are either inside or outside the set.

Fuzzy sets were introduced by Zadeh in 1965 as a means of formalizing vague and uncertain semantics. Fuzzy sets are appropriate for representing *relative constraints* or *relative modifiers,* such as *near, very, almost, small,* or *fast* [4]. Although absolute modifiers are typically insensitive to context, relative modifiers tend to be highly sensitive to context. For instance, fuzzy constraints can depend on the distance from a goal state, the current weather conditions, or the state of another process. *Near* in the context of a division-level force might be quite different than *near* in the context of a company-level force. Since they tend to be insensitive to context, absolute constraints can normally be represented by a single, crisp set. Relative modifiers, on the other hand, often require a family of fuzzy-set membership functions, where each membership function is conditioned on a specific context state.

Although a language that employed only absolute modifiers would be far simpler to emulate than a language that accommodates context-sensitive relative modifiers, the language would also be considerably less expressive.[4] Humans successfully communicate, in part, because of the inherent context sensitivity of relative modifiers in natural language. Even without appealing to the *uncertainty principle,* the

---

3. Adjectives and adverbs are modifiers of nouns, pronouns, verbs, and other modifiers in natural language.

4. This observation partially explains why conventional mathematical approaches provide inadequate support to highly context-sensitive applications.

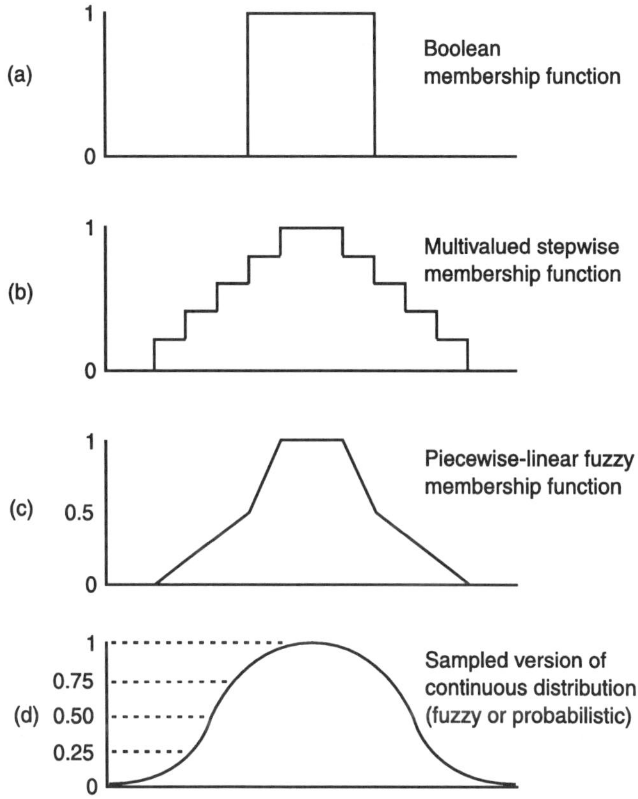

Figure 12.3 Progressive generalization of a Boolean membership function.

reader is likely to agree that as the *precision* of a word, an algorithm, or a concept is increased, its *generality* is decreased.[5]

In general, a modifier can be treated as a *generalized metric.* Although traditional database management systems support the evaluation of absolute metrics, such systems do not directly support the evaluation of relative metrics. Because emulation of human-level reasoning can require the use of both absolute and relative metrics, database management systems should ideally support both crisp and fuzzy queries. The spatial/semantic object database kernel supports the evaluation of both of these metric classes.

5. Although this famous principle [5] stated that the momentum and positions of electrons cannot be simultaneously measured with unlimited accuracy, the general notion appears to be so universal that it has been widely paraphrased (as has been done here).

Consider the simple Boolean query:

Find [n] [objects] (*set function*) [objects] (*set function*) [objects] 12.1

Two examples of this form are

Find [one] [truck] *on* [road A] *within* [spatial region B] 12.2
(intersect) (intersect)

and

Find [all] [(secondary roads) *or* (mobility_3 regions)] *east of* [Rockville] 12.3
(union) (intersect)

For both spatial and semantic data, Boolean query evaluation involves the composition of a crisp constraint set with the database.

In general, semantic data can be adequately represented along relatively independent dimensions, whereas spatial data inherently requires at least two dependent dimensions. Therefore, while the evaluation of fuzzy sets associated with semantic data involves set operations among linear, single-dimensional features, the evaluation of fuzzy sets associated with spatial data inherently involves *areal-based* set operations.

Consider a specific example of a fuzzy semantic query:

Find all emitters whose center frequency is *near* 1 GHz 12.4

As can be observed in Figure 12.4, an emitter with center frequency of 0.95 GHz is *near* 1 GHz to degree 0.5, whereas an emitter with a center frequency that is lower than 0.90 GHz is *not near* 1 GHz.

Next, consider the fuzzy spatial query:

Find all vehicles that are *near* region A 12.5

where region A is depicted in Figure 12.5(a) and *near* is the two-dimensional distance metric described by the fuzzy membership function shown in Figure 12.5(b).

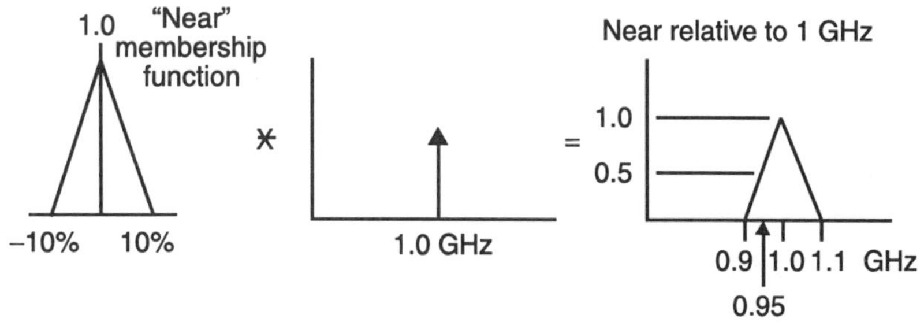

Figure 12.4 Fuzzy membership function "near" convolved with 1 GHz center frequency yields membership function for "near 1 GHz."

Consider an example of the evaluation of query (12.5). Clearly, any vehicle further than 4 km from region *A* is *not near* region *A*. If, as shown in Figure 12.6, a 4-km-resolution search window is constructed around a 4-km-resolution-level representation of region *A*, all known vehicles that are *near* region *A* will be located within this search window. In this particular case, the search space consists of four quadtree nodes that cover region *A* and 12 quadtree nodes that represent the 4-km region growth about these four nodes. Therefore, any vehicle associated with these 16 quadtree nodes is a candidate for being within 4 km of region A. For the data set depicted in Figure 12.6, three candidate vehicles are discovered.

In order to determine each vehicle's "degree of nearness," the actual distance between the vehicle and region *A* must be determined. Because of the multiple-resolution nature of the spatial representation, highly efficient distance computa-

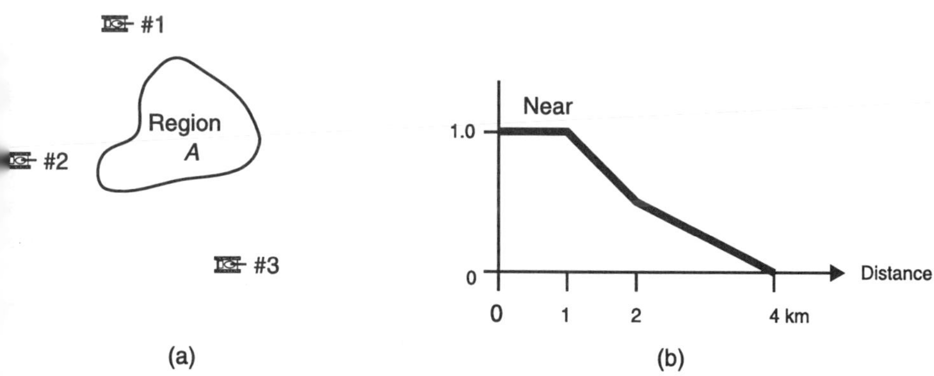

Figure 12.5 (a) Spatial windows about point *A* associated with the two break points in the fuzzy-set function in Figure 12.4. (b) Linear piecewise fuzzy-set membership function "near."

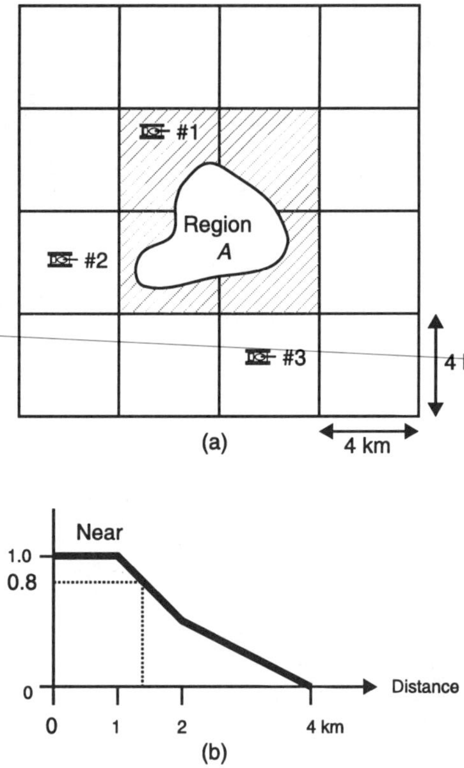

Figure 12.6 (a) Sixteen-node, coarse-resolution search space used to identify candidate vehicles that are "near" region *A*; (b) fuzzy-set membership function that defines the metric "near" for high-resolution evaluation of all candidate vehicles.

tion (as described in Section 8.3) is possible. When the accuracy requirement of a particular application exceeds the highest resolution quadtree node size, the high-resolution vector boundary representation of the region must be utilized. In this case, Vehicles 1 and 2 are found to be *near* region *A* with degree 1.0 and Vehicle 3 with degree 0.8. Although only a very simple top-down evaluation approach was presented, more sophisticated approaches can be employed.

Complex constraint sets can be constructed by concatenating sets of constraints. For instance, a "concealed road" database might be developed as the union of:

1. The intersection of the *road network* database with the *heavy foliage* data set;
2. The intersection of the *road network* with the complement of a *field-of-view* data set.

12.2.2 Spatially Anchored Search

As discussed in Part II, *spatially anchored search* represents true two-dimensional spatial search relative to (or anchored by) a point, line, or region object. Consider the query:

Find all lakes that are greater than 1 acre and less than 10 acres that are within 5 to 10 km of city *A* (*point anchor*), south of an interstate highway (*line anchor*), and within 1 km of forested areas (*region anchor*) 12.6

In a conventional database system, this query could require the generation of the exhaustive intersection among all the region-grown sets shown in Figure 12.7. Because the spatial/semantic object database supports true two-dimensional spatially anchored search, the query constraints can be treated as a spatial-search window through which the data is viewed. Because data outside the query window are *not* interrogated, spatially anchored search tends to be highly efficient.

12.2.3 Fuzzy Spatial Templating

Spatial templating involves the recognition of two-dimensional (or three-dimensional) patterns or organizations of objects. Typical data fusion applications of fuzzy spatial templating include the recognition and possible identification of:

1. Individual natural or man-made objects;
2. Patterns of terrain features;
3. Complex composite objects.

Fuzzy templates of manmade objects in aerial imagery, for instance, can employ a set of fuzzy variables that are associated with target size, shape, spacing, orientation, image contrast, and camera observation angle.

In addition to tactical data fusion applications, fuzzy spatial templating supports the analysis of a wide range of other problem classes. For example, consider the development of an automated plan recognition system for football based on the observed pattern of players on the field.[6] In perhaps the simplest approach to auto-

6. The game of football and tactical military operations are metaphorically very similar. In both domains, players possess historical knowledge of their opponent's strategies, strengths, and weaknesses. Numerous physical laws and rules of engagement constrain the confrontation. Deception is a critical element of strategy formation for both sides. However, much like the chess metaphor presented in Chapter 1, due to the relatively well-defined nature of competitive sporting events, plan recognition in football tends to be considerably more straightforward than it is in the military domain.

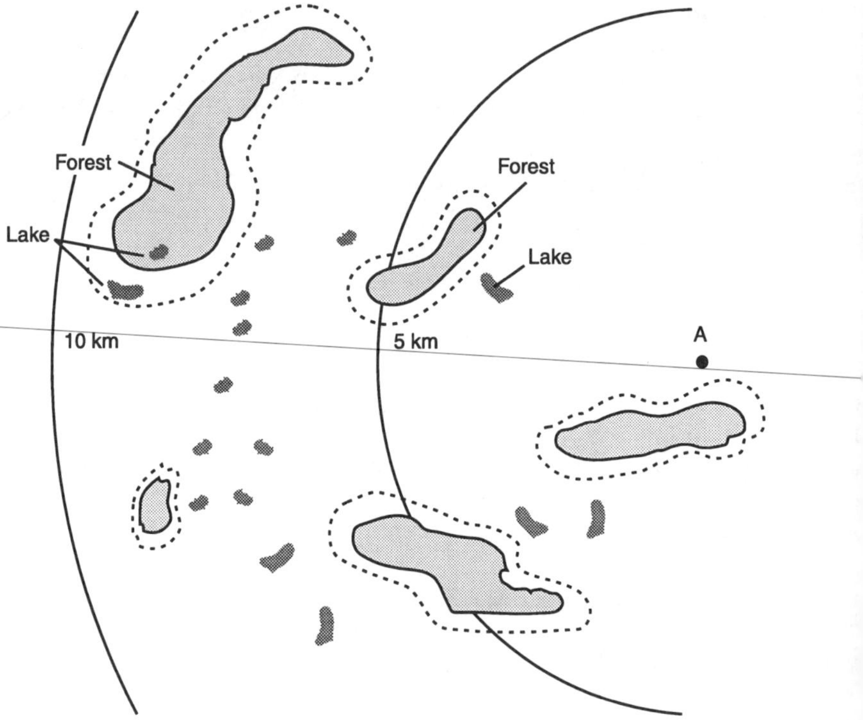

Figure 12.7 Database relevant to query (12.6).

mated plan recognition, a set of rigid templates could be compared to an exhaustive depiction of the possible player lineups associated with each expected behavior pattern. Such rigid templates effectively represent Boolean tests on the semantic features (distance to the goal line must be less than 40 yards) and spatial patterns (precise positions of the front linemen). The template that best matches both the semantic and spatial conditions (the template that exhibits the largest intersection with the observed player lineup) would be selected as the most likely play to be executed. Due to the large number of variations in patterns to be recognized, rigid templating can require a large number of templates.

Sports commentators and experienced fans can often anticipate team tactics, probable running and passing patterns, and field-goal attempts based on factors such as:

1. The relative spatial locations of key players at the time the ball is snapped;
2. The number of remaining downs;
3. The distance to the scrimmage line and goal lines;
4. Time left in the quarter;

5. Individual player strengths and weaknesses;
6. Historical team behavioral patterns.

A fuzzy model-based templating paradigm naturally emulates a human's reasoning process as well as eliminates the need to generate, maintain, and test a large set of rigid templates. Suppose, for example, that one of the key characteristics of a particular play is that the Tight-End is *normally to the right* and *slightly forward* of the quarterback.

To the right and *forward* are absolute (Boolean) modifiers. Given the location of either of the entities, the other elements can be readily located using spatially anchored search. *Forward* is a dynamically evaluated absolute modifier whose meaning is determined based on the current position of the quarterback and the goal direction. *Slightly* is a context-sensitive relative distance metric that can be represented by a conditional fuzzy membership function that depends on key semantic and/or spatial features (the name of the individual playing the Tight-End position, the relative positions of key players in the opposing team's lineup, and the remaining number of time-outs).

Thus, rather than templating based on the correlation between the database and a set of rigid templates, constraints can be imposed by "viewing" the data through the appropriate Boolean and fuzzy spatial and semantic windows. For example, a database search window can be constructed using the expected relative spatial relationships between the quarterback and the rest of the players. The greater the percentage of the data that "can be seen" through the window, the better the data "matches" the template conditions.

Since fuzzy templates can capture relative, context-sensitive relationships that are not easily emulated by more conventional approaches, a fuzzy templating paradigm tends to be much more flexible than an approach requiring that the correlation between the data and an exemplar exceed some numerical threshold. In most cases, if none of the templates were to pass this rigid test, the match criteria would typically be relaxed until a decision was secured. Reducing the decision-function threshold is perhaps the most common form of relaxation. However, because complex relationships may exist within the underlying phenomenon, simple relaxation of the decision threshold often proves unsatisfactory.

With a fuzzy model-based template, on the other hand, the observed player lineup is effectively compared to the flexible models of sets of candidate plays. Since individual problem attributes are modeled, relaxation of *individual* template constraints (as opposed to sensitivity adjustment of the overall template) is possible. The efficiency of the templating process can often be improved by the use of a hierarchical evaluation approach. For instance, if the strongest constraint for a given play depends on the relationship between the "Tight-End" and the "Running Back," the relative spatial relationships of these two positions would be evaluated first. Once the most significant constraint is satisfied, the next strongest constraint would be applied. The fuzzy hierarchical constraint set that best matches the data would be declared the winner. Because fuzzy templating mimics key aspects of human reasoning, the development of fuzzy models tends to be much more natural for humans than attempting to generate a "complete" set of rigid templates that account for all possible data set variations.

12.2.4 Path Planning

In this section, a hierarchical algorithm is outlined that generates admissible paths in a multiple-resolution grid-based domain. The multiple-resolution, pyramid-based path-development algorithm employs a heuristic "next-node selection criteria" that depends on a hierarchical set of mobility constraints. The algorithm first generates a relatively coarse resolution path that satisfies high-level domain constraints; this path is then recursively refined at higher resolution levels using more local spatial constraints and path-selection heuristics. Since subsequent stages of the algorithm simply refine the paths developed in the previous stage, only the four offspring nodes of each node in the path set from the last iteration must be considered. The recursion terminates when an admissible path is developed to the desired resolution. Path refinement below the highest resolution pyramid representation requires use of the hybrid representations of appropriate point, line, and region features.

Because pyramid nodes summarize the characteristics of relatively large regions of space, a given node can possess multiple mobility classes. For example, a particular node might contain roads, rivers, steep terrain, as well as gentle slopes. Thus, while a given node might contain an attribute that satisfies the path selection criteria (for example, the node contains at least one road), mobility through a cell is not assured. For example, if the cell contained a deep canyon that went from one side of the cell to the other, passage through the cell would be blocked.

When a potential path extension node contains a feature that could prevent mobility through the node (boulder field, canyon, river), a simple "look-ahead" algorithm can be used to determine whether passage through the cell is actually possible. Figure 12.8 demonstrates several possible ways that a blocked cell can be identified by investigating its next higher resolution level offspring nodes. In this figure, a node is marked with a large dot if it satisfies the path-extension constraint set and with an *X* if the cell contains any mobility barriers that violate the constraint set. If passage through the cell is discovered to be blocked, the candidate node would be dropped from consideration. This simple "look-ahead" capability effectively adds a depth-first search component so as to avoid the backtracking that would be required when a path would be discovered to be blocked during a subsequent stage of the algorithm.

To illustrate the characteristics of the proposed algorithm, consider the development of a path between points *A* and *B* in Figure 12.9 under the constraints that the values of Mobility Resistance (MR) must be between 0 and 3. Two simple heuristic next-cell selection criteria are used:

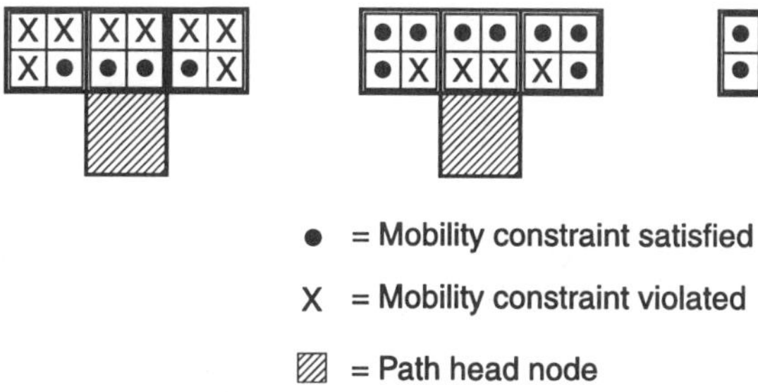

Figure 12.8 Examples of path extensions that are found to be blocked using a depth-first "look-ahead" algorithm. Each candidate extension node is evaluated at twice the resolution of the basic path resolution.

1. Favor minimum MR-valued cells;
2. If more than one neighbor of the current head node of the developing path satisfies all constraints and the first heuristic, select the node that is the minimum distance to the goal state.

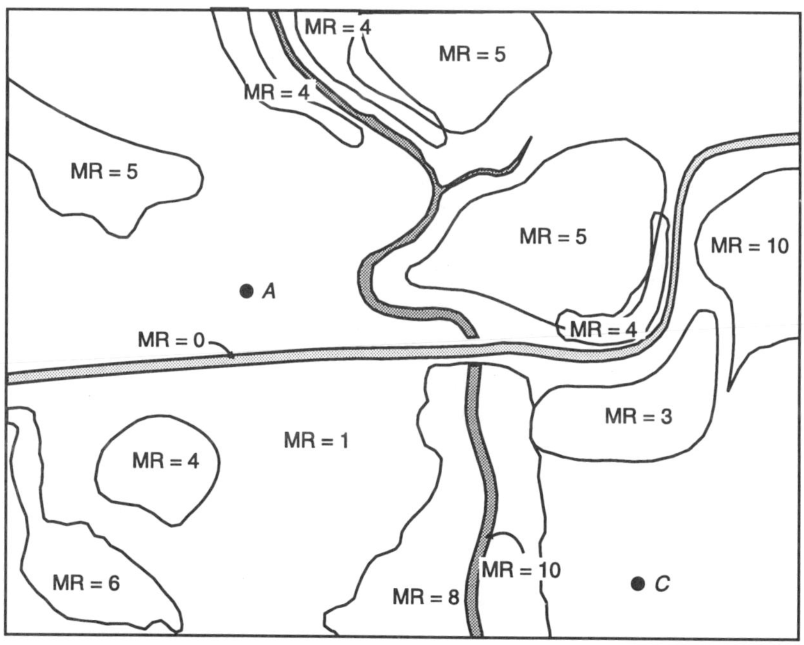

Figure 12.9 Mobility resistance map for hierarchical path-plan example.

The first stage of the algorithm produces the 1-km-resolution connected-cell path shown in Figure 12.10. This path is then recursively refined by applying the same heuristics at a 1/2-km resolution (Figure 12.11), followed by a 1/4-km resolution (Figure 12.12).

By varying the spatial constraint set, as well as the node-selection heuristics, the same algorithm can generate alternate paths between the two endpoints. Each path can be rank-ordered with respect to its *total length, minimum travel time,* or *maximum supported travel speed.* As mentioned in Chapter 8, the path length is approximately proportional to the weighted (weighted by their respective node sizes) sum of the nodes in the path.

A number of extensions and enhancements to the above algorithm are possible. For instance, by simply "closing the loop" between the path-finding and path evaluation processes, an ensemble of paths can be generated and ranked against a selected evaluation criteria. Backtracking can be minimized by using the look-ahead algorithm in conjunction with a list of previously considered path segments. An analysis of the spatial feature set contained within the bounding pyramid node of the start and goal locations may help choose the most appropriate first-pass spatial resolution size. Finally, an evaluation function can be developed to trap paths that show little promise of reaching the goal state.

Not only does a hierarchical path-planning algorithm support efficient automated route development, it naturally supports dynamic replanning as well. Sup-

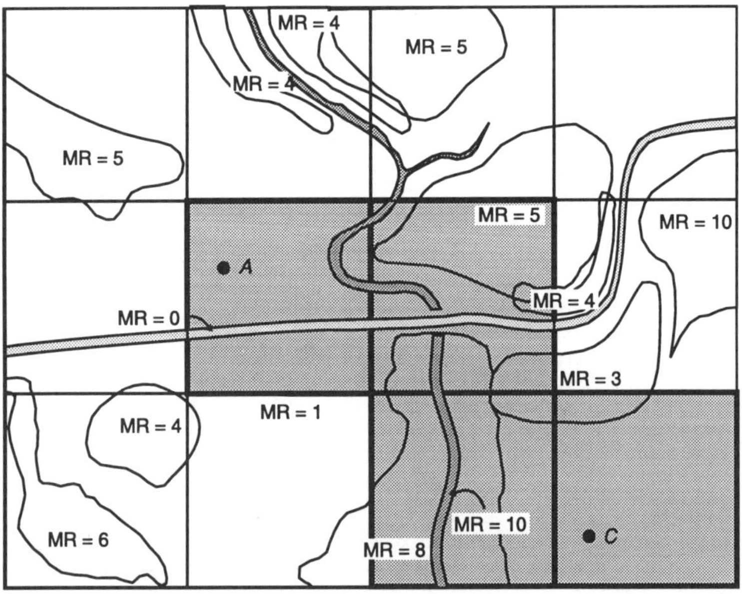

Figure 12.10 Coarse-resolution path from point *A* to point *C* for MR ≤ 3.

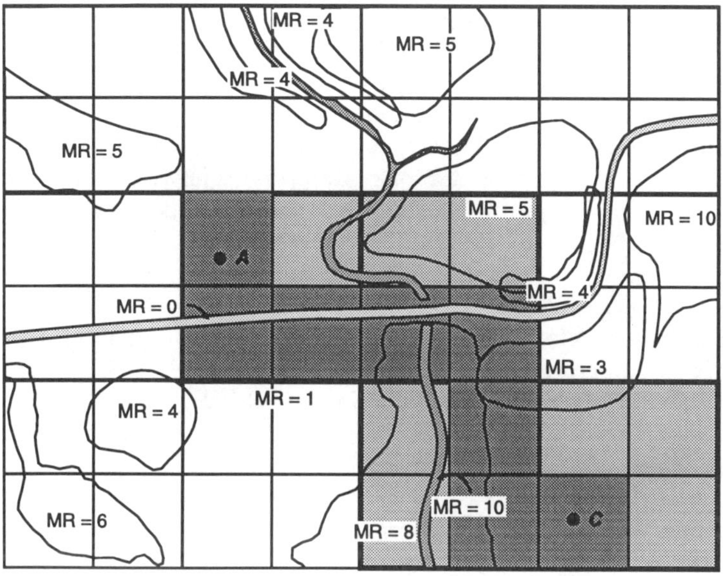

Figure 12.11 Refinement of the path shown in Figure 12.10.

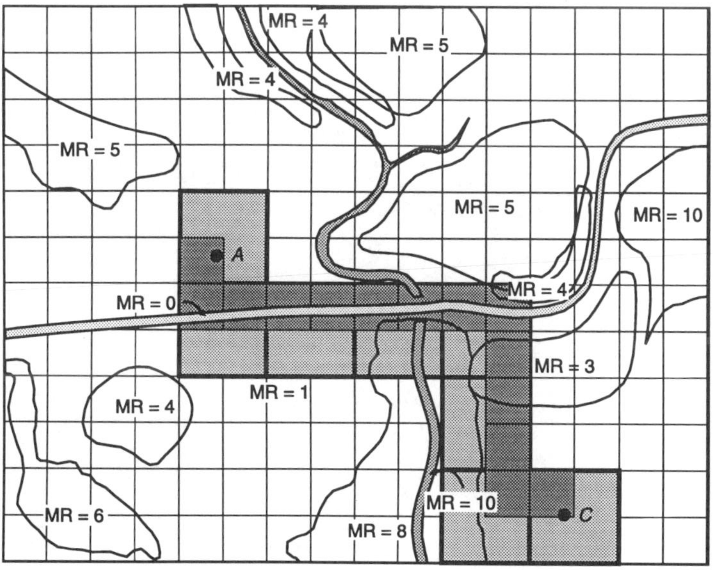

Figure 12.12 Refinement of the path shown in Figure 12.11.

pose that Figure 12.12 represents the pre-mission route plan for a semi-autonomous vehicle. When the vehicle reaches point *B*, rain begins to fall at a rate of 2 inches/hour. If the on-board control system determines that with the degraded weather conditions, travel through regions with MR = 3 is no longer feasible, a new plan must be generated from point *B* to the goal state *C* that excludes regions where MR = 3. Based on a new starting location and the same goal state, the recursive path-finding algorithm can efficiently generate the top-down route plan shown in Figures 12.13 through 12.15.

Since the proposed path-planning algorithm represents a canonical form VII problem-solving approach, no claims of global optimality can be made. However, the algorithm can develop admissible paths for rather complex spatial constraint sets and path-selection heuristics. A more robust, context-sensitive, and efficient version of the above algorithm is outlined in the next chapter.

12.2.5 Terrain Analysis

Terrain analysis involves the transformation of relatively context-insensitive geographical and cultural feature databases into context-sensitive products, such as vehicle mobility corridors, choke points, land use, optical line of sight (LOS), and fields of view (FOV).

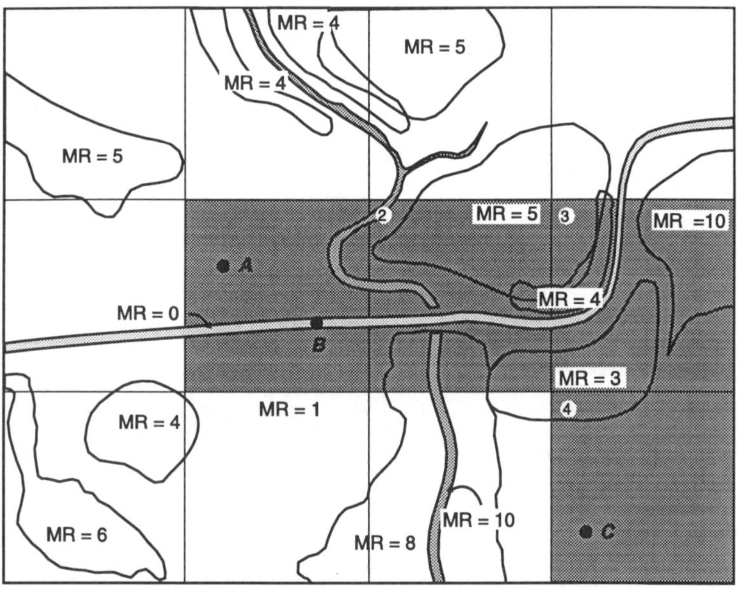

Figure 12.13 Coarse-resolution path from point *B* to point *C* for MR < 3.

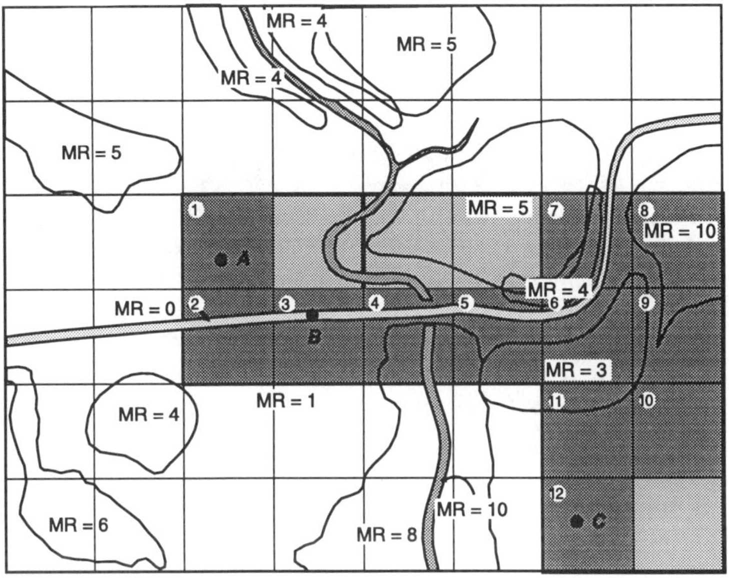

Figure 12.14 Refinement of path shown in Figure 12.13.

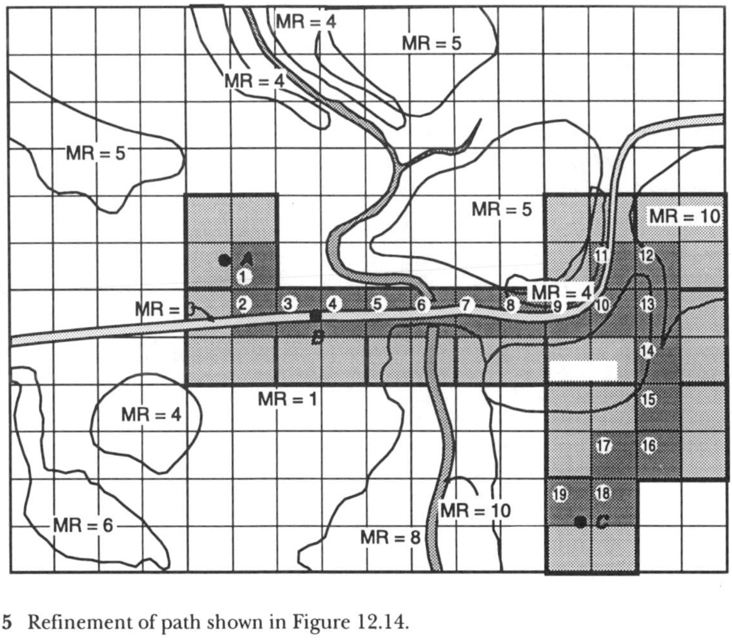

Figure 12.15 Refinement of path shown in Figure 12.14.

Mobility Maps

Ground-based mobility depends on numerous problem and domain characteristics, such as the vehicle class, terrain slope, vegetation, surface material, soil type, and weather. Mobility maps are typically generated as the result of intersections and unions of natural features (terrain, waterways, foliage, soil type) and cultural features (roads, telephone lines, air ports, cities), which effectively decompose the earth's surface into disjoint regions of similar mobility characteristics. Mobility maps coupled with other environmental information (recent rainfall amounts, temperature) can be used to evaluate the domain's impact on ground-based vehicle movement. A somewhat hilly region with little vegetation, for example, might offer little resistance to a wheeled vehicle in dry weather yet be impassable in wet weather. While a four-lane paved highway offers very low resistance to wheeled vehicles, it would represent a somewhat higher resistance for metal-tracked vehicles.

Mobility Corridors/Choke Points

Mobility corridors are connected, trafficable regions that are wide enough to permit the movement of a particular force structure. Mobility corridors can be generated as a special case of hierarchical route planning where the maximum required resolution is equal to the size of the minimum required corridor width.

Choke points are locations within mobility corridors that restrict the doctrinal flow of units. Such flow restriction can be caused by geographical features, such as mountains, ravines, boulder fields, forests, or bodies of water. The choke points within a mobility corridor are readily located using an adaptation of the rolling ball image-processing algorithm. When a round ball (a disk in two dimensions) is rolled along one edge of the mobility corridor, the center of the ball as well as the outside perimeter generate a locus of points that follows the original contour but are displaced by the ball radius and diameter, respectively. At all times, the disk is normal to the surface point of contact. Figure 12.16 depicts an example of a quadtree-based "rolling" disk.

By choosing the disk diameter that is equal to the minimum doctrinal width of a particular unit class, the doctrinal flow of the unit would be restricted at all locations where the disk extends beyond the mobility corridor. A quadtree-based rolling ball algorithm can be efficiently implemented using moving window, spatially anchored search. Depending on the inherent pyramid base resolution size and the disk diameter, significant hierarchical search-space pruning might be possible. In addition, only the boundary nodes of the disk that are near the edge of the corridor may actually require testing.

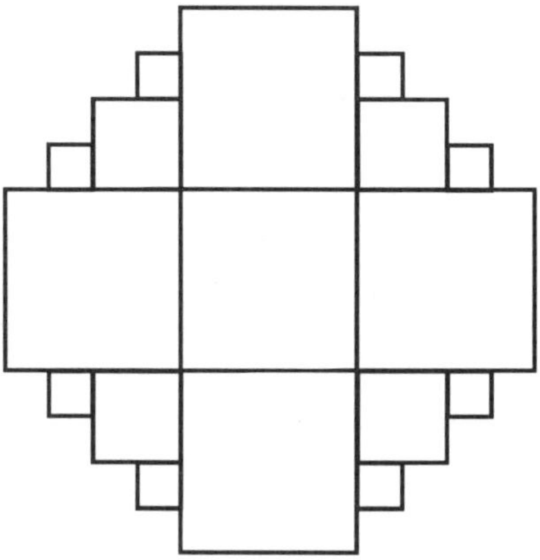

igure 12.16 Example of quadtree-based "rolling" disk.

bject Abstraction

patial objects that are not explicitly maintained in the database can often be de-ived from existing database objects. For example, the identification of regions that ould be considered "high meadow" can be developed as the intersection of (1) re-ions of a certain range of elevations, (2) regions with certain slopes, and (3) regions at are grasslike. In a similar fashion, "erosion-prone regions" can be determined by in-rsecting data for certain slope, soil type, and vegetation characteristics.

The development of open line features that are not explicitly represented in e database requires first locating an arbitrary point on the desired linear feature, llowed by feature extension through a series of nearest neighbor node moves. A vine feature, for example, can be generated by first searching for a global mini-um of the region of interest using top-down directed search. Once a node associ-ed with a ravine is located, the balance of the feature can be discovered by cursively searching for the minimum elevation neighbor nodes using simple gra-ent descent. A more robust gradient test can be developed using higher order dif-rences [6]. If a more global solution is required, the recursive path-finding proach discussed above can be employed.

A wide range of conventional image-processing and image-understanding chniques can be used to support database query and manipulation operations. Al-ough a detailed discussion of such techniques is beyond the scope of this book, a w basic operations relevant to database abstraction are discussed. *Spatial filtering* is e two-dimensional analog of a one-dimensional signal filter. Just as linear signal tering involves the convolution of a one-dimensional transfer function with a one-mensional data set, spatial filtering involves the convolution of a two dimensional transfer function (or spatial window) with a two-dimensional data set. *Edge detectio* and *edge enhancement* are generally performed using high-pass spatial filters; con versely, *image smoothing* for noise removal is generally performed using low-pass spa tial filters.

Erosion and *dilation* operations are used to smooth and restore images follow ing edge-detection and other image-processing operations. The rolling disk men tioned above is the traditional operator used in such operations. In an erosio process, as the disk is rolled along the inside of a segmented feature, all data that fal within the disk are removed; during a dilation operation, the "eroded" area is re stored by rolling the disk along the outside edge of the previously eroded region Since the erosion process effectively removes small-scale phenomenon, such a noise and fractal-like detail, the result of erosion, followed by dilation, is smoother, less detailed data set. This operation is often employed to eliminate bot noise and small holes in a large object. Since the new object has reduced comple ity, it therefore may have reduced storage requirements.

Fields of View (FOV)

Military applications of field-of-view maps include equipment siting, locating poss ble positions for forward observers, and identifying routes that provide cover an concealment. The most common method of generating these products, ray tracin using gridded elevation data sets, tends to be highly combinatoric. Since it is ofte impractical to pregenerate all FOV maps that might ultimately be required by given application, an efficient real-time FOV generation capability is highly desi able. This section outlines a simple, highly efficient multiple resolution FOV alg rithm that is well-supported by the semantic/spatial object database kernel.

Masked areas (or its complement, *unmasked areas*) are regions that are lower (o higher) than the optical LOS from a given vantage point. As illustrated in Figur 12.17, for a low-depression angle large areas can potentially be masked. If each pyr mid node maintains the minimum and maximum elevations associated with th particular region of space, a simple straight line variant of the recursive path pla ning algorithm can be used to test for masked cells.

A cell is considered *masked* if the maximum elevation of that cell could b blocked by the minimum elevations of any intervening cells to the vantage point V Since grid-based elevation data sets do not support true radial search, visibility tes ing requires the construction of the cell set that could potentially block LOS (i practice, various approximations and thinning algorithms are often employed to r duce the computational requirements). Figure 12.18 depicts the cell sets that ca potentially block the visibility from V_p to any part of the cell of interest.

The top-down FOV algorithm begins by identifying all *masked* cells at a rel tively coarse resolution level. The visibility to all *unmasked* cells is then re-evaluate at twice the resolution of the previous stage. This process is recursively repeated u til all masked regions (or their complement, unmasked regions) have been gene

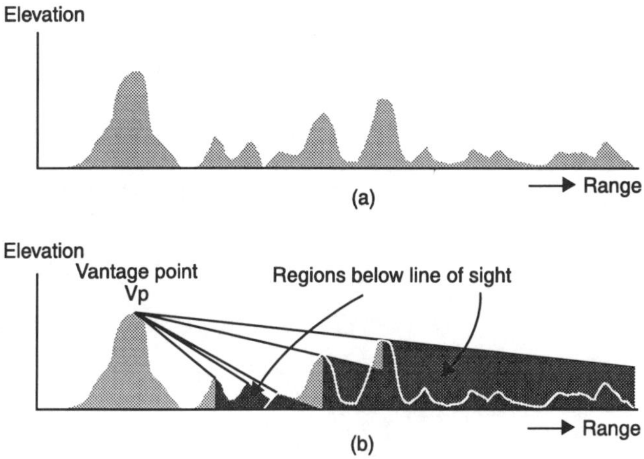

igure 12.17 (a) Example of a two-dimensional terrain-elevation profile, and (b) the optically masked areas from vantage point V_P (shown in dark shading).

ated to the appropriate resolution. Cells are marked as *masked* only if the *highest* lo-ation in that cell could possibly be blocked by the *lowest* elevation of any interven-ng cell. Thus, at the early stages of the algorithm, few pyramid nodes are likely to be dentified as "masked." However, as the resolution increases, progressively larger umbers of masked cells will be discovered.

Since the identification of masked nodes at the early stages of the algorithm liminates the subsequent testing of large numbers of base-level pyramid nodes, a ultiple-resolution ray-tracing algorithm can be much more efficient than the tradi-ional single-resolution approach. With the addition of spatial constraints, spatially indowed FOV maps can be generated. Earth curvature can be accounted for by ad-usting the LOS masking angle as a function of the distance from the vantage point.

2.3 TEMPORAL REASONING

ince many problem domains exhibit a dynamic, time-varying character, temporal easoning is an underlying requirement of many data fusion applications. In a tacti-al application, for example:

1. Targets might be at a particular location at a certain time or during a given time interval;
2. Bursts of RF energy occur as an illuminator sweeps past the observer;
3. Weapon system emissions related to "acquisition," "lock-on," and "end-game" each have a time-varying and/or time-sensitive aspect;
4. Both independent and coordinated activities begin, evolve, and terminate.

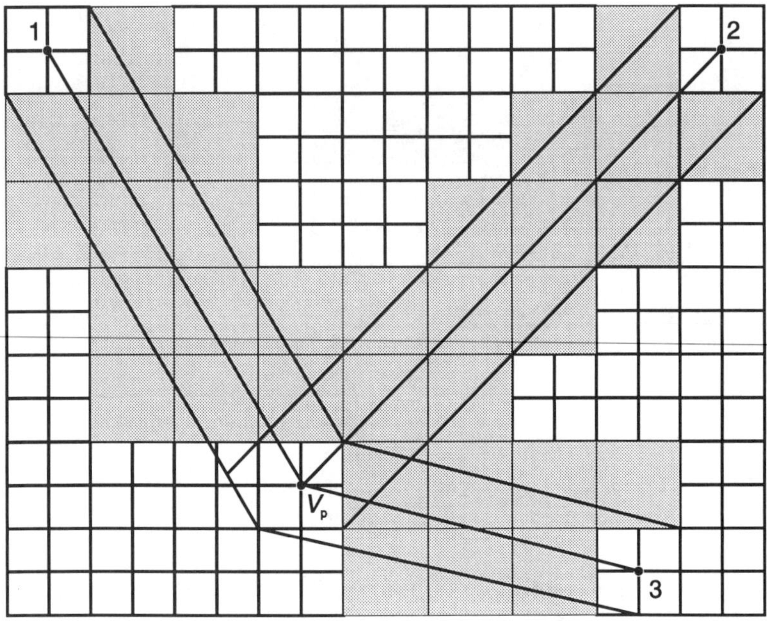

Figure 12.18 "Blocking cells" for line-of-sight algorithm to points 1, 2, and 3.

The identification of static and dynamic objects that are close both in spac *and* time is a key element of the data fusion process. Chapter 6 discussed the advan tages of a spatially indexed and temporally tagged database. Although collapsin three-dimensional information (two spatial dimensions and time) into two spatia dimensions might appear to violate the representation principle that data should b maintained along its most natural search dimensions, *time* can be viewed as an *attri ute* of data that *is* maintained along its natural spatial dimensions. By subordinatin *time* to another natural search dimension, the efficiency of a wide variety of data fu sion applications can be dramatically enhanced.

Because discrete and sampled events are of considerable interest in machin reasoning applications, time is routinely treated as a discrete variable. In reality however, time is both inherently *continuous* and *monotonically increasing*. The latte characteristic imposes an *order* on dynamic events. Signal filtering, target tracking and time multiplexing are simple examples of time-ordered reasoning. At a slightl higher level of abstraction, the sequence of actions involved in activities such a crossing a river or building a house imposes a particular order on the events. For ex ample, the foundation of a house must be built before the walls can be erected; an the walls must be erected before the roof can be constructed.

In addition to event *order*, temporal reasoning involves analysis that is sensitive o event *duration*. Activities, such as travel between points *A* and *B*, aircraft refueling, nd changing a flat tire are all accomplished over an interval of time. In some cases, he temporal duration of an activity or an event might be appropriately described by single number (2.6 seconds), a uniform interval ([2.0, 3.2]), a fuzzy membership unction, or a probabilistic distribution function.

Two classes of temporal reasoning will be considered: *implicit* and *explicit*. *Im-licit* temporal reasoning is associated with processes where time is an intrinsic ele-nent of both the data and the analysis process (filtering of time series data, range ating of radar return). Implicit temporal reasoning normally involves data-driven rocess flow and a simple control element. A tracking radar provides geopositional nd parametric measurements that are an implicit function of time. Conventional irget classification is developed by sequential processing of uniformly time-sam-led radar backscatter. *Explicit* temporal reasoning, on the other hand, typically equires consideration of *both* time order and time duration. The following discus-ons focus on explicit forms of temporal reasoning.

2.3.1 Plan Development/Plan Recognition

Iost planning tasks involve explicit temporal reasoning. In a tactical bridging op-ration, for example, a sequence of events necessarily precedes and follows the ac-ıal crossing. Units become associated, engage in a range of activities, cross the ver, and are subsequently realigned when the crossing operation is completed. ver time, sensor-derived evidence either confirms or refutes the individual stages f the operation. Since each of the individual activities in such a script requires a ertain amount of time to execute, both prediction/estimation and templating re-uire consideration of the *order* and *duration* of the individual observed or inferred tivities and events.

A simple path-development example is discussed next that illustrates both mporal order and temporal duration. Figure 12.19 depicts a set of script frag-ents (or *subscripts*) that can be used to *construct* a number of alternative travel plans r the morning commute to the office. While plan generation involves a concatena-on of a set of individual script fragments, the *observation* of a sequence of such ript fragments supports *plan recognition*.[7] Thus, planning is the constructive proc-s of building a composite plan from a set of possible subordinate plans, while plan cognition is the process of recognizing higher level plans based on the observa-on of their component elements. Thus, planning and plan recognition are duals.

Plan recognition, as discussed here, presupposes that the observed activities represent intentional or directed behavior. In fact, virtually all forms of expectation-based reasoning carry important, yet often unstated, assumptions.

The plan-recognition process can be considered a form of generalized templating. Both the plan-development and the plan-recognition process can involve complex multiple level-of-abstraction reasoning.

The likely *duration* of the various script fragments are shown in Figure 12.18 In some cases, a uniform distribution of time is judged to be most appropriate ("travel time to interstate highway"), while in other cases, experience or task analysis might suggest a fuzzy distribution ("travel to exit 19 conditioned on possible road construction delays"). Several of the script fragments are clearly optional elements of any plan ("stopping for coffee and doughnuts"), while others are not required for every instance of the plan ("stopping for gas" might be required only once per week). Although several alternative commute plans can be generated from these script fragments, only a single plan would actually be implemented. In general, plan recognition can require dealing with both the local as well as the global aspects of behavior.

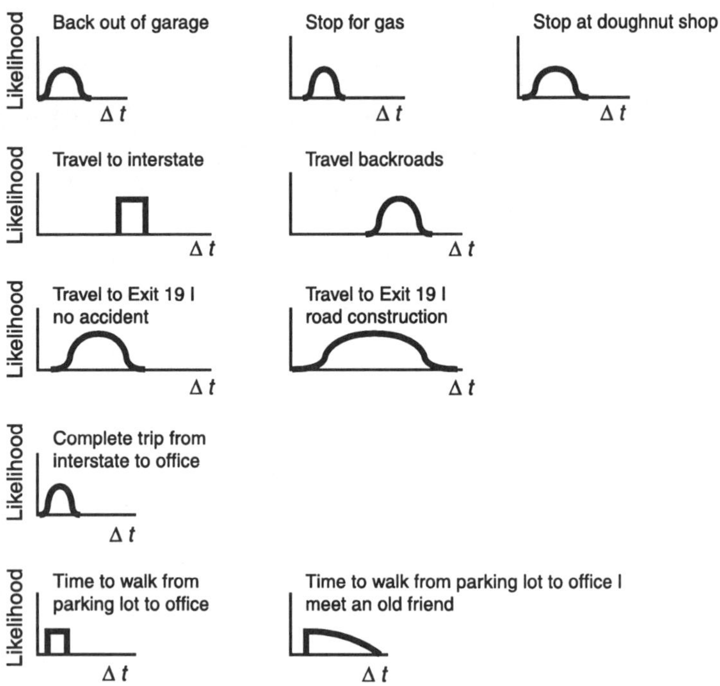

Figure 12.19 Fuzzy membership functions defining the expected times to complete various script fragments used to either plan or recognize a "driving to work" script.

12.3.2 Multiple-Hypothesis Reasoning

Many numeric-based problem-solving approaches (signal processing, statistical classifier, statistical-based tracker) process the available input data and produce a single decision. Gradient descent, minimum-distance, nearest-neighbor, and maximum likelihood-based approaches identify the "best" solution to a given problem. Similarly, although symbolic-based approaches may evaluate a large number of candidate solutions, typically only a single decision or a rank-ordered set of decisions is produced. Rigid templating paradigms, the Euclidean TSP problem, and conventional raster-based path-development algorithms all generate and then select from multiple hypotheses. However, since they do not maintain multiple competing hypotheses over time, traditional problem-solving approaches employ *single-hypothesis reasoning.*

Multiple-hypothesis reasoning methods, on the other hand, develop and maintain multiple competing hypotheses for the purpose of deferring final decision making until adequate information has been obtained to arrive at a single decision. When applied to dynamic problem solving, multiple-hypothesis reasoning represents an important form of temporal reasoning. A classical application of multiple-hypothesis reasoning is that of target tracking [7]. Due to the dynamic, time-varying character of both sensor data and the situation development process, a much wider range of data fusion tasks benefit from the maintenance of multiple competing hypotheses.

A significant problem with multiple-hypothesis reasoning is that the number of competing hypotheses tends to increase combinatorially. Thus, to be practical, multiple-hypothesis reasoning paradigms must:

1. Generate only highly plausible new hypotheses;
2. Provide for the efficient removal of all "invalidated" (incorrect) or "under-supported" (stale) hypotheses.

The use of robust, context-sensitive fusion algorithms tend to minimize the development of implausible, inappropriate, or unnecessary hypotheses. The maintenance of data and hypotheses along natural database search dimensions can facilitate highly efficient hypothesis removal.

Hypotheses are created or destroyed as the result of either the *acquisition of new evidence or* the *absence of supporting evidence.* Both *hypothesis creation* and *hypothesis destruction* are critical elements of multiple-hypothesis reasoning. Eight classes of evidence/hypothesis interaction are of interest. There are four ways *new evidence* can affect problem hypotheses.

1. If *new evidence supports an old hypothesis,* the status of the old (or existing) hypothesis is updated. Thus, class 1 supports *strong constructive* reasoning (track file update).

2. If *new evidence refutes an old hypothesis,* the old hypothesis must be re-evaluated in light of the evidence. Thus, class 2 supports *strong destructive* reasoning (the existence of a staging area in a certain region is refuted by the movement of all units out of that region).
3. If *new evidence supports a new hypothesis,* a new hypothesis is posted. Thus, class 3 supports *strong constructive* reasoning.
4. If *new evidence refutes a new hypothesis,* this knowledge must be preserved so that it can subsequently be utilized. Class 4 evidence evaluation represents a strong destructive process that argues against establishing a future hypothesis.

There are four ways that the *absence of new evidence* regarding an old hypothesis or an anticipated hypothesis can affect the hypothesis set. While the existence of new evidence supports relatively strong inferences, the absence of new evidence supports much weaker inferences.

5. *No new evidence supporting an old hypothesis* can *weakly suggest* that:
   a. An event continues although the sensors have failed to observe it (due to ECM, deficiencies of the available sensor suite, or current sensor tasking);
   b. The event has been, at least temporarily, discontinued (command post "off the air" may suggest a technical problem, "emissions control" has been ordered or the command post is being moved to a new location);
   c. The event has been completed (the river has been crossed).
6. *No new evidence refuting an old hypothesis weakly supports* the old event. Thus, classes 5 and 6 represent *weakly destructive* and *weakly constructive* processes, respectively.
7. *No new evidence supporting an anticipated event* supports *very weak destructive* inferences.
8. *No new evidence refuting an anticipated event* supports *very weak constructive* inferences. Lacking additional information, classes 7 and 8 tend to hold each other in check.

The existence of either new evidence that refutes an old hypothesis or the absence of anticipated evidence that supports an old hypothesis can lead to the removal or the downgrading of existing hypotheses. Since by refuting one hypothesis the support for existing hypotheses can change, or entirely new hypotheses can be created, such operations can generate multiple side-effects. For example, information that establishes the new location of an enemy unit on the friendly side of a river might provide both a force element location update and a refutation of the existing situation hypothesis that a "possible aggressive river-crossing operation is in progress." The collision between the eight evidence/hypothesis classes tends to drive a fusion system to the most plausible interpretation of the currently available information. Although some hypotheses will stabilize over time, others may continue to evolve due to the time-varying, nondeterministic nature of the problem domain.

Table 12.2 summarizes the eight general classes of evidence–hypothesis interactions. Combined with the data representations discussed in Part II, this table provides the foundation for a powerful truth-maintenance capability. In effect, data fusion system update and maintenance involves the interaction among eight functional agents: four constructive and four destructive. Four agents evaluate *new evidence* from either the sensors, the current situation description, or the long-term factual and reasoning knowledge, while four agents evaluate the impact of the *lack of new evidence.*

Table 12.2
The Eight Principal Blackboard Interaction Classes

Input		*BB Data*	*BB Effect*	*Inference*
New evidence	Supports	Old event	Update new event	s. constructive
New evidence	Refutes	Old event	Update new event	s. destructive
New evidence	Supports	New event	Create event	s. constructive
New evidence	Refutes	New event	Suppress event	s. destructive
No new evidence	Supports	Old event	Decay/special	w. destructive
No new evidence	Refutes	Old event	Maintain/special	w. destructive
No new evidence	Supports	Anticipated event	Special	vw. destructive
No new evidence	Refutes	Anticipated event	Special	vw. destructive

When a particular hypothesis is retracted, two classes of side-effects must be considered:

1. The evidence that previously appeared to support the hypothesis might require redistribution;
2. All hypotheses that relied on the retracted hypothesis might require re-evaluation.

Both such adjustments tend to lead to complex and highly combinatoric operations. However, if all relevant evidence associated with a hypothesis is explicitly preserved, hypothesis retraction can be both relatively simple and rather efficient.

A simple notation (H_i, E_j) is used here to capture the explicit relationship between a hypothesis H_i and its supporting evidence E_j. The *confidence* in a hypothesis $C(H_i, E_j)$ is the evidential support for (H_i, E_j). When the confidence of an hypothesis reaches a specified level, explicit reference to its supporting evidence is dropped. For example,

$$\text{if } C(H_1, E_1, E_2, E_3) = 99\%, \textit{ then } (H_1, E_1, E_2, E_3) \rightarrow (H_1) \text{ and } C(H_1) = 99\%$$

Once such a decision is made, retraction of that hypothesis and redistribution of evidence is no longer possible.

An example of a multiple-hypothesis target tracker is presented next to demonstrate important characteristics of such a paradigm. To simplify the example, an ad hoc evidence aggregation model is used. Assume that at time $t = t_a$, evidence A arrives representing a radar detection at (x_1, y_1, t_1).[8] With the semantic/spatial object database, all candidate tracks that are spatially near (x_1, y_1, t_1) can be efficiently located via true two-dimensional spatial search. The time tags associated with all targets that match the spatial query must then be examined to identify candidate tracks.

Suppose that two validated tracks, T_1 and T_2, are found to be both spatially and temporally *near* (x_1, y_1, t_1). Thus, the new detection might be associated with:

1. The existing Track T_1;
2. The existing Track T_2;
3. A new target, T_3.

Suppose that evidence A generates the following hypotheses:

$$C(T_1, A) = 40\% \quad C(T_2, A) = 40\% \quad C(T_3, A) = 20\%$$

Next, assume that at time $t = t_b$, evidence B (a radar detection at (x_2, y_2, t_2)) arrives that strongly supports hypothesis (T_2, B) and weakly supports the presence of a new target, T_4, such that

---

8. Although we assume t_3 t_2 t_1, there need not be any particular relationship between message arrival times (t_a, t_b, t_c) and actual target detection times (t_1, t_2, t_3).

$$C(T_2, B) = 90\% \quad C(T_4, B) = 10\%$$

as well as strongly refutes the existing hypothesis (T_2, A). On the strength of evidence B, evidence A support for T_2 must be redistributed between support for T_1 and T_3. Assuming the support is equally divided, the confidence associated with the hypotheses that exist at time $t = t_b$ are:

$$C(T_1, A) = 60\% \quad C(T_2, B) = 90\%$$

$$C(T_3, A) = 40\% \quad C(T_4, B) = 10\%$$

At time $t = t_c$, suppose evidence C (a radar detection at (x_3, y_3, t_3)) arrives that strongly supports the hypothesis (T_1, A, C), weakly refutes the hypothesis (T_3, A), and weakly supports hypothesis T_5. Thus, at time $t = t_c$,

$$C(T_1, A, C) = 95\% \quad C(T_2, B) = 90\%$$

$$C(T_3, A, C) = 20\% \quad C(T_4, B) = 10\% \quad C(T_5, C) = 20\%$$

Because track extensions associated with T_1 have reached the 95% confidence level, the hypothesis is accepted as valid and T_1 is extended to include points (x_1, y_1, t_1) and (x_3, y_3, t_3). Thus, the hypotheses that were being carried for T_1 are dropped (that is, (T_1, A, C) is set to (T_1) with $C(T_1) = 95\%$). The active hypotheses following the introduction of each piece of evidence are summarized in Figure 12.20.

12.4 SUMMARY

This chapter has demonstrated that hierarchical, spatial, and temporal reasoning are underlying elements of the data fusion process. In addition, the chapter has illustrated the practical advantages of the proposed semantic/spatial object database kernel in terms of its support to the development of highly intuitive as well as highly effective and efficient fusion algorithms. As we have seen, machine-reasoning paradigms can often exploit the hierarchical character of many real-world problems. Hierarchical reasoning facilitates the development of more sophisticated algorithms for at least three different reasons.

1. Multiple level-of-abstraction algorithms tend to naturally match the character of complex, real-world problems.
2. Top-down reasoning approaches tend to generate globally optimal solutions to real-world problems that exhibit complex or multimodal behavior.

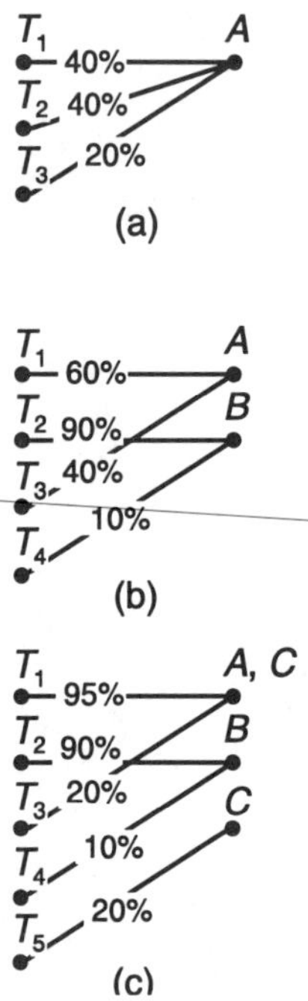

Figure 12.20 Active hypotheses following the introduction of evidence *A*, *B*, and *C*.

3. Top-down reasoning often leads to more efficient algorithms than single level-of-abstraction approaches.

Efficient search and manipulation of extensive spatially organized information is a critical element of many data fusion applications. Especially in the case of the higher order canonical form algorithms, the database requirements imposed by spatial reasoning tend to eclipse other application requirements. As demonstrated by numerous examples in this chapter, spatial reasoning efficiency can generally be enhanced by top-down, multiple-resolution, true two-dimensional spatial representations.

Due to the dynamic nature of the world, temporal reasoning underlies the entire situation development process. Two distinctly different classes of temporal reasoning were discussed. *Implicit* temporal reasoning includes those processes where time is an implied element of both the data and the reasoning process (the correlation of time series data). *Explicit* temporal reasoning, on the other hand, requires the analysis of time order and duration of processes and behaviors. The duality between plan development and plan recognition was demonstrated using an example of script-based reasoning. The maintenance of multiple lines of reasoning was shown to be an important aspect of dynamic problem solving.

References

[1] Fu, K.S., "Syntactic Pattern Recognition," *Applications of Pattern Recognition,* K.S. Fu (ed.), Boca Raton, FL: CRC Press, Inc., 1982.

[2] Chang, S.K., *Principles of Pictorial Information Systems Design,* Englewood Cliffs, NJ: Prentice-Hall, 1989.

[3] Zadeh, L.A., "Fuzzy Sets as a Basis for a Theory of Possibility," *Fuzzy Sets and Systems,* Vol. 1, 1978, pp. 3–28.

[4] Zadeh, L.A., "Common Sense Knowledge Representation Based on Fuzzy Logic," *Computer,* October 1983, pp. 61–65.

[5] Gamow, G., *Thirty Years that Shook Physics: The Story of Quantum Theory,* Garden City, NY: Doubleday, 1966.

[6] Skidmore, A.K., "A Comparison of Techniques for Calculating Gradient and Aspect from a Gridded Elevation Model," *Int. J. Geographical Information Systems,* Vol. 3, No. 4, 1989, pp. 323–334.

[7] Reid, D., "An Algorithm for Tracking Multiple Targets," *IEEE Trans. on Automatic Control,* Vol. AC-24, No. 6, December 1979, pp.843–854.

CHAPTER 13

Comprehensive Examples

Chapter 3 illustrated the spectrum of possible canonical form problem-solving approaches for a number of typical data fusion applications. As we observed in those examples, the highest order canonical form problem-solving approaches supported the development of sophisticated algorithms that could mimic the reasoning of human analysts. This chapter describes detailed canonical forms XV and XVI algorithms for several familiar fusion applications. The examples demonstrate that the application of the algorithm and database principles summarized in Chapter 11 permit the development of highly intuitive, as well as robust, context-sensitive, and efficient data fusion algorithms.

All the algorithms employ domain-sensitive reasoning models as well as a flexible, top-down, multiple level-of-abstraction problem-solving strategy. Because the algorithms emulate human problem-solving approaches, the algorithm-development process tends to be highly intuitive. Although database issues associated with both *algorithm development* and *algorithm performance* were largely ignored in Chapter 3, the critical relationships between algorithms and the supporting data management system are stressed in this chapter. The objective this chapter is to demonstrate that it is possible to develop algorithms that exhibit the following properties:

1. Highly intuitive formulation;
2. Robust performance;
3. Sensitivity to both static and dynamic domain context;
4. Computationally efficiency;
5. Efficient algorithm/database interaction.

Four specific data fusion applications are highlighted. The first and second sections present canonical form XV approaches to *target tracking* and *target classification*, respectively. The third section discusses *route planning* and introduces a path-planning algorithm that employs a true global control element. Although presented as a tactical route planner, the algorithm is equally applicable to path development for nonmilitary terrestrial, undersea, or extraterrestrial robotic vehicle applications. The fourth section provides a general discussion of the strengths and weaknesses of various approaches to *templating.* A robust and context-sensitive doctrinal templating algorithm is outlined in some detail that is applicable to a wide range of problems, including image understanding, fault identification, and land-use evaluation.

13.1 TARGET TRACKING

Because an active radar can support both real-time surveillance and targeting while being relatively insensitive to environmental conditions (time of day, weather conditions, battlefield obscurants), radar-derived information supports dynamic situation development and maintenance.[1] In this section, an algorithm is presented for tracking targets based on the return from a pulse-Doppler radar. A pulse-Doppler radar system offers three important capabilities. First, by employing moving target indication (MTI) processing, the relatively small cross-section return of moving targets can be separated from the backscatter due to relatively stationary objects, such as the ground, buildings, nonmoving vehicles, and foliage. Second, the return can be "range-gated" to estimate the target range. Third, by recovering coherent Doppler important target information, such as the target's radial velocity, direction, and modulation characteristics, can be obtained.

The requirement for context-sensitive, class-dependent target-tracking models was discussed several times in earlier chapters. We begin our present discussion with a few comments on a closely related problem: premature *track termination.* For a target that remains within the field-of-view of the radar, a target track will terminate either when the target stops moving (the target return can no longer be separated from the stationary target backscatter based on its Doppler shift) or when any of the following conditions occur.

1. The target moves tangential to the radar.[2]
2. The target becomes occluded by foliage or terrain.
3. The signal-to-noise or signal-to-clutter ratio drops below the detection threshold.

---

1. Certain microwave frequencies do exhibit significant attenuation due to aerosols and other atmospheric phenomenon.

2. Since the radar directly measures only relative motion toward or away from the radar, a target on tangential trajectory exhibits no velocity component toward the radar. In reality, a nonzero cutoff frequency is required to separate targets and clutter. Thus, very low speed targets and targets with only a small component of motion toward (or away) from the radar will be undetectable as well.

Track termination generates two primary alternative hypotheses.

1. The target stopped moving and therefore remains near its last detected position.
2. The target continued to move and is therefore no longer at the location of the last track update.

If the first hypothesis is true (the target actually stopped moving), the target database must be updated to reflect the presence of a stationary object at the last track update point. If multiple target tracks appear to terminate within a relatively small geographic region, a staging area or other concentration of equipment would be suspected. Similarly, the detection of a considerable amount of undirected movement within a relatively small area may also suggest a staging area. Future tracks would be expected to originate from the vicinity of such equipment concentrations.

If the second hypothesis is true (the target did *not* stop moving), the present location of the target is unknown. In this case, it would be desirable to extrapolate the track so that future detections of the target would be properly associated with the existing track. We will consider each of the above explanations for an anticipated but missing detection.

The first explanation for a "missing" detection is that the target began moving tangential to the radar. In this case, the track could be tentatively extrapolated in the most likely tangential direction. If a target is subsequently detected in the vicinity of the extrapolated track, the tangential track extension hypothesis would be supported. The inability of a Doppler radar to detect nonradial motion can be mitigated by netting two or more radar systems that possess significantly different vantage points. With more than two noncolinear radar systems, a given target trajectory will be nonradial with respect to at least one radar.[3]

Next, consider the case when a target under track moves into a region that is masked from the radar. Since a sensor's field of view can be computed as a function of the location of the radar, the location of the target, and the knowledge of the intervening terrain and vegetation, masked regions can be predicted using the appropriate electrical line-of-sight model. When a target moves into regions where detection is unlikely, the vehicle track would be extended using the most likely target behavioral model or models. For example, if the target had been following a road, the most obvious strategy is to extend the target track along that road. Once the target re-emerges from the masked region, the model that provided the best explanation for the new detections would be accepted and the hypotheses based on alternative behavioral models would be retracted. Just as in the case of nonradial motion, multiple netted radars would increase the likelihood that a given moving target would be visible by at least one sensor.[4]

In the case of only two radars, targets moving tangential to the line connecting the radars will not be detected.

Although only radar sensors are considered to simplify the problem, this example illustrates one of

Finally, suppose the track terminates in a region that is determined to be visible by the radar but that the target continues to move in a nontangential trajectory. In this case, the target might not be detectable due to signal interference (electronic attack), low signal-to-clutter ratio, or some atmospheric phenomenon (chaff). Once again, the target could be extrapolated using its most recently validated context-sensitive motion model. If the target had been following a minimum terrain gradient path, for example, the most reasonable assumption is that this behavior would continue. On the other hand, if a large swamp appeared to block continued "minimum terrain gradient" motion at the point where the track terminated, the support to the hypothesis that the target actually stopped is increased. Alternatively, a new motion model could be hypothesized based on the extent of the swamp and other characteristics of the region. Because considerable uncertainty is associated with any form of prediction, track extension models are most appropriate for "gluing together" relatively closely spaced track fragments.

As demonstrated in the last chapter, the maintenance of *multiple lines of reasoning* can help ensure that information is effectively utilized as it becomes available. Decisions that might be totally unresolvable at a particular point in time often clarify as the situation evolves. Thus, multiple-hypothesis reasoning adds a *temporal dimension* to uncertainty management. On the other hand, multiple-level-of-abstraction problem-solving approaches can add a *hierarchical dimension* to decision making. For example, a target's local behavior can be expected to be influenced most strongly by relatively local domain characteristics and constraints, while the target's higher level-of-abstraction behavior can be expected to be influenced by both local as well as a more global context.

Figure 13.1 depicts a target trajectory that has been overlaid on the geographical and cultural feature database. When viewed at a relatively low level-of-abstraction, trajectory fragment 1 appears consistent with a "minimum terrain gradient-following" behavior, fragments 2 and 3 appear consistent with a "road-seeking" behavior, and fragments 3 to 6 are each consistent with a "road-following" behavior. Viewed from a slightly higher level perspective, trajectory fragments 1 to 4 are consistent with a "river-crossing objective" hypothesis, while fragments 5 and 6 support a "road-following" hypothesis. These medium level-of-abstraction interpretations can themselves be viewed with respect to more global, goal-oriented objectives, such as to "reinforce a unit" or "reach a particular staging area."

Based on the above discussion, three important observations are offered.

13.1.1 Observation 1

Although traditional level-1 fusion algorithms have relied almost exclusively on lower order canonical form paradigms, there exists no theoretical justification (al-

---

the many advantages of fusing the output from multiple distributed sensors.

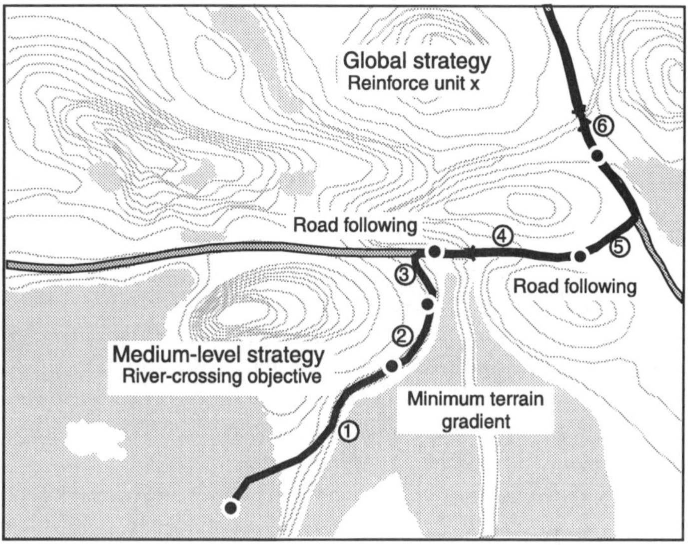

Figure 13.1 Interpretations of target detections within a hierarchy of contexts.

though there might be valid practical reasons) for reserving higher order canonical form problem-solving forms for the higher level fusion processes (levels 2 to 4).[5] Human analysts performing level-1 tasks certainly do not employ such a limitation.

13.1.2 Observation 2

Although target tracking is often treated as a signal-processing problem, viewed abstractly, tracking represents a special case of *plan recognition.* While the lowest level interpretations of the track in Figure 13.1 are associated with level-1 analysis, the higher level interpretations are associated with level 2 and level 3.

In general, plan recognition seeks to explain observed behavior in terms of objectives or intent. Whether performed by man or by machine, plan recognition generally carries an implicit assumption that the observations are the result of intentional or purposeful actions. Behavior that exhibits a "cause-and-effect" relationship between actions and observations is referred to as *causal.* Since statistical-based tracking models depend only on past behavior and limited target knowledge (maximum velocity, maximum acceleration), statistical trackers rely on weak causality assumptions. When strong causality assumptions are appropriate, more powerful expectation-based target-tracking models can be utilized. Target tracking based on

5. The complexity of the operations required by some of the algorithms presented here are currently impractical for highly time-critical decision making (for example, for decision processes that operate in the millisecond time frame.)

an evaluation of potential tactics and doctrine in full domain context can represent sophisticated plan-recognition algorithms. In general, local interpretations of target behavior support the recognition of relatively low level-of-abstraction plans, while more global interpretations of behavior support the recognition of higher level plans.

Although constraints afforded by causality assumptions can enhance both manual and automated reasoning, a cautionary note is in order. Expectation-based reasoning, especially in the case of adversarial problem domains, is inherently problematic. The famous story of Hannibal crossing the Alps using elephants, the failure of the Maginot Line to protect France during World War II, and the successful large-scale deception operations by the Allies prior to and during the Normandy invasion demonstrate the limitations of causal-based reasoning. Thus, while expectation-based reasoning is a critical underpinning of situation development (whether the analysis is performed by man or machine), intentionally deceptive and even apparently irrational behavior may be observed.

When novel or unexpected situations arise, conventional expectation-based reasoning tends to be inadequate. *Nonmonotonic reasoning* [1] attempts to deal with situations where causality assumptions are violated. Because nonmonotonic reasoning remains a rather open-ended and somewhat controversial subject, our focus remains on data fusion problems where causality assumptions are appropriate.

13.1.3 Observation 3

Holistic approaches to data fusion seek a synergistic combination of all available evidence. Traditional data fusion approaches, on the other hand, treat the fusion subtasks as relatively independent functional operations. Target identification, for instance, is often treated as a series of independent sequential processes (detect, centroid, track assignment, target classification, target identification). Although problem decomposition is a valuable problem-solving strategy, significant degrees of freedom can be lost if target centroiding is assumed to be independent of the target-detection process, for instance. Similarly, treating target tracking and target classification as entirely independent functions can potentially handicap the overall analysis process. A uniform approach to the combination of both sensor-derived and non-sensor-derived information has been recommended with the goal of generating information products whose overall value is greater than the sum of the individual parts.

The synergistic combination of both sensor-derived and non-sensor-derived information can be illustrated by a simple example. To avoid being located and subsequently targeted, tactical ballistic missiles (TBM) and their associated launchers are often moved from storage locations to a launch position and then returned to storage or moved to a new location following missile firings. Launch vehicles can be hidden from direct observation in buildings, under tree canopies, or beneath bridges. Locating TBMs is an inherently difficult task because of both their small size and mobile character. However, important domain constraints do exist that can reduce the size of the search space and increase the likelihood of identifying such systems prior to launch. For example, if it is known that missile launchers require relatively open, flat launch sites, wheeled vehicle access, favorable weather conditions, and specific communications requirements, these constraints can be used to focus the reasoning process.

When coupled with the natural and cultural feature database, such doctrinal knowledge supports the development of overlays that identify potential TBM storage and launch locations. Sensor-derived information can then be evaluated with respect to these constraints. Level-4 fusion (whether manual or automated) could direct the level-1 through level-3 processes to attempt to validate this potentially high-threat target as well as generate collection requests for supporting information. For instance, if a radar-derived target track begins from within a region deemed suitable for TBM storage and terminates within a region suited to missile launch, a potential mobile missile launcher would be suspected. If SIGINT identifies characteristic prelaunch transmissions from the vicinity of the last detected track position, the TBM hypothesis would be even more likely.

In the next section, the practical advantages of more sophisticated approaches to target tracking are explored by investigating relationships between target tracking and target classification.

13.2 TARGET CLASSIFICATION

In the mid-1970s, the author was involved in the development and evaluation of a prototype ground-based site-security system [2,3]. The system consisted of a UHF pulse-Doppler radar that provided 360° day–night, all-weather coverage, and a dual magnification passive optics system to be used to validate radar-generated tracks.[6] Because of the high value of the assets to be protected, the system specifications required a high probability of detection of humans at a range of 10 km. Since the guard force would be required to respond to all unidentified alarms, only a single system-level false alarm per day was considered acceptable.

Although the system specifications were driven by justifiable operational requirements, achieving such specifications represented a significant technical challenge. To achieve a high probability of detection, the radar system was required to process thousands of range-azimuth cells using matched filter signal processing to maximize the target signal-to-clutter ratio (S/C) and signal-to-noise ratio (S/N). To control false alarms due to wind-blown foliage and wind-blown manmade objects, such as power lines and chain-link fences, sophisticated constant false alarm rate (CFAR) processing was required to dynamically adjust the detection threshold

6. The use of both in-phase and quadrature return permitted true directional Doppler processing.

based on spatial, spectral, and temporal clutter estimation techniques.[7] In addition to wind-generated false alarms, the backscatter from large mammals and high-flying birds can generate false target detections.[8]

Although the track-association algorithms that were employed significantly reduced the detection-level false alarm rate, the objective system-level false alarm rate was not achieved.[9] As suggested by the operational requirements, reducing the probability of false alarm by lowering the probability of detection was an unacceptable tradeoff. Thus, to meet both the detection and false alarm requirements, post-detection processing was required to attempt to eliminate both nontarget false alarms and alarms due to nonthreat targets.

To this end, a set of target classification algorithms were developed to attempt to discriminate between true threat target detections, nonthreat target detections, and nontarget false alarms. Ground-based vehicles, low-flying helicopters, and humans represented the target classes of interest. Large quadrupeds and birds represented nonthreat targets, and swinging power lines, fences, blowing foliage, and random noise represented false alarms. As mentioned in Chapter 3, model-based target classifiers were developed for each of these targets.[10] Summary-level results of the single-cell classification study were discussed in Chapter 3 and are replicated in Table 13.1. For example, the classifier achieved a 92% probability of correct classification against more than a hundred samples of a single man walking at various step sizes, step rates, and trajectories relative to the radar. The classifier correctly identified 78% of the samples of multiple humans walking both in-step and out-of-step with each other.

Because of the relatively good performance of the single-cell classifiers for both threat and nonthreat targets, a simple sequential target classifier was evaluated. Table 13.2 shows the predicted probability of correct classification (based on assumed cell-to-cell independence of the return) for both threat targets and nonthreat targets. For example, while the probability of correctly identifying a single walking human in a given range-azimuth cell was 92%, the probability of correctly identifying a track of a walking man approached 100% for tracks as short as three contiguous range-azimuth cells. Based on these predictions, high-confidence track classification appeared to be achievable.

---

7. As clutter return tends to be spatially nonhomogeneous and wind velocity and direction are often very local phenomena, the estimation of clutter backscatter in a specific range of cell benefits from a cell-by-cell characterization (rather than the spatial integration over many range-azimuth cells).

8. Because there exists a ground-reflected wave that is out-of-phase with the direct signal return, the range law for a ground-based target is proportional to R^{-8}, while for a free-space target is proportional to R^{-4}. Thus, at a given range, a large flying bird can have a radar return as large as that for a walking human.

9. Nearest neighbor track assignment and "m-out-of-n" track initiation criterion were employed.

10. Although helicopter data was not collected as part of this program, because there exists a strong rotor blade modulation component in the return, a helicopter classifier would be relatively straightforward to develop.

Table 13.1
Single-Cell Classifier Performance Matrix for High Signal-to-Interference Ratio Target Return

	Classification				
	Clutter	*Vehicle*	*Human*	*Animal*	*Bird*
Clutter	1.0	0	0	0	0
Vehicle	0	0.95	0	0.5	0
Single man	0	0	0.92	0.08	0
Multiple men	0	0	0.78	0	0
Dog	0	0	0.04	0.88	0.11
Pheasant	0	0	0	0.11	0.89

Table 13.2
Probability of L Correct Classifications Given Tracks of Length K
Probability of Correct Decision = 1 − P(error) where P(error) = $[K!/L!(K - L)!]p^{L}q^{K\ L}$ (Cell-to-Cell Independence Assumed)

Actual Target	*Probability of Correct Decision*			
	$L = 1$, $K = 1$	$L = 2$, $K = 2$	$L = 2$, $K = 3$	$L = 3$, $K = 3$
Vehicle	0.95	1.00	0.99	1.00
Single man	0.92	0.99	0.98	1.00
Dog	0.88	0.99	0.96	1.00
Bird	0.89	0.99	0.97	1.00

Although a simple sequential classifier may be adequate for several of the tar-et classes, the approach proved inadequate for bird targets. Figure 13.2 depicts the oppler signature of a wild duck flying through three contiguous range-azimuth ells. In the first cell, the radial component of the duck's velocity was 42 ft/s (which within the expected range of duck flight speeds indicated in Figure 13.3), and its ing-beat rate was approximately 6 beats/s (which is reasonably close to the nomi-al rates for ducks shown in Table 13.3). In the second cell, the bird set its wings nd glided through the range cell. Because wing-beat modulation was the principal feature in the bird target classifier, the absence of a detectable wing-beat harmonic resulted in target misclassification.[11] Finally, in the third range cell, the target velocity decreased and the duck landed in a dry lake bed just beyond the range cell. Because the backscatter from the third range cell was not characteristic of any target or nontarget signature, the detection was classified as "due to an unknown source."

Table 13.3
Typical Bird Wing-Beat Rates

Species	*Rate (Beats/s)*
Sparrow	13
Duck	9
Crow	3 to 4
Stork	2
Pelican	1.1
Hummingbird	10 to 80

Since three consecutive assessments of a single target produced three substantially different "high-confidence" interpretations, a conventional sequential classifier would be expected to perform poorly on this data set. The usual recommendation for enhancing classifier performance is to collect more data to help refine the parameters of the discrimination function. In this view, the shortcomings of the sequential classifier could ultimately be overcome by obtaining a more complete training set. However, for targets that exhibit nondeterministic and time-varying behavior, the simplifying assumptions required by the conventional sequential classification model (cell-to-cell independence, unimodal target characteristics) are simply not valid.

Target modulation characteristics (jet engine turbine, ship propeller, engine room mechanical noise, rotating antennas, and vehicle track modulation) are important elements of many target classification and identification algorithms. Apart from the high signal-to-noise ratio requirements of such algorithms, detection of the harmonics that are relied on by such algorithms can be a function of the geometrical relationships between the target and the sensor as well as a function of

11. Bird wing-beat rates have been used by ornithologists to differentiate migrating bird species. Models exist that predict wing-beat rate based on bird weight and wing length.

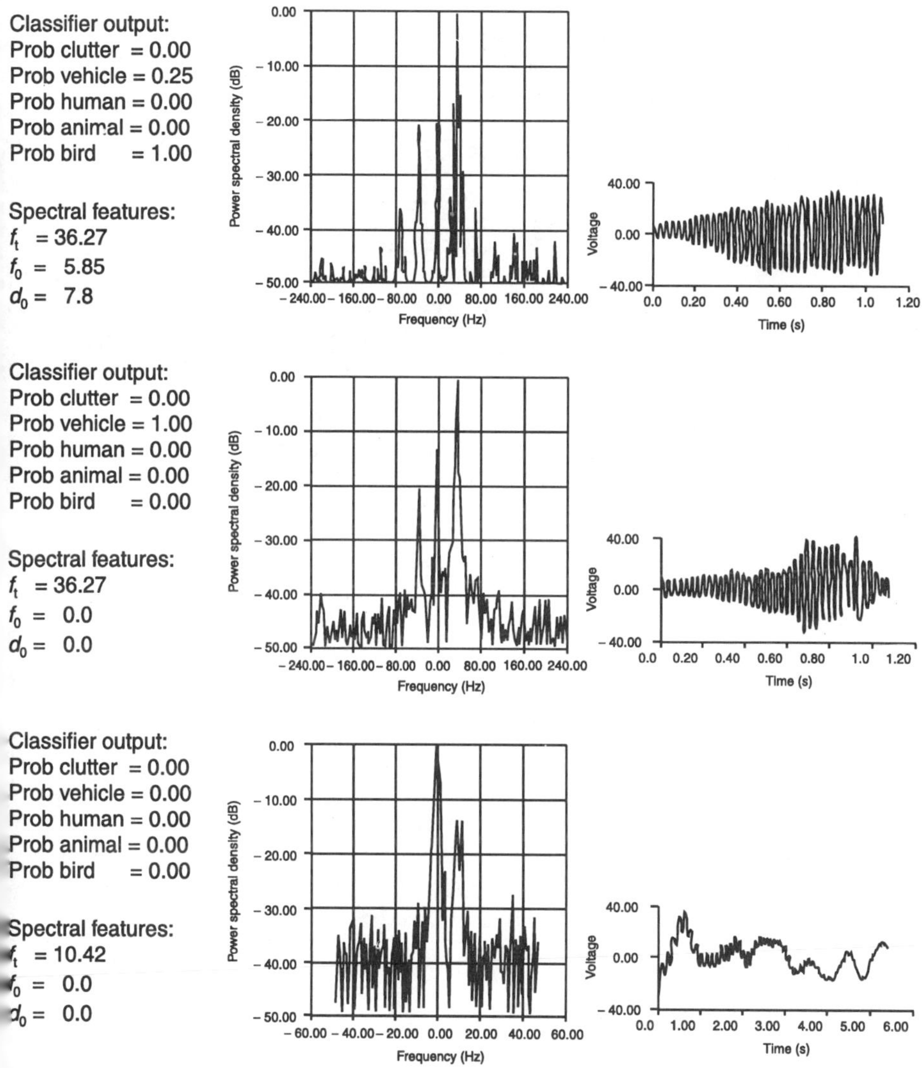

igure 13.2 Doppler waveforms, power spectral density functions, and classifier results for backscatter from a wild duck flying through three contiguous range-azimuth cells.

me-varying characteristics of the target itself. For instance, track modulation from tank is generally only detectable from certain view angles. In addition, if the upper racks are covered, the second harmonic of the target's radial velocity might not be bserved from any angle.

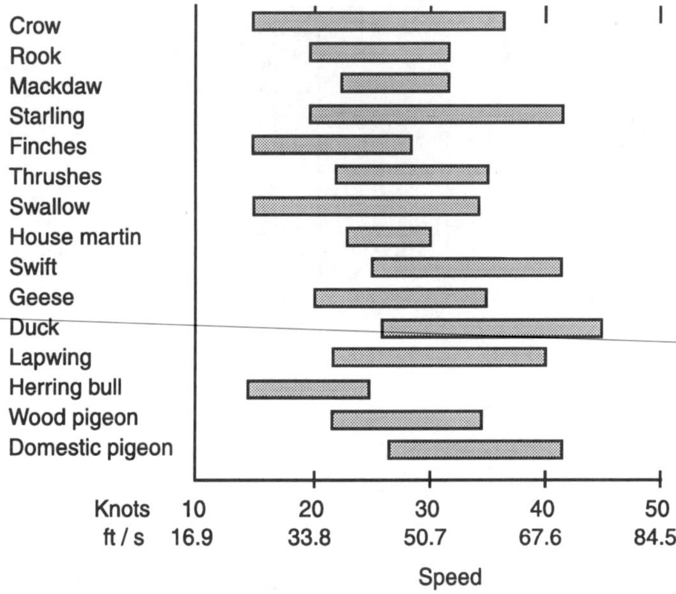

Figure 13.3 Typical flight speeds of various bird species.

In terms of the site security application, if birds beat their wings constantly (a smaller birds must do) and exhibited slow angular acceleration, classification base on estimates of target velocity, wing-beat rate, and cross section would adequatel characterize bird tracks. However, because bird behavior and flight patterns ar complex and nondeterministic, gross classification errors can occur.

In the case of the duck track shown in Figure 13.2, the tri-modal spectral char acteristics observed in the radar backscatter are due to:

1. The bi-modal behavior of large birds in flight (wing beating/wing set);
2. Transients associated with takeoffs, landings, and high-angular accelerations.

Radar operators can readily identify the tracks of large birds once they becom familiar with this multimodal pattern. The traditional sequential classifier, on th other hand, will likely fail to produce a global interpretation of the target track sinc it merely combines a sequence of highly local interpretations.

While the conventional sequential *detect–then–track–then–classify* paradigm i entirely adequate for certain problems (single-sensor, unimodal target return), more sophisticated approach is required to handle targets that exhibit time-varyin or highly context-sensitive behavior. For example, although the traditional sequen tial classification paradigm was satisfactory for a constant-velocity human target, more sophisticated approach would be needed if the target speed and travel direction changed dramatically over the spectral feature estimation window.

Based on the data fusion principles that have been offered, several potential enhancements to the traditional single-sensor sequential classifier are possible.

13.2.1 Enhancement 1

When the information content of the radar signature alone does not permit reliable classification, the dimensionality of the relevant information can be increased by the use of additional sensors that provide at least partially independent information. For example, laser radar and IR imaging systems could be queued by the UHF radar system to provide high spatial resolution and thermal characterizations of targets, respectively.[12]

13.2.2 Enhancement 2

Algorithm robustness can be improved by employing target class–dependent track classification models that interpret the radar backscatter with respect to appropriate arget behavior models. The development of a bird track classifier that is based on a multimodal behavior model would be capable of significantly more robust and conext-sensitive performance than a simple sequential classifier.

3.2.3 Enhancement 3

The classification algorithm can made dynamically sensitive to historical, physical, nd logical constraints. For example, since ducks regularly flew from, and returned o, the same large partially dry lake-bed, the comparison of the observations to hisorical behavior could enhance the algorithm performance. Likewise, because the etection occurred during daylight hours, all night-flying bird species could be precluded from consideration. Seasonal dependence of bird species (due to migration) nd consideration of the present weather conditions (water fowl often fly during eavy rains, while other bird species generally roost) could further refine the analysis process. In a similar fashion, since range-azimuth cells that contain power lines nd chain-link fences routinely generate false alarms for wind gusts over 20 miles er hour, a more sophisticated constant false alarm rate system could model this argely deterministic phenomenon.

The inherent complexity of the bird classification problem demonstrated the ortcomings of simple numerical-based algorithms. The principle shortcoming of

---

12. The use of an active optics system was of considerable interest as an adjunct to the radar system. Since funding and technology limitations precluded this approach, a passive optics system (with a man-in-the-loop) *was* evaluated in parallel with the radar backscatter classification study as a "final filter" on the automated process output.

the conventional track classifier is the information-limited nature of the problem-solving paradigm. A more robust and context-sensitive bird classifier could be developed by employing an effective combination of three classes of information:

1. Target-dependent behavior models;
2. Non-sensor-derived domain knowledge;
3. Nonconventional problem-solving constraints.

Thus, in addition to the adequacy of target signal-to-noise and signal-to-clutter ratios, algorithm developers must be concerned about the adequacy of the total *information content* used by the decision process.

13.3 PATH PLANNING

Path planning is an underlying requirement of many real-world fusion applications involving such tasks as mobility corridor determination, autonomous vehicle control, and robotic assembly operations. While humans plan many routine actions almost subconsciously (walking from one room to another, picking up a glass of water), other tasks can require considerable conscious deliberation (the selection of a route for a multiple-day automobile trip). As an example of the latter, suppose that a route must be developed for travel from Washington, DC, to Portland, OR. Although the road maps for all possible counties through which such a route migh pass could be acquired and carefully scrutinized, most people would employ a muc more efficient top-down strategy. For example, a coarse-resolution map of th United States would likely be sought to help identify all interstate highways tha might be appropriate to use. Once the interstate highway network had been se lected, state maps would be acquired to help select the portions of the route tha cannot use the interstate highway system and to locate tourist attractions that ar not far from the interstate highway that will be used. Finally, in order to locate an in dividual residence in Portland, a street map of the city would be used to refine th final stage of the plan. Thus, the top-down path-development strategy begins by con structing a skeletal plan that is recursively refined to the required level of detail.

This same top-down reasoning approach supports *dynamic replanning* as wel Replanning can involve only relatively local changes or much more global change to an existing plan. For instance, consider a detailed travel plan that called for dri ing down the right-hand lane of Route 32. If during plan execution, a doubl parked truck is encountered, the plan must be locally adjusted to avoid the vehicl On the other hand, much more global modifications of a plan can be required. Fo example, suppose that a traffic report announces that a serious water main brea has caused Route 32 to be closed for several hours. In this case, the original rou plan might require extensive revisions in order to entirely avoid the use of this roa Because it tends to minimize the size of the search and problem-solving spaces, mu tiple level-of-abstraction planning can be significantly more efficient than single level-of-abstraction planning paradigms.

A hierarchical algorithm is outlined next for planning wheeled vehicle routes for mixed "on-road" and "off-road" paths. Domain features (geographical and cultural) that have the most impact on vehicle mobility tend to represent the most significant constraints on route selection. If the domain constraints are ordered from the most to the least significant, path development can be based on a top-down, multiple-stage, constraint-satisfaction paradigm. In stage 1, all potential *extended barriers* to ground mobility (barriers, such as rivers, canyons, neutral zones, ridge lines that cannot readily be circumnavigated) between the "start position" and the "goal position" are identified based on top–down spatial search of the natural and cultural feature database. If an extended barrier is detected, candidate barrier-crossing locations are determined (bridges or fording locations for a river, passes for a mountain range). One or more of these barrier-crossing sites are selected as potential high-level path subgoals. This global search strategy is reapplied at each high-level path subgoal. Once the final goal state is attained, one or more candidate path sets will have been identified that satisfy the highest level domain constraints.

Because roads generally have the next most significant impact on route selection, stage 2 establishes road connectivity between all subgoal pairs that are "near" roads. Top-down, multiple-resolution true two-dimensional spatial search permits the efficient discovery of all roads that are "near" existing path subgoals. If two consecutive subgoals are discovered to be on or near a road, the shortest road network between the subgoal pair is developed; otherwise overland travel between the subgoals is required. Upon completion of stage 2, the coarse-resolution paths developed in stage 1 have been refined with the appropriate road-following segments.

In stage 3, paths between all subgoal pairs that have not been linked by roads are developed. Just as *extended barriers* are naturally associated with the global constraints on mobility, *nonextended barriers* (barriers whose physical extent is small enough so that circumnavigation is feasible) play an important role in more local route-planning decisions. Ordering the set of barriers from the largest to the smallest permits stronger constraints (larger barriers) to be considered before weaker constraints (smaller barriers). The efficiency and the effectiveness of the algorithm can be maintained by employing progressively higher resolution spatial representations as the barrier size under consideration decreases. The path-refinement process terminates when the objective path resolution is achieved. By establishing a mobility resistance threshold that constitutes a vehicle barrier (for example, MR = 5), standard mobility resistance maps can be used to identify both the extended and nonextended barriers.

Although generally more complex, the higher order canonical form approaches tend to have significant advantages over lower-order approaches. For example, a hierarchical constraint-satisfaction-based path-development paradigm can support the development of coarse-resolution paths that satisfy rather global objectives, such as to provide "maximal concealment from a particular vantage point" or "minimal travel time to a goal state." A path that meets these requirements might consist of a set of road segments that are not visible from the vantage point, plus high mobility off-road path segments for those sections of the roadway that *are* visible from the vantage point. Because hierarchical constraint-satisfaction-based reasoning captures the character of human problem-solving approaches, algorithm development tends to be highly intuitive. In addition, the problem solutions tend to be intuitively satisfying as well. Because each subgoal pair is known to be a potential member of the final solution set, recursive path refinement tend to be much more efficient than algorithms that directly develop high-resolution candidate solutions. Once all candidate path segments are developed, individual paths can be rank-ordered against path-evaluation metrics, such as minimum travel time, minimum path length, or maximal concealment.

To explore the computational implications and database issues associated with such an algorithm, consider the development of a route for a wheeled vehicle from the "start" to the "goal" state depicted in Figure 13.4. In stage 1, spatially anchored search of the cultural and geographic feature database (the lower resolution repre sentation of these features is maintained by the object representation of two-dimen sional space) reveals a potential extended ground-mobility barrier (a river) between the start state and the goal state. As demonstrated by Figure 13.5, top-down search within the pyramid reveals that the river does, in fact, lie between the start and the goal nodes. Because this extended barrier must be breached if a path developmen plan is to be successful, bridges or suitable fording locations must be located.

As shown in Figure 13.6, semantic search of the object *River* 1 reveals a smal number of candidate bridges and no known fording locations. To simplify the ex ample, only the closest bridge is selected as a subgoal (denoted $SG_{1,1}$). Althoug the closest bridge could have been located via spatial search along the river, seman tic search tends to be more efficient because bridges are explicit attributes of the se mantic object *River* 1.

The first-stage algorithm is then reapplied as each high-level subgoal is deve oped until the goal state is reached. In this case, spatial search from $SG_{1,1}$ towar the goal node reveals another *potential* river barrier. Top-down spatial search verifie the existence of a second extended barrier (*River* 2). Just as before, the closes bridging location, denoted as $SG_{1,2}$, is determined by selecting the closest bridg maintained by the semantic object *River* 2. Spatial search from *Bridge* 2 toward th goal state reveals that no extended barrier exists that blocks ground travel betwee $SG_{1,2}$ and the goal state. As depicted in Figure 13.7, the first stage of the path-deve opment algorithm generates a single path consisting of the three subgoal pai (start, $SG_{1,1}$), ($SG_{1,1}$, $SG_{1,2}$), ($SG_{1,2}$, goal) that satisfies the global objective "reac the goal state by crossing all extended barriers." At the conclusion of stage 1, the pr mary alternatives to path flow have been efficiently identified for a problem th could easily overwhelm a high-resolution, raster-based canonical form I algorithm.

In stage 2, all adjacent subgoals that are either *on* or *near* a road with a degre greater than 0.25 are connected by the shortest road network. Figure 13.8 defin

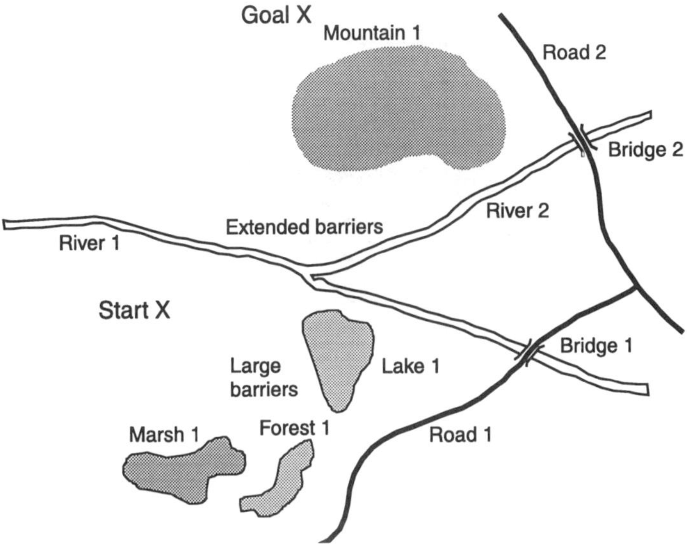

Figure 13.4 Domain mobility map for path-development algorithm.

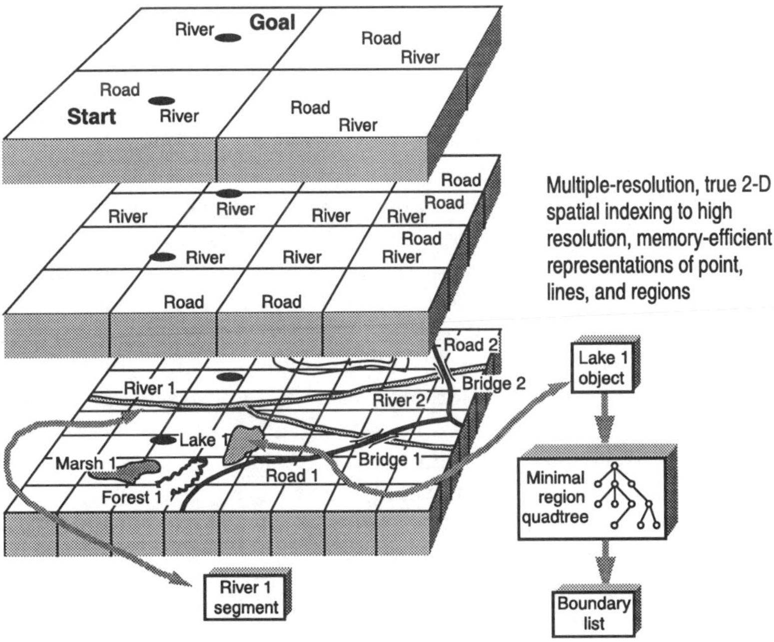

Figure 13.5 Top-down multiple-resolution spatial search from the start toward the goal node reveals the existence of a river barrier.

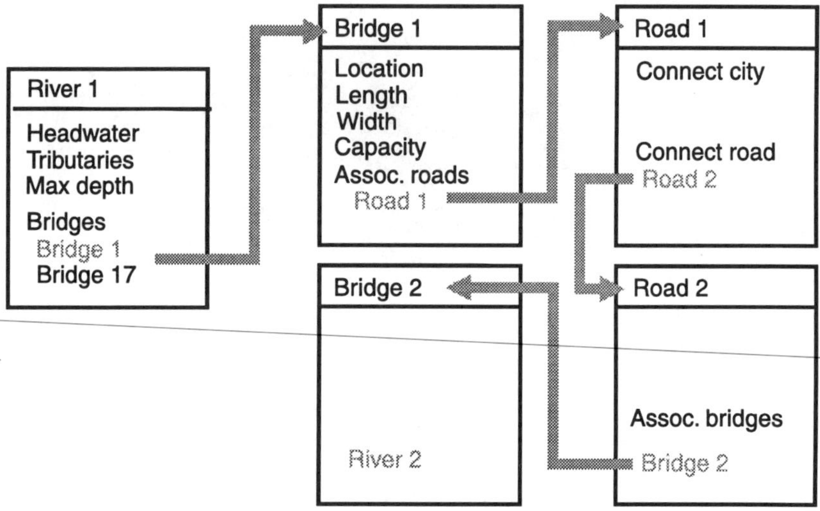

Figure 13.6 Semantic object database for the path-development algorithm.

the fuzzy qualifier *near* in terms of a Euclidean distance metric. Since most bridges support vehicle traffic, it is quite likely that the subgoal pair ($SG_{1,\,1}$, $SG_{1,\,2}$) is traversable over a road network. In this case, since both *Road* 1 and *Road* 2 are known to be members of the candidate solution set (by virtue of their association with *Bridge* 1 and *Bridge* 2, respectively), a road network solution can be efficiently developed using the semantic object representation.

The *start* and *goal* nodes are investigated next. Top-down, multiple-resolution spatial search in the vicinity of both the *start* and the *goal* nodes (Figure 13.5) reveals that the *start* node is *near* a road with degree 0.5, while the *goal* node is *not near* a road. These assessments are made by computing the Euclidean distance from both nodes to the nearest road within 1 km. Figure 13.8 is then employed as a weighting function to evaluate the "degree of nearness."

Because the *start* node *is* near a road, in addition to direct overland travel toward *Bridge* 1 ((start, $SG_{1,\,1}$)), an alternative strategy is to travel overland to the nearest road (($SG_{2,\,1}$)), which would then be followed along the minimum length road path to *Bridge* 1 (($SG_{2,\,1}$, $SG_{1,\,1}$)). Although a spectrum of variants exist between direct travel to the bridge and direct travel to the closest road segment, at this level of abstraction only the primary alternatives must be identified. The solution state at the conclusion of stage 2 is shown in Figure 13.9.

In stage 3, all existing off-road path segments are refined based on the mobility constraints imposed by progressively smaller-sized barriers. First, large barriers such as lakes, marshes, and forests, are considered. Spatial search from the *start* node to $SG_{1,\,1}$ reveals a large lake. Since circumnavigation of the lake is required, two subgoals are generated ($SG_{3,\,1}$ and $SG_{3,\,2}$) as shown in Figure 13.10. In a similar

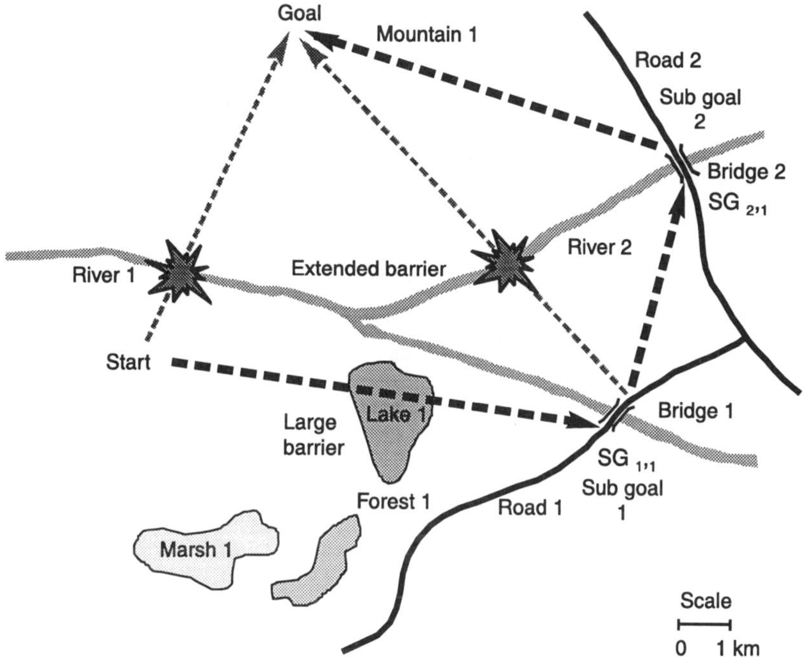

'igure 13.7 Results of the first stage of the path-development algorithm.

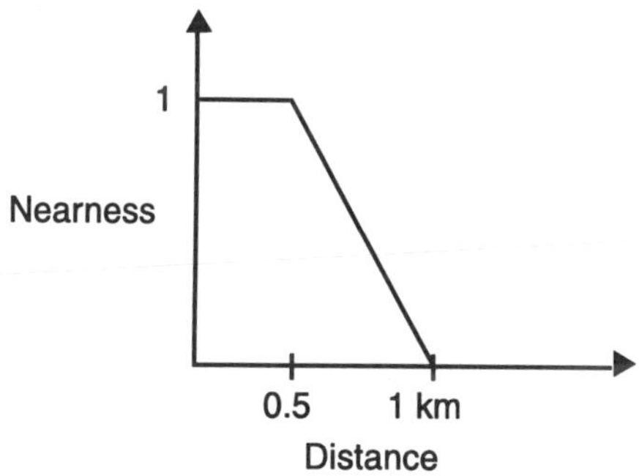

igure 13.8 Membership function for fuzzy metric "near."

ıanner, spatial search from the start state toward $SG_{2,\,1}$ reveals a large marsh, gen-rating, in turn, two alternative subgoals ($SG_{3,\,3}$ and $SG_{3,\,4}$).

Spatial search from both $SG_{3,\,3}$ and $SG_{3,\,4}$ toward $SG_{2,\,1}$ reveals a forest obsta-e (*Forest* 1). Assuming that the forest density precludes wheeled vehicle travel, two ıore subgoals are generated representing a northern route ($SG_{3,\,5}$) and a southern

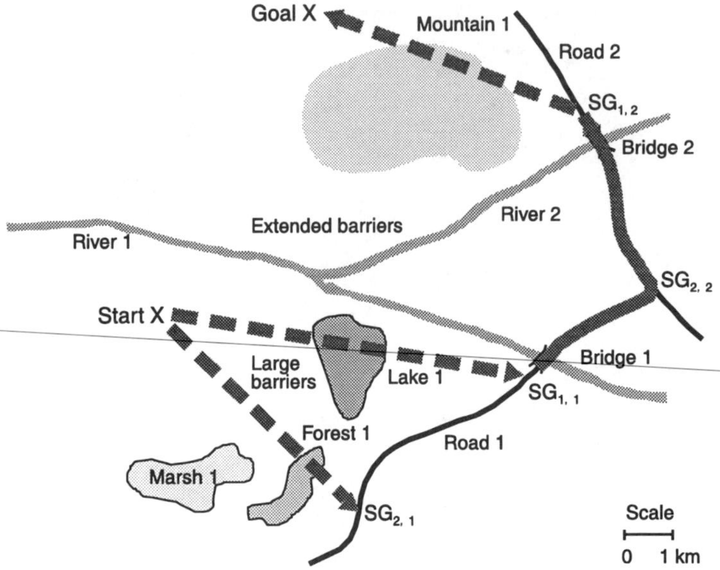

Figure 13.9 Path-development following stage 2 (dotted path implies that future refinement i required).

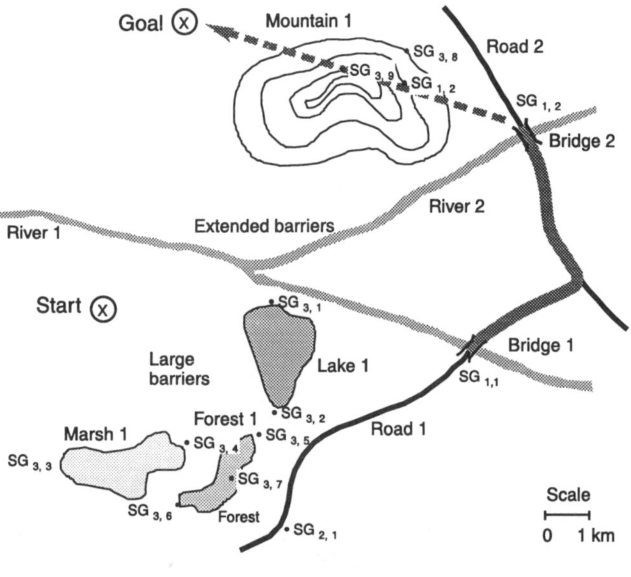

Figure 13.10 Path-development subgoals based on the consideration of "extended mobility barriers road connectivity of subgoals, and "large mobility barriers."

route ($SG_{3,\,6}$) around the forest. Because there might exist a road that passes through the forest, the possibility of a third subgoal (road travel through the obsta-cle) can be investigated by intersecting the region representation of *Forest* 1 with the road database.

With strictly vector-based representations of roads and regions, such intersec-tion operations could require interrogation of a significant portion of the road data-base. However, the semantic/spatial object database kernel permits direct spatial search of that portion of the road database that is within *Forest* 1. In this example, suppose a dirt road is discovered that intersects the forest. Since at this level of ab-straction, no objective criterion exists for evaluating the "best"[13] subpath(s), an additional subgoal, ($SG_{3,\,7}$) is established. For example, although the road length hrough the forest might be shorter than the travel distance around the forest, the road might not enter and exit the forest at locations that satisfy the overall con-straint set or it might be too narrow or too rough to support the vehicle class of nterest.

Next, spatial search from $SG_{1,\,2}$ to the goal state identifies a mountain obsta-le. Because of the inherent flexibility of a constraint-satisfaction-based problem-olving paradigm, a wide range of local path-development strategies can be onsidered. For example, the path could be constrained to:

1. Circumnavigate the obstacle ($SG_{3,\,8}$);
2. Remain below a specified elevation ($SG_{3,\,9}$);
3. Follow the minimum terrain gradient from $SG_{3,\,10}$.

Figure 13.11 shows the path-plan subgoal graph following stage 1, stage 2, and stage 3. In subsequent stages of the algorithm, the detailed path between all sets of subgoals would be recursively refined based on the evaluation of progressively more ocal domain-constraint criterion, such as minimum mobility resistance paths or ninimum terrain gradient paths.

Whereas typical path-planning algorithms directly generate a high-resolution ath based on a single level-of-abstraction constraint set, the multiple-stage algo-thm just discussed begins by first satisfying a set of global objectives and con-raints. Subsequent stages of the algorithm focus on progressively more local bjectives and refined constraints. The high-level constraints can be evaluated using oarse-resolution spatial representations, while highly local constraints require uch higher resolution data.

---

13. Objectives like "best" and "optimal" tend to be highly context-sensitive, often requiring a complex, multiple level-of-abstraction perspective. In addition, although optimal solutions *can* be developed for highly combinatoric problems, the finite state-space formulations themselves may require significant simplfying assumptions. Thus, while combinatoric optimization may generate solutions that are technically optimal, they may not generate truly optimal solutions.

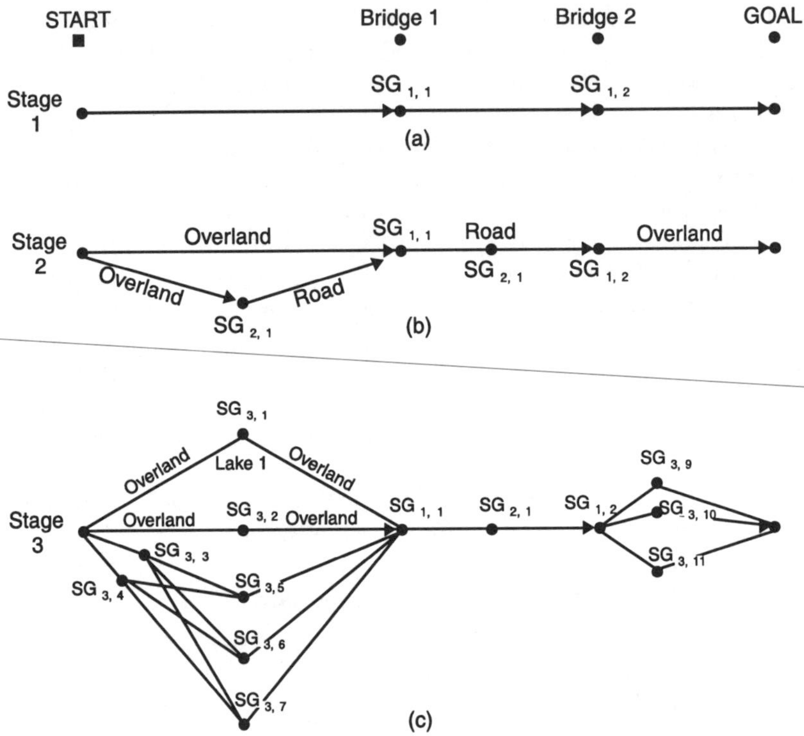

Figure 13.11 Path-development graph following (a) stage 1, (b) stage 2, and (c) stage 3.

Figures 13.12 to 13.14 illustrate the efficient top–down refinement of path segments (start, $SG_{1,1}$) and ($SG_{1,2}$, goal). In contrast, a canonical form I path-planning algorithm uses specific declarative knowledge and rigid, single level-of-abstraction control knowledge generates highly local, nearest neighbor path-extension strategies. Because the semantic/spatial database kernel supports multiple-resolution, true two-dimensional spatial search, the top–down path-planning algorithm requires significantly reduced search-space size compared to a single-resolution algorithm. For example, the imposition of a high-level constraint may eliminate a single 1-km × 1-km resolution cell from further consideration, which avoids the search-and-test of as many as ten thousand 10-m × 10-m resolution cells.

The A^* algorithm is often employed to attempt to control the size of the search space. Although such algorithms employ a flexible control element that supports general heuristic evaluation criteria, its declarative element typically represents specific declarative knowledge and the control is generally of a single level of abstraction. Thus, the A^* algorithm is fundamentally a canonical form V problem solving approach. Lacking global control, a typical algorithm may attempt to de

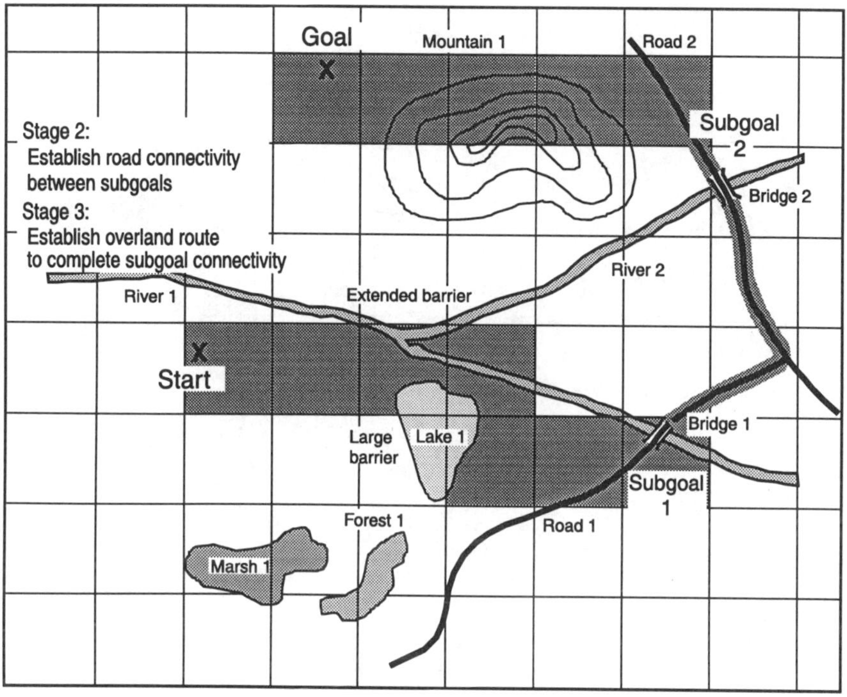

Figure 13.12 Coarse-resolution path development between "start" and $SG_{1,1}$ and between $SG_{1,2}$ and "goal" state.

velop a straight line path from the start node to the goal state. At the point the river barrier is encountered, two alternative paths (moving up-river and down-river) would be explored until a river-crossing location is discovered. In general, path-planning algorithms that are based on movement along the edge of barriers tend to generate nonoptimal, as well as counterintuitive, paths. If the domain mobility map is highly fragmented, the plan produced by such algorithms may "wander"excessively.

The canonical form XV path-development algorithm, on the other hand, emulates the efficient problem-solving style of a human analyst. The highest level strategy effectively exploits potential "weaknesses" in extended mobility barriers. All candidate paths that satisfy the global constraints are recursively refined using progressively higher resolution domain knowledge and progressively more local domain constraints. The constraint sets can include both default constraints (wheeled vehicles are not found in a lake) and problem-specific constraints (favor highly concealed routes; pass through region *A*; prefer road travel to off-road travel unless an alternate off-road path exists that is less than 1/3 as long and has a maximum slope of 15%). In general, default constraints represent generic, context-insensitive facts

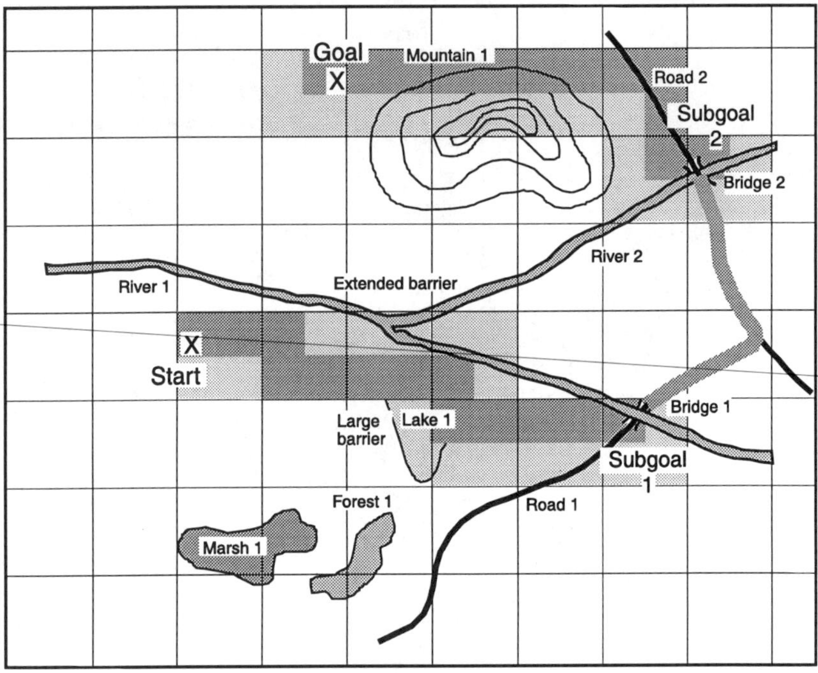

Figure 13.13 Refinement of path shown in Figure 13.12.

or rules, while context-sensitive problem-specific constraints refine or over-ride default constraints.

The hierarchical path-development algorithm can be seen to achieve the five objectives outlined in the chapter introduction.

1. The algorithm employs a reasoning approach that is quite natural for humans.
2. Since the constraint sets can be tailored to a specific vehicle class, the algorithm tends to be highly robust.
3. The algorithm is dynamically sensitive to the current domain context.
4. Due to the top-down nature of the constraint set, the algorithm tends to generate efficient global solutions.
5. The spatial/semantic database kernel provides critical algorithm support, including efficient, top-down, spatially anchored search and fully integrated semantic and spatial object search.

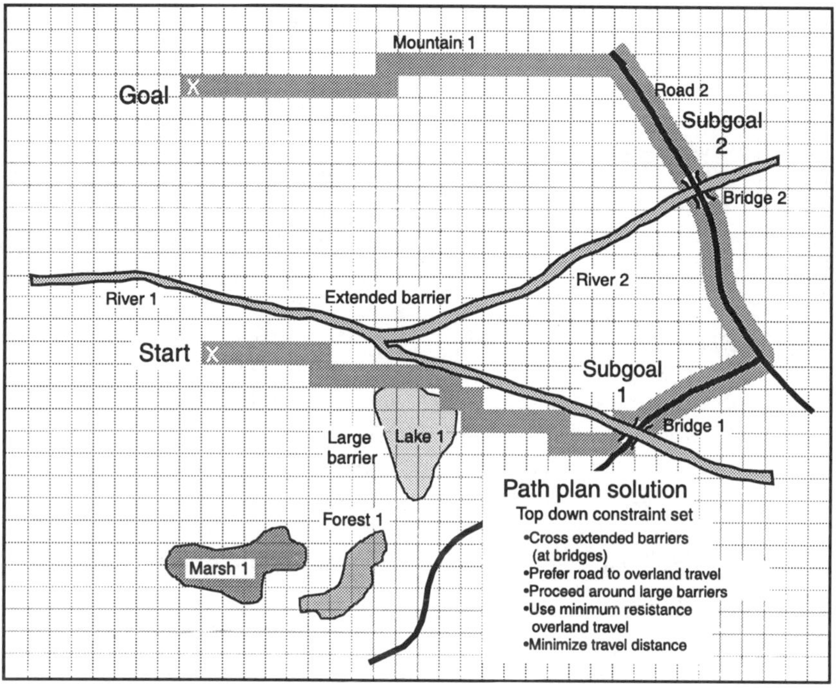

Figure 13.14 Refinement of path shown in Figure 13.13.

13.4 TEMPLATING

Templates are stereotypical or prototypical representations of classes of data, objects, events, or activities. Both biological and artificial situation-understanding systems attempt to categorize data sets based on their similarity to prototypical patterns. Since templates represent a priori expectations, templating represents a form of pattern recognition. Conceptually, templates can be treated as either *explicit* or *implicit*. Explicit templates depict a fixed, rigid pattern, while implicit templates are more flexible representations. The latter includes both model-based and constraint set–based templates.

Several examples of templating have been presented in earlier chapters. A template for scoring multiple-choice questions is an example of a highly rigid explicit template. If the student's answers align with the holes punched in the template, the answers are correct; otherwise they are wrong. Although optical character recognition for machine-printed characters rely on rigid templates of all letters and symbols, a rigid template approach tends to be inadequate for recognizing handwritten characters. In fact, explicit templating approaches have been found to be inadequate for numerous real-world problems [4].

Templates can be used to recognize both individual objects (a specific ship) or composite objects (a carrier-battle group). Templates can operate on strictly spatially organized information (the detection of tanks in imagery), nonspatially organized information (the identification of radar-type based on ELINT parameters), or a combination of both spatial and nonspatial information (doctrinal templating based on both the spatial and the functional organization of observed military units).

13.4.1 Event Template

An *event* will be treated loosely as any observable action that has a time-varying character. Typically an event begins and ends at specific times or during certain time intervals. Bridge construction, a frontal attack, and a training maneuver are examples of events. In this section, the potential threat posed by an incoming aircraft track (a specific event) is evaluated using both a rigid, noncontext-sensitive templating paradigm and a context-sensitive, model-based paradigm. The available sensor-derived information is assumed to be:

1. The estimate of the current aircraft speed;
2. The mean target speed for the full track history;
3. The target's altitude above sea level;
4. The detailed spatial representation of the track history.

Figure 13.15 shows an example of a rigid, frame-based template representation of the event "a likely hostile intelligence gathering mission;" similar templates would exist for other event classes. Individual template parameters can be either single-valued or multivalued. Parameters can be either alphabetic or numeric and can possess an associated certainty. If observed data match all (or most) of the paramters associated with a particular template, the data is considered to represent an instance of that template. For templates that use a frame-based approach, parameters represent slots as described in Part I.

By contrast, Figure 13.16 illustrates a high-level view of a hierarchical, model-based template that involves the consideration of five principal subtemplates:

1. Early detection avoidance;
2. Penetration of the friendly air-space;
3. No weapon system signatures or weapons release;
4. No jamming of air defense radar;
5. No nearby hostile activity.

Current speed	450 MPH
Average speed	400 MPH
Altitude	4000 ft above sea level
Track file history	

(a)

Mission 1 template	
Average speed	(300–500 MPH)
Altitude	(1000'–1500')
Track history—approaching Region A	

Mission 1 template	
Average speed	(500–700 MPH)
Altitude	(20 Kft–30 Kft)
Track history—approaching Region A	

• • •

(b)

Figure 13.15 (a) Track file attributes and (b) set of rigid templates with elastic parameters.

Each subtemplate would be evaluated based on the context-sensitive interpretation of the observables. For instance, an aircraft that attempts to avoid early detection by ground-based radar will probably fly just above tree-top level. If the area of interest is mountainous, the absolute altitude of the aircraft used by the rigid template approach is considerably less relevant than the target's relative altitude above the terrain or vegetation. While both templating approaches can compare observations to individual template parameters, model-based approaches can evaluate complex relationships among problem attributes with respect to the domain context.

Similarly, the evaluation of the second subtemplate in Figure 13.16 requires that the target track be evaluated within the context of friendly force assets that might be the objectives of a hostile reconnaissance flight. If the enemy aircraft is on an intelligence gathering mission, the aircraft's flight path will most likely have been selected so that air defense radar systems of interest will be activated, yet at the same time minimizing the probability that the penetrating aircraft will be intercepted. The third subtemplate is satisfied if the target employs self-protection measures and releases no weapons. The fourth subtemplate is activated as long as the potential "targeted" radars are not jammed. Finally, the intelligence gathering nature of the flight is further suggested if only a single aircraft is involved and if no nearby hostile activity is observed. Had a large number of aircraft been active in the vicinity, a strict intelligence-gathering mission would have been unlikely.

Thus, a model-based template that evaluates input data with respect to the typical behavior of an "intelligence-gathering overflight" would be expected to be considerably more effective than a rigid template that cannot evaluate key relationships between the flight profile and the mission characteristics.

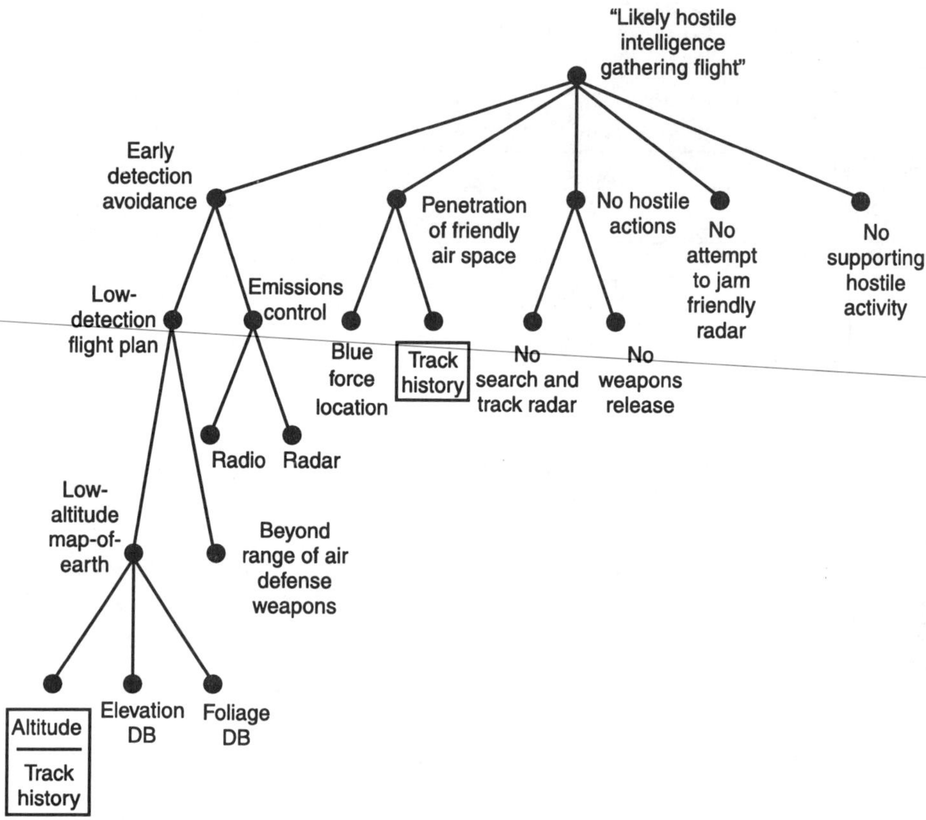

Figure 13.16 Characteristics of a flexible model-based template.

13.4.2 Network Template

Next, consider the identification of a communication network based on the observations over time of individual communications. Suppose a model for a particular radio network states that:

1. Control nodes are typically located within 500m of primary roads and within lightly forested regions;
2. Subordinate nodes must maintain LOS with the control node and are typically located within 10 km of primary nodes.

Figures 13.17 and 3.18 illustrate the set operations that are required to evaluate this two-stage template. The first stage generates the intersection of the 500-m region-grown primary road network, the lightly forested region map, and all target error ellipses that are associated with primary node locations. The small black regions in Figure 13.7 represent the uncertainty regions associated with control nodes that satisfy the stage 1 constraints.

Relative to the centroid of the two candidate primary node locations, the line-of-sight and distance constraints of the second-stage template are applied to the secondary node database as shown in Figure 13.18. The resulting black regions represent the possible locations of secondary nodes. As compared to the large uncertainty regions provided by the uncertain regions associated with the individual radios, the two-stage spatial template permits the location uncertainty of both the primary and secondary nodes to be significantly reduced.

Rather than treating the template-evaluation process in terms of exhaustive set operations, template evaluation can be handled using highly efficient optimal clause-order, constraint-based search. While a conventional set-intersection approach might require the generation of complete stage 1 and stage 2 overlays for potentially large regions of space, template evaluation based on spatially anchored search permits constraints to be directly evaluated using highly efficient database search. Since roads generally possess less areal-extent than either forests or terrain-elevation information, the region-grown road database could be used as the spatial window through which to view the forest database. The resultant product would then become the spatial window within which to view the elevation database. The final spatial filter would represent the search window for locating all known targets that exhibit the characteristics of a primary node. Alternatively, if the uncertainty re-

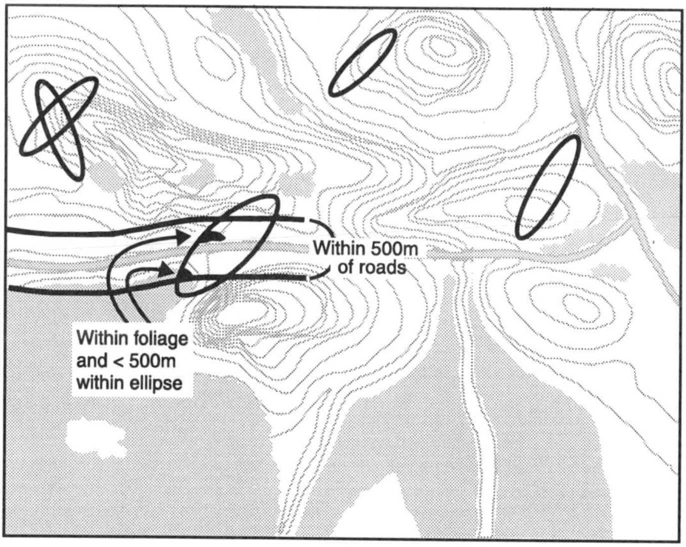

Figure 13.17 Stage-1 template for a particular communications network.

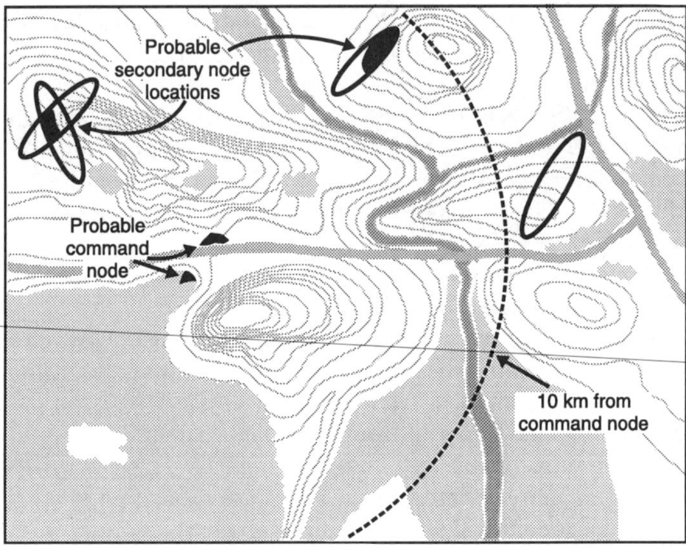

Figure 13.18 Stage-2 template for communications network.

gions associated with all candidate emitters are relatively small, it might be more efficient to treat the individual emitter uncertainty representations as the spatial window with the smallest extent, which would then be the initial filter used to search the data set possessing the next smallest extent.

In the second stage, a 1-km spatial window centered on all candidate primary node locations would be used to search for targets possessing the characteristics of secondary nodes. The use of the hierarchical line-of-sight (LOS) algorithm discussed in Chapter 12 supports the efficient generation of a spatial window for locating secondary nodes that possess LOS to candidate primary nodes. Just as in stage 1, the templating efficiency can be enhanced if the set of secondary node uncertainty representations are treated as the spatial window for evaluating the stage 2 constraints.

The semantic/spatial object database kernel provides highly efficient evaluation for complex constraint sets that involve either semantic or spatial attributes. More sophisticated templates can be developed by simply adding relevant constraints to the hierarchical constraint set. The ability to construct rich and highly intuitive mixed symbolic/numeric templates represents a critical requirement of general machine reasoning.

13.4.3 Doctrinal Template

Doctrinal templating seeks to recognize a group of entities that are organized according to a set of rules or principles. In a military application, a particular unit or activity must be identified from limited observables using a priori knowledge about the composite object as well as its constituent parts. In certain special cases, a military organization can be uniquely identified by a specific set of sensor measurements. For instance, if radar class *X* and radio type *Y* are always associated with unit class *Z* and no other unit class, the detection of these two emissions from a relatively small geographic region would provide direct evidence of the existence of such a unit. Since no ambiguity exists in the decision process, a simple canonical form I templating approach would be adequate.

In general, however, military units are complex and dynamic organizations of objects that may not be uniquely differentiated based on the available sensor measurements. In addition, the spatial organization of the various elements of a unit can be affected by local geographic and cultural features and can vary significantly over time, taking on radically different forms during a staging operation, a forced march, or a contingency mission. Although military analysts who perform doctrinal templating use field manual depictions of doctrinal organizations merely as guidelines for their analysis, many early attempts to automate doctrinal templating treated such prototypes as rigid templates that were overlaid on the situation database in much the same way as a grading template is used to score multiple-choice questions. A "template match" was judged to have occurred if the degree of correlation between a specific template and the perceived situation was above a fixed threshold. When multiple templates achieve a degree of match above the decision threshold, the template receiving the highest score would be normally selected as the "best" match. Although relatively simple to implement, the exhaustive correlation between a sizable template set and a large data set can lead to highly combinatoric computational requirements.

If no template receives a degree of match that exceeds the decision threshold, the match criterion is normally relaxed. For example, the templates can be allowed to stretch or contract (for example, rubber sheet templating), a statistical match criteria might be used, or various distance metrics can be utilized. In general, match criteria relaxation can lead to even higher order computational combinatorics.

Quite apart from the computational issues, a rigid templating paradigm rests on rather tenuous assumptions. As has been done several times before, these assumptions will be explored using a simple biological metaphor. In particular, military *organizations* will be compared to biological *organisms.* In biological systems, an *organism* consists of multiple *organs*; *organs*, in turn, consist of multiple classes of *tissue*, and so forth. Military organizations possess a very similar form of hierarchical specialization. Biological organs perform specialized functions required to support an organism, just as armor, surveillance, and communication assets perform specialized functions to support a military organization. Strong dependency can exist between certain biological organs (the stomach and the liver) in much the same way as strong dependency can exist between certain elements of a military organization (fire-control center and weapon launchers). On the other hand, some organs are relatively independent of each other (the stomach and the lungs) just as are certain elements in a military organization (motor pool and the mess hall).

Although there is considerable symmetry in the metaphor, there do exist significant differences between the recognition of biological organisms and the recognition of military organizations. While biological organisms are spatially and organizationally well defined, military organizations exist only as a collection of associated objects. Many biological organisms are identifiable from a small number of distinguishing features (tigers are large, four-legged animals with stripes, claws, and a long tail). Military organizations, on the other hand, must be deduced or inferred based on observable-level attributes of the individual entities and the apparent logical, functional, and organizational relationships among the entities. Thus, doctrinal templating is analogous to the problem of identifying biological organisms based on observations of their internal structure. Since many species of animals possess similar organs that perform very similar functions, specie identification at the suborganism level tends to be considerably more difficult than at the organism level.

Just as *organism identification* based on limited organ-level observables can lead to ambiguous results, so can *organization identification* based on the observables from individual constituent entities. As stressed throughout this book, ambiguity resolution requires the acquisition and utilization of additional relevant information. This additional information can be supplied by additional sensor measurements, a priori reasoning knowledge, or non-sensor-derived domain knowledge.

The development of models of individual biological organisms that depend on the characteristics of the individual organs as well as the functional relationships among the organs could potentially support specie identification. For example, since a mammal's heart-beat rate tends to be inversely related to the animal's physical size, heart-beat rate measurements would help identify the organism's approximate size and weight. Similarly, a measurement of the blood flow rate may help establish the volume of the organism; measuring the temperature of the blood would discriminate between mammals and reptiles. These individual measurements are analogous to the disparate, highly fragmented information gathered by tactical sensor systems.

A model-based approach to doctrinal templating would have the same advantages as a model-based approach to specie templating. Models could be developed that capture the key organizational principles of individual military units, such as:

1. Subordinate elements are controlled by higher level elements;
2. Units typically have a well-established reporting protocol;
3. Each subordinate element possesses certain organic capability, while depending on other elements for various types of support.

In addition, elements of a military unit possess functional responsibilities and tactical, doctrinal, and logistical constraints. For example, subordinate staff elements might be expected to be less than 5 minutes travel time from their command post, line-of-sight communications must be maintained between certain elements, mobility requirements must be satisfied, and cover and concealment requirements must be met.

Clearly, the effectiveness of either an organ-level specie identification algorithm or a doctrinal templating algorithm relies on the effective utilization of available information. The limited success to date of rigid doctrinal templating algorithms can be attributed to both the inherent difficulty of the problem and the tacit assumption that simple observable-level attributes (the relative spatial relationships among key elements of a unit) can be directly used to differentiate organizations. Since only a subset of the total number of elements in a given unit might actually be both observed and identified, additional uncertainty exists. Although conventional rigid templating can provide an acceptable discrimination model when the relationships among attributes are relatively simple, rigid templates cannot adequately capture complex *relationships* between the observables and the domain constraints, such as:

1. Units are typically within 1 to 5 km of each other;
2. One unit may typically be forward of another;
3. There should be no maneuver barrier between two elements of a single unit.

Additionally, rigid templates may not be capable of effectively utilizing potentially relevant *domain knowledge*, such as:

1. The unit status and readiness level;
2. The implications of current and predicted weather conditions;
3. The apparent tactical mission;
4. Recent reorganizations;
5. Logistical support requirements;
6. Apparent force objectives;
7. The commanding officer's historical preferences.

While the organization of real military units is multidimensional, dynamic, and highly sensitive to its current mission, as well as its operating domain, most rigid doctrinal templates capture only a single static aspect or dimension of a unit's organization (for example, nominal spatial organization). In addition, since relationships and functions among elements of a unit can change with time (an engineer battalion may construct a bridge one day and clear a mine field the next), an enormous number of static, explicit templates can be required to accommodate the range of expected unit variations.

To illustrate the potential of a canonical form XVI approach to doctrinal templating, consider the development of a model-based algorithm for recognizing a composite object W (for example, a division-level force), based on observations of its constituent elements. The observables are assumed to be radio intercepts, radar intercepts, and radar detections of moving (Doppler MTI radar) and stationary vehicles (synthetic aperture radar). Suppose that *W* is known to be comprised of a single *X*, between one and three *Y*, and either one or two *Z*. Since the unit commander is known to be associated with node *X*, both *Y* and *Z* represent subordinate nodes. Just as *W* is a composite entity, *X*, *Y*, and *Z* are composite entities, as well, consisting of multiple objects, attributes, and constraints as described in Table 13.4. The joint attributes that describe the relationships between the individual elements are shown in Table 13.5.

Table 13.4
Attributes of the Primary Elements of Unit *W*

	Object W
Contains	1 Object X 1–3 Object Y 1–2 Object Z
	Object X
Contains	1 Class 1 vehicle 2–5 Class 2 vehicles Type 1 radios Type 2 radios
Preferred geographic/cultural features	Within 0.5 Km of secondary roads Light concealment
Disallowed geographic/cultural features	Lakes, rivers Elevations > 2000 m Populated areas
	Object Y
Contains	5–20 Class 2 vehicles 2–5 Type 1 radar 0–1 Type 2 radar Type 1 radios
Preferred geographic/cultural features	Low mobility resistance Primary and secondary roads
Disallowed geographic/cultural features	Lakes, rivers Mobility resistance > 0.8

Object Z	
Contains	5–10 Class 3 vehicles 0–10 Class 4 vehicles Type 1 radios
Preferred geographic/cultural features	Low mobility resistance Primary and secondary roads
Disallowed geographic features	Lakes Mobility resistance > 0.8

Table 13.5
Joint Attributes Among Elements of Unit *W*

	Object Y and Object Y	
(a)	Communication	Radio Type 1, frequency f_1
	Separation	0.5 to 2.5 km
	Mobility requirements	No maneuver barrier
	Object Z and Object Z	
(b)	Communication	Radio Type 1, frequency, f_4
	Separation	1 to 3 km
	Mobility requirements	No maneuver barrier
	Object X and Object Y	
(c)	Communication	Radio Type 1, frequency, f_2
	Separation	1 to 5 km, no LOS, Y typically forward of X
	Mobility requirements	No maneuver barrier
	Object X and Object Z	
(d)	Communication	
	Separation	2 to 10 km, no LOS
	Mobility requirements	No maneuver barrier

Table 13.5 (continued)

		Object Y and Object Z
(e)	Communication	Radio Type 1, frequency f_2
	Separation	4 to 10 km
	Mobility requirements	No maneuver barrier

The hierarchical constraint set described in Tables 13.4 and 13.5, in effect, represents a highly intuitive "model" that characterizes key elements of the logical, physical, organizational, logistical, and doctrinal infrastructure of the abstract object *W*. The model is both robust (because it provides a very general characterization of object *W*) and context-sensitive (because it depends on a wide range of non-sensor-derived domain knowledge). The hierarchical and spatial relationships among the physical objects are depicted graphically in Figure 13.19.

Template evaluation requires that the observables be evaluated with respect to organizational constraints (control versus primary communication node characteristics, communication constraints) and doctrinal constraints (maximum permitted spatial separation of units, mobility requirements) that depend on the appropriate underlying domain knowledge (terrain/elevation, mobility maps). Based on the degree of match between the data and the constraint set, a certainty factor is assigned that identifies the likelihood that the observations represent an instance of unit *W*.

The database issues associated with the templating process are considered next. Figure 13.20 illustrates the problem data set. Table 13.6 summarizes all relevant constraints that must be evaluated for object *X*. As shown in Figure 13.20, at the highest level of abstraction, four possible elements (a single *X*, a single *Z*, and two *Y*) of *W* have been identified that satisfy the highest level template constraints. The individual elements were identified by testing both their object-specific constraints (does object *X* possess both a type 1 and a type 2 radio?) and the joint characteristics among the candidate elements of *W* (do Object *X* and Object *Y* appear to be communicating over radio type 1 at frequency f_1?).

Consider the identification of the unit control node *X*. Region *X* is the region that encloses all physical objects associated with node *X*. We assume that both radio types used by unit X have been detected within region *X*. No class-1 vehicles, but several class-2 vehicles within region *X* have been identified from SAR imagery. Next, region *X* must be evaluated with respect to both preferred and disallowed conditions. In terms of the preferred conditions, *high ground* will be interpreted as "elevations that are greater than 100m above the lowest point within 1 km" and "light concealment" is taken to be "light foliage." Based on the support provided by the semantic/spatial object database kernel, region *X* is determined to satisfy all three

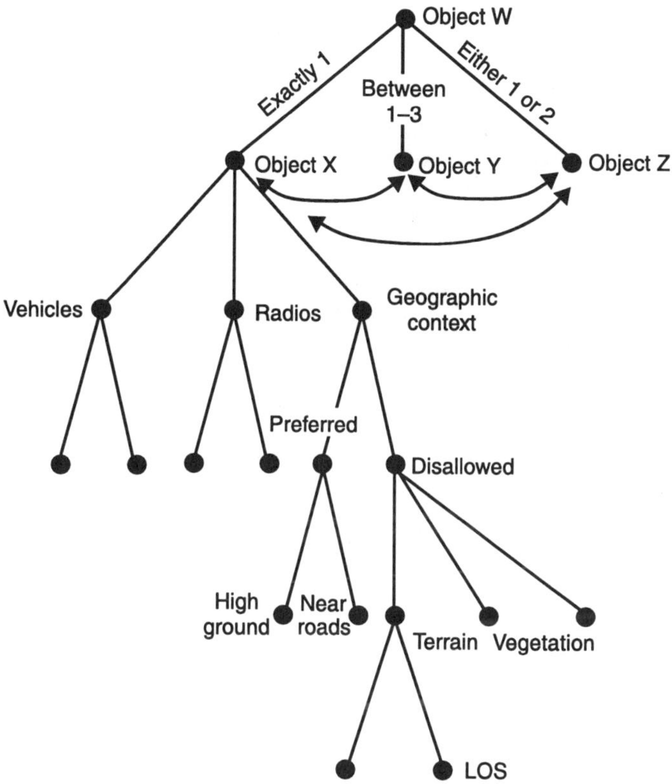

Figure 13.19 Graph representation of the constraints for the doctrinal templating example.

characteristics. Next, the disallowed conditions must be evaluated. In this case, it is determined that region *X* (1) does not intersect a water feature, (2) possesses an elevation less than 2,000m, and (3) is not in a populated area. Thus, the candidate elements of unit *X* satisfy all of the template constraints.

The association between unit *X* and units *Y* and *Z* can be investigated in much the same manner. In this case, it is determined that *X* and *Y* appear to be communicating over a type 1 radio at frequency f_2, region *X* and region *Y* (the region that contains all elements apparently associated with unit *Y*) are separated by approximately 2 km, no LOS or maneuver barrier exists between them, and one of the two candidate *Y* units is forward of *X*. Since the qualifier "typically" in the template permits the condition that "not all Y must be forward," once again, the subtemplate constraint is satisfied. Because the semantic/spatial object database supports top–down true two-dimensional spatial search, the required set operations can be computed using highly efficient spatial search.

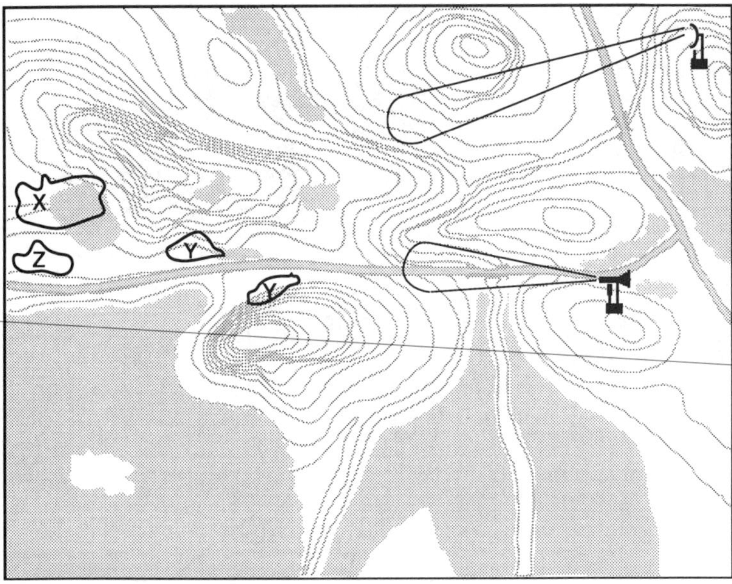

Figure 13.20 Underlying domain database for the doctrinal templating example.

Table 13.6
Constraints Relative to Object *X*

Radio type 1, radio type 2
Prefers high ground AND < 0.5 km to secondary roads AND light concealment
Not in water, low mobility resistance
Communicate to Y_i on radio type 1, frequency f_2
$1 < \lvert X - Y \rvert < 5$ km
X, Y LOS required
X, Y no maneuver barrier
Y typically forward of X
Communicates to Z on radio 2, frequency f_3
$2 < \lvert X - Z \rvert < 10$ km
X, Z LOS required
X, Z no maneuver barrier

A constraint-based canonical form XVI templating paradigm can be readily ex-
ended to handle both a larger number of constituent elements and arbitrarily com-
plex constraint sets. For instance, the typical separations between objects might be
est characterized by a nonuniform fuzzy-set membership function. Even though
wo objects that require continuous, direct communication do not possess radio
OS, communication over fiber optics (for separations that are less than 1 km) or
ndirect communication via a relay station could satisfy the communication require-
nents. In forested areas, due to both the concealment potential and the electro-
nagnetic attenuation of foliage, objects might tend to deploy closer together than
n a flat, relatively open setting. Object *X* might typically be on higher ground than
ither *Y* or *Z*. Since *Y* and *Z* don't communicate directly, their relationship can be
uite arbitrary.

Although the top-down, model-based doctrinal templating paradigm just de-
cribed represents a considerable improvement over rigid, single level-of-abstraction
emplating paradigms, as noted earlier in this chapter, all forms of expectation-
ased reasoning possess definite limitations. In recent years, the highly rigid Soviet-
tyle doctrine that had been relied on to support expectation-based reasoning
Indications & Warnings, threat assessment) is rapidly being replaced by the doc-
rine of much more loosely organized military forces. Thus, many analysis tech-
iques, both automated, as well as manual, that rely on rigid tactical and doctrinal
nowledge must be re-evaluated as the military embraces its new roles in low-inten-
ty conflicts and other nontraditional military operations. However, because the
eneralized constraint-based templating approach outlined above can handle a
ide range of constraints, the problem-solving methodology is not as dependent on
ighly formalized doctrine as are the more rigid templating paradigms.

References

] Ginsberg, M.L. (ed.), *Readings in Nonmonotonic Reasoning*, Palo Alto, CA: Morgan Kaufmann, 1987.

] Antony, R., "Target Classification for the Installation Security Radar System," U.S. Army Electronics Research and Development Command, Harry Diamond Laboratories, Technical Report No. HDL-TR-1976, November 1981.

] Antony, R., "Target Classification Using Spectral Features of Doppler Radar Return," *Proc. Twenty-fifth Annual Tri-Service Radar Symp.*, U.S. Air Force Academy, 1979.

] Coombs, M.J., and R.T. Hartley, "Design of a Software Environment for Tactical Situation Development," *U.S. Army Symp. on Artificial Intelligence Research for Exploitation of the Battlefield Environment*, El Paso, TX, November 15–16, 1988, pp. 1–15.

CHAPTER 14

Conclusions and Future Directions

14.1 TOWARD A UNIFIED INFORMATION-PROCESSING MODEL

In the beginning, there were no artificial data fusion algorithms; all information fusion was performed by humans. As sensors, such as radar and sonar, that produce vast quantities of measurements became available, processing techniques were developed to help "reduce the data" before it was analyzed by humans. Over time, progressively more powerful algorithms have evolvedto help automate many of the lower level data fusion processes. Humans, however, continue to perform most of the higher level fusion functions. Because the amount of sensor-derived information has tended to steadily *increase* over the years, while the required decision-making timelines has tended to *decrease,* many of the higher level functions demand increased automation. For many years, conventional wisdom has held that numerically based algorithms were fully capable of supporting level-1 processing, while more symbolic-based reasoning appeared to be required to automate level-2 and level-3 tasks.

The renaissance of artificial intelligence in the early 1980s offered the promise of sophisticated machine-based reasoning that would support the automation of much of the higher level situation analysis. Despite extensive R&D efforts, significant shortfalls remain in level-1 processing, while few operational level-2 and level-3 fusion applications exist. In general, at least four somewhat independent reasons for such shortfalls seem apparent:

1. The lack of adequate sensor-derived information;
2. Inadequate use of problem domain knowledge;
3. Fundamental limitations of traditional problem-solving paradigms;
4. The inherent difficulty of automated reasoning that relies on dynamic, incomplete, ambiguous, inaccurate, and uncertain remotely sensed information.

Improvements in sensor technology, more effective control of the collection process, and wider distribution of sensor-derived information seek to overcome the first limitation. The use of deep problem knowledge; non-sensor-derived domain knowledge; and relevant physical, logical, and logistical constraints can overcome the second shortfall. The development of more powerful and more efficient problem-solving approaches that synergistically combine all relevant information can overcome the third shortfall. Although overcoming the first three shortfalls can significantly enhance fusion products, the elimination of all uncertainty may not be feasible.

Traditional data fusion algorithms *combine* multiple-sensor information using a variety of technical approaches. In many of these approaches, sensor-derived data from the various sensors are effectively treated as elements of a composite feature vector. In a pattern recognition–based approach to decision making, for example, a linear discriminant function provides a direct mapping from an n-dimensional *feature space* to an m-dimensional *decision space.* Templating paradigms that map a set of ELINT parameters (frequency, pulse width) and a set of COMINT parameters (call sign, frequency) into a decision space is a common example. Most such algorithms carry the implicit assumption that the available features provide adequate discrimination among the set of possible decisions. For relatively static problems that do not exhibit either strong internal dependencies (which demand the use of general domain knowledge) or strong domain dependencies (which requires reasoning in context), the direct combination of sensor-derived features can be entirely adequate.

On the other hand, many data fusion problems violate such assumptions. The current reliance on narrow-discipline paradigms, unrealistic simplifying assumptions, and highly combinatoric algorithms reflects a bottom–up view of problem solving. As observed in numerous examples, humans successfully perform many level-1 fusion tasks that have not been successfully automated. Since humans excel at symbolic reasoning but possess rather limited computational capabilities (relative to a digital computer), it seems clear that many level-1 tasks may benefit from the use of symbolically oriented reasoning to complement traditional numerical-based approaches. Similarly, restricting level-2 and level-3 algorithms to symbolic-based reasoning may effectively handicap the higher level processes as well. As discussed in Part I, the guiding principle should be that individual problem requirements must dictate the selected problem-solving methodology.

This book has emphasized the potential benefits of intuitive model-based algorithms that synergistically combine available sensor-derived and non-sensor-derived information. When both possible and practical, algorithms should employ a global control element. Since the development of algorithms that can achieve more than simple "sum-of-the-parts" improvements requires a clear understanding of the strengths and weaknesses of existing paradigms, the problem-solving form taxonomy developed in Part I offers important insights into the similarities and differences of individual problem-solving paradigms. Although in a simple "side-by-side" comparison, two problem-solving approaches might appear to be entirely unrelated, striking similarities may become apparent when the paradigms are viewed from a more top-down perspective.

The primary elements of information analysis are problem domain knowledge and control knowledge; the nature of these two elements and their interaction differentiates the traditional problem-solving approaches. While conventional deterministic and statistical-based algorithms tend to tightly couple both problem domain knowledge and control knowledge, AI approaches tend to separate them. Neural networks are most similar to conventional approaches in that reasoning and control knowledge are tightly coupled. However, whereas conventional approaches use an explicit form of problem domain knowledge, neural nets employ an implicit form that is acquired during the weight adaptation phase of network training.

Long before concerns, such as:

1. "Does Dempster–Shafer or Bayes represent the better evidence accrual method?"
2. "What Expert System shell offers the best performance?"
3. "What variant of back-propagation learning requires the least training time?"

he fusion task requirements must be fully understood by both the algorithm designer and the system developer. Since no problem-solving paradigm can compensate for fundamental inadequacies in the information content of the data to be used, relevant sensor-derived information, non-sensor-derived domain knowledge, and expectation-based reasoning models must be identified and fully exploited. Once the information requirements of a fusion task have been established, a problem-solving methodology can be selected that is capable of effectively combining he various classes of information.

14.1.1 Heterogeneous and Homogeneous Problem-Solving Paradigms

As has been observed, robust, context-sensitive, and efficient algorithms can often be developed by mimicking the style of human information analysis and decision making. Although all conventional problem-solving paradigms (production rule–based paradigms, gradient descent–based methods, artificial neural systems, Bayesian inference methods) possess fundamental limitations, by effectively combining two or more paradigms the weakness of one approach may be at least partially offset by the strength of another approach. Because the canonical form taxonomy supports the identification of *similar* and *complementary* problem-solving approaches, it can provide insight into the development of both *heterogeneous* and *homogeneous* reasoning paradigms.

In the former approach, multiple, distinct problem-solving approaches are utilized. One subproblem might employ a finite state space, generate-and-test-based reasoning approach, while another subproblem might employ top-down, model-based reasoning that depends dynamically on the result derived for the first subproblem. A heterogeneous approach to target recognition could use various procedural algorithms to perform feature preprocessing and evaluate relevant context states, while a rule-based algorithm validates candidate target behavior models. A neural net signal classifier could be sensitive to the terrain, elevation, and vegetation along an intercept line-of-bearing that was provided by a procedural-oriented preprocessor.

Many reasoning paradigms (including blackboard-based and multiple hypothesis–based approaches) rely on some form of problem decomposition. In general, any logically decomposable problem is a candidate for a heterogeneous problem-solving paradigm. For example, a target-tracking system might employ both a statistical reasoning element (Kalman filter) as well as a context-sensitive rule-based element. The statistical-based component would represent a canonical form I problem-solving approach, while the context-sensitive road-following tracking component might represent a canonical form IX approach. Tracks would be updated based on which of the competing algorithms best explained the observations.

Rather than matching individual subproblems to specific problem-solving paradigms, in a *homogeneous* problem-solving paradigm, the tools, concepts, and methodologies from multiple disciplines would be integrated. For instance, in such a paradigm, a hierarchical set of procedural-based models that relied on object-oriented representations of physical and spatial objects could be employed. The local and global problem-solving context could be established using temporally sensitive semantic constraint sets that control the parametrization of the relevant procedural algorithms. The overall control might be provided by a rule-based process manager.

Similarly, a homogeneous problem-solving approach could consist of a set of functions implementing both numeric-based algorithms and symbolic-based reasoning that are controlled by a hierarchical set of production rules that are, in turn, conditioned on the outputs of a fuzzy-based domain-constraint analyzer. The domain constraint analysis could depend on the output of a bank of artificial neural systems that are dynamically monitoring critical states of the current situation description. Thus, in a heterogeneous approach, individual functional elements would employ a distinct problem-solving paradigm, while in a homogeneous approach, individual functions would be implemented within a composite problem-solving paradigm.

14.1.2 Production Rule Extensions

Extending an existing problem-solving paradigm offers a straightforward approach for developing more homogeneous paradigms. Consider, for example, the extension of the basic production rule paradigm.

Four Rule Classes

Although, in recent years, IF/THEN rules have become closely associated with non-procedural-based reasoning, the rule

IF (X) THEN (Z)

fundamentally represents a form of procedural knowledge. When the left-hand side condition set X is fully satisfied, a set of actions Z is taken (state-change, numeric computations, database search). Both X (*condition set*) and Z (*action set*) can represent arbitrarily complex numeric and/or symbolic expressions. Since both condition sets and action sets can be either *certain* or *uncertain*, there exist four general classes of IF/THEN rules as identified in Table 14.1.

Table 14.1
Four Classes of IF/THEN Rules

Class	*X*	*Z*
1	Certain	Certain
2	Uncertain	Certain
3	Certain	Uncertain
4	Uncertain	Uncertain

In class 1, both X and Z are *certain*. Examples of this class include:

IF ($X > 10$ and Transmitter = "ON") THEN Receiver = "ON"

and

IF ($X < 5.5$) THEN Go to 100.

Procedural programming applications rely on this form of the IF/THEN rule.

In a class-2 rule, the left-hand side conditions are *uncertain* (fuzzy, probabilistic, certainty factor), while the right-hand side conditions are *certain*. For example,

IF (X = "very large" AND H = "high") THEN (target is an aircraft)

and

IF ($Y \cap$ error ellipse E_1 with $P = 0.9$) THEN (assign Y to target 1).

In a class-3 rule, the condition clauses are *certain* but the action clauses are *uncertain*. For example,

IF [(Frequency = 108 MHz) AND (pulse repetition frequency = 50 kHZ)] THEN (either RADIO-1 with confidence 60% or RADIO-4 with confidence 40%).

In a class-4 rule, both the conditions and the conclusions are *uncertain*. For instance,

IF [(target velocity is "low") AND (target is a wheeled vehicle with $P > 0.85$)] THEN (Target is on a reconnaissance mission with confidence = 0.75).

The support for individual clauses can take on numerical values ranging:

1. From 0 to 1 for probabilistic relations;
2. From strong negative evidence (for example, −1.0) to strong positive evidence (for example, +1.0) for fuzzy relations;
3. Over a finite set of ad hoc semantic certainties (*highly likely, likely, somewha likely, not very likely, highly unlikely*).

In either Boolean or fuzzy rule–based system applications, the condition se generally represents a template or constraint and the action side generally perform some nonnumeric operation, such as changing the problem state in a finite-stat problem domain. In rule-based reasoning, the order of rule firing is relatively unim portant, whereas in more procedural-based applications, rule order tends to be cru cial. Although only a small number of IF/THEN rules may be required to contro the reasoning in procedural-based applications, such rules represent the bulk of th knowledge base in an Expert System or fuzzy rule–based system.

Decision Rule Capture

One or more of the four classes of IF/THEN rules can be utilized in a generalize rule-based system. For rule classes 2 to 4, a certainty calculus must be adopted tha permits the combination of uncertain information. Several popular certainty calcul were discussed briefly in Chapter 4. For the default fuzzy conjunction (the *minimu* operator) and the Bayesian product of probabilities (the joint probability is as sumed to be the product of the individual univariate probabilities), the feature pos sessing the lowest membership value (or probability) tends to dominate th decision. For example, in the case of the "product of the probabilities rule," if a features have probability 1.0 except a single feature that possesses probability 0.1 the probability of the decision is 0.1. Thus, the feature with the lowest value effe tively *captures* the decision rule. If the feature with the lowest probability (or mem bership value) is also the least significant feature in the decision process, high counterintuitive results are possible.

By selecting a decision model that supports the use of feature-dependent co ditional weights, decision rule capture by both low confidence and/or relatively u important features can be avoided. Perhaps the simplest such decision function the linear weighted-sum evidence aggregation model C_j defined as

$$C_j = \frac{\sum \alpha_{i,j} K_{i,j}(f_{i,j}(x))}{\sum (\alpha_{i,j})} \qquad 14.1$$

where C_j is the confidence in the jth decision state, $\alpha_{i,j}$ are weights that represent the relative importance of feature i to decision state j, $K_{i,j}$ are deterministic, probabilistic or fuzzy membership functions that represent conditional evidence or support for decision j due to feature i, $f_{i,j}$ is the value of the ith feature for decision state j, and x is the data.

Partial Condition Set Match

Permitting IF/THEN rules to fire on partial condition set match represents yet another possible generalization of the basic rule-based paradigm. In this case, by designating rule conditions as either *required* or *optional*, rule firing can occur when either (1) all clauses, (2) all required clauses, or (3) a normalized weighted sum of all satisfied clauses achieves some minimal level of support for a rule conclusion. Assuming a linear weighted sum evidence aggregation model, the confidence in decision states Z_1 and Z_2 can be expressed as

$$CZ_1 = \alpha_{11} C_1 + [\alpha_{13} C_2] + \alpha_{13} C_3 + [\alpha_{14} C_4] + \alpha_{15} C_5 \qquad 14.2$$

$$CZ_2 = \alpha_{21} C_1 = \alpha_{22} C_2 + \alpha_{23} C_3 + [\alpha_{24} C_4] + [\alpha_{25} C_5]$$

where the unbracketed terms represent required conditions and the bracketed terms represent optional conditions. With this generalization, rule firing depends on the available evidence.

Explicit Control

The advantages of top-down control of reasoning has been emphasized throughout this book. In order to provide a measure of top-down control and yet preserve the nonprocedural-based character of the production-rule paradigm, hierarchical knowledge (sometimes referred to as metaknowledge) is often included within the rule set ("apply a more general rule before a more specific rule"). However, a generalization of the rule-based paradigm that permits explicit top-down control of rule firing could support more sophisticated reasoning approaches.

Condition Set Evaluation

In the traditional production-rule paradigm, rule applicability is determined by directly comparing rule conditions (or conclusions, in the case of backward-chained reasoning approaches) to the current problem state. Many data fusion applications, however, require considerably more powerful rule-instantiation methods. For instance, consider the rule condition set

("the target is on a road AND the target is a member of a convoy").

In this case, constraint-set satisfaction requires context-sensitive analysis of the current situation description. Since the evaluation of context-sensitive conditionals within a strictly rule-based paradigm can be both awkward and inefficient, functions or procedures are required to evaluate conditions that are not directly testable in "working memory."[1]

The extension of the conventional production-rule paradigm to support non-Boolean condition and decision functions, more intuitive evidence aggregation methods, global control, partial condition set match, and a more procedural-oriented character would yield a paradigm that possessed most of the characteristics of the highest order canonical problem-solving forms. Classical production rules clearly represent a special case of this generalized rule-based reasoning paradigm.

14.1.3 Object-Oriented Reasoning

As discussed earlier, the object-oriented reasoning paradigm has become extremely popular in recent years. Because humans are accustomed to reasoning about physical objects, the notional extension to reasoning about abstract objects tends to be quite natural. Since individual objects can maintain arbitrarily complex procedural and/or nonprocedural-based knowledge, the paradigm inherently supports both procedural-oriented and symbolic-oriented reasoning. By encouraging problem decomposition and domain-specific property inheritance, object hierarchies facilitate multiple level-of-abstraction procedural-oriented reasoning as well as multiple level-of-abstraction nonprocedural-oriented reasoning. Thus, the object-oriented reasoning paradigm exhibits the key characteristics of the higher order canonical form problem-solving approaches. In addition, the paradigm embraces important software engineering concepts, such as functional modularity, encapsulation, reusability, and interface standards.

14.2 TECHNICAL CHALLENGES

Chapter 8 outlined a house-building metaphor to help illustrate the scope and the complexity of building large-scale, automated situation awareness systems. The construction of a high-quality house within both budget and time constraints requires a proper design, a solid foundation, adequate building materials, and the proper

1. As suggested in Chapter 9, a substantial portion of such functionality could be provided by a high-level database system shell.

ools. Rather than being required to make all the bricks needed, rip the framing umber, and forge their own nails, contemporary builders rely on well-proven, ighly efficient construction methods and the considerable infrastructure that was leveloped to supply both labor-saving tools and high-level building materials. Aodular construction techniques can significantly speed on-site construction time s well as potentially improve the quality of the finished product. Various levels of nspection ensure that proper standards of structural design, materials, and worknanship are maintained.

A similar approach is needed for building large-scale information-processing ystems. Much like a house, an information-processing system requires a proper deign, a solid foundation, adequate building blocks, and the proper tools. Regardless f how well managed, well staffed, or well funded, no single R&D program can dequately deal with the myriad of technical issues (spatial reasoning, temporal easoning, uncertainty management, parallel implementations, massive databases, nultilevel security, human computer interaction) required to develop sophistiated, real-world fusion applications from the bottom-up using low-level programning languages and relatively primitive database management systems.

Without an adequate infrastructure investment, the progress in any field tends o be limited. Unless telephone companies invest in the construction of switching enters and terrestrial and extraterrestrial communication networks, rapid, worldvide communications would not be possible. However, historically, individual reearch and development programs have often "taken on" the entire fusion problem. n many cases, such efforts have contributed little to the information-processing inrastructure. Rather than attempting to tackle the entire data fusion problem, more esearch and development effort must be directed at improving the information rocessing tools and the building blocks of complex information-processing ystems.

Although much remains to be accomplished, considerable progress has been nade in recent years. Many of the tools and building blocks required by data fusion pplications are the same as those required by other information analysis applicaons. Thus, the data fusion community is the beneficiary of hardware, software, and rchitecture developments that support a wide range of other applications as well. Iardware improvements include faster central processing units, increased parallelm, enhanced size and speed of random access memory, massive high-speed offoard memory, and new display technologies. Software improvements include dvances in software engineering and design, object-oriented environments, persistnt object databases, geographic information systems, enhanced user interfaces, nd visualization tools. Architecture improvements include improvements to blackoard-based reasoning models and other distributed problem-solving paradigms. As result, the outlook for the development of sophisticated data fusion applications as never been brighter.

In addition to general infrastructure support, the data fusion algorithm develoment process requires consideration of the following issues.

1. Given adequate experience and analysis time, can a human perform th task that is to be automated? If not, task automation is high risk and furthe analysis of the problem may be indicated.
2. Assuming that the task is relatively well understood, what sensor-derived in formation can be expected to be available? What domain constraints migh be relevant? What a priori reasoning knowledge is required to construct a algorithm that is both robust and context-sensitive? In other words, doe adequate information exist to achieve the problem-solving objectives?
3. What canonical problem-solving form best matches the inherent problen characteristics and requirements? What are the most appropriate problem solving techniques that support the selected canonical form? What are th algorithm efficiency considerations?
4. Can the data storage, access, and manipulation requirements of the task b satisfied?

Many of the algorithm and database issues that have been discussed in thi book revolve around issues of efficiency. One "school of thought" argues that futur hardware advances alone will overcome most of these efficiency concerns. Howeve another "school" takes the opposite position (for example, "hardware advances i the next decade will not make brute-force solutions economical because the scale c the prospective applications is simply too great" [1]).

Humans routinely solve intractable problems by applying various forms of ap proximations and simplifications. Demanding machine-generated optimal solution to many complex, real-world problems may be:

1. *Unrealistic* if there exists no known approach for generating provably opt mal solutions;
2. *Impractical* if the computational complexity of an optimal solution is to high.

For example, it may be unrealistic to demand that an "optimal" weather-forecastin algorithm be developed. Weather is a complex, large-scale, nonuniform, and highl dynamic phenomenon. Currently, weather forecasting employs a judicious comb nation of both art and science. By imposing certain finite-state space assumptior (only certain states are predicted, such as wind speed, precipitation, and temper ture) and closed-world assumptions (only a fixed set of relevant domain measu ables are used by the prediction model), the problem dimensionality is reduced to computationally manageable size. Even when provably "optimal" solutions to pro lems can be formulated, solutions that demand combinatorial optimization may b infeasible when the state space size is simply too large. Thus, true global optimali may need to be sacrificed in favor of tractability and performance efficiency.

Numerous suggestions have been offered regarding both the *paradigm selectic* and the *algorithm-development* process. The wide range of problem-solving paradign vailable to the algorithm developer can be thought of as small, relatively isolated islands in a vast sea of technology. For a given application, developers have traditionlly selected an island (*an individual technical paradigm*) and relied on the resources t provided. As demonstrated in Part I, since some islands possess inherently very diferent resources (*paradigms, techniques, tools*), the selection of the island with the "opimal" set of resources (*best problem-solving paradigm, most sophisticated techniques, most nature tools*) can involve complex tradeoffs.

On the other hand, an algorithm developer with a canoe and some basic naviational charts could potentially travel from one island to another to gather addiional resources (*multiple heterogeneous problem-solving paradigms*). However, since the anoe is small and the trip is dangerous, few (*narrow discipline specialists*) will be inlined to risk the journey. Cross-cultural fertilization and resource exchange would e greatly improved if solid bridges were built between the islands (*formal mappings etween all problem-solving paradigms*). However, not only is the construction of such ridges a formidable task, a large number of bridges would be required.

Rather than actually developing formal mappings between the numerous echnical paradigms (neural networks and statistical pattern recognition, expert sysems and decision theory), the development of a unified information-processing heory would permit individual problem-solving approaches to be treated as merely pecial forms of a generic algorithm. Although the data fusion model proposed in his book falls short of such a formal theory, the past 14 chapters have attempted to y some of the required groundwork. In addition, general guidelines have been ofered for mapping individual tasks into appropriate classes of algorithms as well as r evolving more capable problem-solving paradigms. Although the development f a "unified electromagnetic field theory" has been sought (without success) for enerations, the development of a more formal theory of data fusion appears to be realistic goal. Because of its holistic approach to data analysis, such a formalization ould provide direct benefit to the pursuit of general machine-based reasoning.

References

] Silberschatz, A., M. Stonebraker, and J. Ullman, "Database Systems: Achievements and Opportunities," *The "Lagunita" Report of the NSF Invitational Workshop of the Future of Database Systems Research,* Palo Alto, CA, February 22–23, 1990, Report No. TR-90-22, University of Texas at Austin, Austin, TX.

Afterword

On numerous occasions during the last decade, the question has arisen as to why I was investigating subjects such as spatial reasoning and database management systems when I was supposed to be doing research in tactical data fusion. Unfortunately, I was never able to come up with a concise, "hard-hitting" reply that seemed satisfactory. After receiving an invitation to become a research fellow at the George Mason University during the academic year 1991–1992, I decided that I might be able to resolve this dilemma by preparing a short monograph linking together several of my research papers. To the disappointment of my management, the scope of the effort broadened somewhat and the project took a little longer than I originally envisioned (two years longer, in fact!). Although I never did come up with the "short answer" I sought, I hope that this book has provided useful insights and perspectives that would not have been possible in a shorter format.

In the early stages of its preparation, a number of friends and associates referred to the manuscript variously as a *monologue,* a *monolith,* and a *monogram.* Although I was not always certain whether the slip was intentional or merely accidental, after a bit of introspection, I decided that none completely missed the mark. A monograph is an extensive treatise on a particular subject. Since a monologue is a prolonged discourse by a single speaker, a monograph could be considered a written form of a monologue. A large stone column that supports the roof of a building is an example of a monolith. This monograph attempts to build a foundation and associated framework that supports the construction of more sophisticated data fusion systems. Finally, a monogram is a single character consisting of one or more letters that are combined. Analogously, this monograph represents the combination (synergistic, of course!) of a number of separate research papers into a single monolithic monologue.

Because data fusion embraces the processing of numerous kinds of information using a wide variety of technical problem-solving approaches, it is not possible to provide a comprehensive treatment of the subject in a single volume. For example, consider the implications of attempting to include an adequate treatment of imagery. In recent years, there has been an explosive growth in the availability of high-resolution, multispectral imagery and powerful, relatively inexpensive analysis software. In addition to direct fusion of multiple registered images, the analysis (either manual or automated) of multidimensional imagery can generate valuable contextual knowledge that supports the analysis of other classes of sensor-derived information. With the increasing availability of real-time tactical imagery and video, these information sources have begun to play a role in both surveillance and targeting of moving objects. Entire books, in fact, have been dedicated to recent advances in image analysis and image understanding. Thus, rather than addressing the details associated with specific sensors and specific information combination approaches, this book has presented a top-down view of the fusion process that can complement a more detailed study of these two topics. In most cases, the general principles that have been identified do not rely on a detailed understanding of particular information sources or information processing technologies.

Bibliography

CHAPTER 1: INTRODUCTION AND PERSPECTIVES

Object Refinement—Level 1

Aggarwal, J.K. (ed.), *Multisensor Fusion for Computer Vision*, New York: Springer-Verlag, 1993.

Blackman, S.S., *Multiple Target Tracking with Radar Applications*, Norwood, MA: Artech House, 1986.

Hovanessian, S.A., *An Introduction to Sensor Systems*, Norwood, MA: Artech House, 1988.

Kadar, I. (ed.), *Proc. SPIE Symp. Signal Processing, Sensor Fusion, and Target Recognition* III, Orlando, FL, April 4–6, 1994.

Luo, R.C., and M.G. Kay, "Data Fusion and Sensor Integration: State-of-the-Art 1990's," *Data Fusion in Robotics and Machine Intelligence*, M.A. Abidi and Gonzalez (eds.), San Diego, CA: Academic Press, 1992.

McKendal, R., and M. Mintz, "Robust Fusion of Location Information," *Proc. 1988 IEEE Conf. Robotics and Automation*, Philadelphia, PA, April 24–29, 1988, pp. 1,239–1,244.

Schenker, P.S. (ed.), *SPIE Proc. Sensor Fusion: Spatial Reasoning and Scene Interpretation*, Vol. 1003, Orlando, FL, 1989.

Thomopoulos, S.C.A., "Sensor Integration and Data Fusion," *J. Robotic Systems*, Vol. 7, 1990, pp. 337–372.

Torrieri, D.J., *Principles of Secure Communication Systems*, 2d ed., Norwood, MA: Artech House, 1992.

Tou, J., and R. Gonzalez, *Pattern Recognition Principles*, Reading, MA: Addison-Wesley, 1974.

Waltz, E.L., "Data Fusion in C3I Systems," *Defense Electronics C^3I Handbook*, 1st ed., Palo Alto, CA: EW Communications, Inc., 1986, pp. 217–226.

Willner, D., C.B. Chang, and K.P. Dunn, "Kalman Filter Algorithms for a Multi-Sensor System," *Proc. IEEE Conf. Decision and Control*, December 1976, pp. 570–574.

Situation Assessment—Level 2

Ball, C.R., "AI Technique Synthesis for Situation Assessment," *Proc. 1989 Tri-Service Data Fusion Symp.*, Laurel, MD, May 16–18 1989, pp. 335–343.

Ben-Bassat, M., and A. Freedy, "Knowledge Requirements and Management in Expert Decision Support Systems for (Military) Situation Assessment," *IEEE Trans. on Systems, Man, and Cybernetics,* Vol. SMC-12, No. 4, July/August 1982, pp. 479–490.

Kadar, I., "Data Fusion by Perceptual Reasoning and Prediction," *Proc. 1987 Tri-Service Data Fusion Symp.,* Laurel, MD, June 1987, pp. 215–224.

Noble, D., "Schema-Based Knowledge Elicitation for Planning and Situation Assessment Aids," *Proc. IEEE Trans. on Systems, Man, and Cybernetics,* Vol. 19, 1989, pp. 473–482.

Threat Assessment—Level 3

Blackman, S., "Theoretical Approaches to Data Association and Fusion," *SPIE Proc. Sensor Fusion,* Vol. 931, 1988.

Chubb, D.W.J., "An Automated Tactical Situation Assessment Methodology: Its Limitations and Impact on Tactical Command and Control," *Proc. 1988 Tri-Service Data Fusion Symp.,* Laurel, MD, May 1988, pp. 257–267.

Garvey, T.D., "The Integration of Multi-Sensor Data for Threat Assessment," *Proc. Fifth Joint Conf. Pattern Recognition,* December 1980, pp. 343–347.

Process Refinement—Level 4

Addison, E.R., and B.D. Leon, "A Blackboard Architecture for Cooperating Expert System to Manage a Distributed Sensor Network," *Proc. 1987 Tri-Service Data Fusion Symp.,* Laurel, MD, 1987, pp. 669–675.

Denton et al., "Towards Modern Sensor Management Systems," *Proc. Sixth Joint Services Data Fusion Symp.,* Laurel, MD, June 14–18, 1993, pp. 659–678.

Hager, G., *Task Directed Sensor Fusion and Planning,* Boston, MA: Kluwer Academic, 1990.

Haskins, T.G., "Sensor Cueing Performance Analysis," *Proc. NAECON,* IEEE, 1984, pp. 262–265.

Steinberg, A.N., "Adaptive Data Acquisition and Fusion," *Proc. Sixth Joint Services Data Fusion Symposium,* Laurel, MD, June 14–18, 1993, pp. 699–710.

Waltz, E.L., and D.M. Beude, "Data Fusion and Decision Support for Command and Control," *Proc. IEEE Trans. on Systems, Man and Cybernetics,* Vol. SMC-16, November/December 1986, pp. 865–879.

Sensory Fusion

Albus, J., *Brains, Behavior and Robotics,* Peterborough, NH: Byte Publications (subsidiary of McGraw-Hill), 1981.

Clark, J.J., and A.L. Yuille, *Data Fusion for Sensory Information Processing Systems,* Boston, MA: Kluwer Academic Publishers, 1990.

Fischler, M.A., and O. Firschein, *Intelligence: The Eye, the Brain, and the Computer,* Reading, MA: Addison-Wesley, 1987.

Lettvin et al., "What the Frog's Eye Tells the Frog's Brain," *IRE Proc.,* Vol. 47, November 1959, pp. 1940–1951.

Spence, C.D., and J.C. Pearson, "Elimination of Ghosts in the Visual/Auditory Target Localization System of the Barn Owl," *Proc. 1989 Tri-Service Data Fusion Symposium,* Laurel, MD, 1989, pp. 153–158.

BB Systems

Craig, I.D., *The Cassandra Architecture: Distributed Control in a Blackboard System,* New York: Halsted Press, 1989.

Engelmore, R., and T. Morgan, (eds.), *Blackboard Systems,* Reading, MA: Addison-Wesley, 1988.

Jagannathan, V., R. Dodhiawala, and L.S. Baum, (eds.), *Blackboard Architectures and Applications*, Boston, MA: Academic Press, 1989.

Llinas, J., and R. Antony, "Blackboard Concepts for Data Fusion Applications," *Int. J. of Pattern Recognition and Machine Intelligence*, Vol. 7, No. 2, 1993, pp. 285–308.

Nii, H.P., "Blackboard Systems: Blackboard Application Systems, Blackboard Systems for a Knowledge Engineering Perspective (Part Two)," *AI Magazine*, August 1986, pp. 82–106.

Nii, H.P., "Blackboard Systems: The Blackboard Model of Problem-Solving and the Evolution of Blackboard Architectures (Part One)," *AI Magazine*, Summer 1986, pp. 38–53.

Speech Recognition/Understanding

Erman, L.D., F. Hayes-Roth, V. R. Lesser, and D. R. Reddy, "Hearsay-II Speech Understanding System: Integrating Knowledge to Resolve Uncertainty," *Blackboard Systems*, R. Engelmore and T. Morgan (eds.), Reading, MA: Addison-Wesley, 1988, pp. 25–86.

Furui, S., and M.M. Sondhi, (eds.), *Advances in Speech Signal Processing*, New York: Marcel Dekker, 1992.

Huang, X.D., Y. Ariki, and M.A. Jack, *Hidden Markov Models for Speech Recognition*, Edinburgh: Edinburgh Press, 1990.

Laface, P., and R. De Mori (eds.), *NATO Advanced Study Institute on Speech Recognition and Understanding: Recent Advances, Trends and Applications*, New York: Springer-Verlag, 1992.

Lee, K.-F., *Automatic Speech Recognition: The Development of the SPHINX System*, Boston, MA: Kluwer Academic, 1989.

Lesser, V.R., and L.D. Erman, "A Retrospective View of the Hearsay II Architecture," *Blackboard Systems*, R. Engelmore and T. Morgan (eds.), Reading, MA: Addison-Wesley, 1988, pp. 87–121.

Rabiner, L.R., *Fundamentals of Speech Recognition*, Englewood Cliffs, NJ: Prentice-Hall, 1993.

Rudnicky, A.I., A.G. Hauptmann, and K.-F. Lee, "Survey of Current Speech Technology," *Comm. ACM*, Vol. 37, No. 3, March 1994, pp. 52–57.

CHAPTER 2: DATA FUSION MODELS

Abidi, M.A., and R.C. Gonzalez (eds.), *Data Fusion in Robotics and Machine Intelligence*, San Diego, CA: Academic Press, 1992.

Antonisse, J.H., "Intelligence Fusion Modeling—A Proposed Approach," MITRE Corp., Bedford, MA, 1983.

Antony, R.T., "Eight Canonical Forms of Fusion: A Proposed Model of the Data Fusion Process," *Proc. 1991 Joint Services Data Fusion Symp.*, Vol. 3, Section 7, Laurel, MD, October 7–11, 1991.

Antony, R.T., "A Generalized Data Fusion Process Model and Database Representation Implications," *Proc. 1989 Tri-Service Data Fusion Symp.*, Laurel, MD, May 16–18, 1989, pp. 59–79.

Antony, R., "A Framework for Automated Tactical Data Fusion," *Proc. 1987 Tri-Service Data Fusion Symp.*, Laurel, MD, June 9–11, pp. 380–396.

Antony, R., "A Distributed Machine Intelligence Architecture for Battlefield Signal and Information Processing," Report No. HDL-TR-2067, The Harry Diamond Laboratories, Adelphi, MD, 1985.

Antony, R.T., "A Framework for Knowledge-Based Battlefield Signal and Information Fusion," DELHD-RT-RD-85-4, The Harry Diamond Laboratories, Adelphi, MD, 1985.

Dasarathy, B.V., *Decision Fusion*, Los Alamitos, CA: IEEE Computer Society Press, 1994.

Fine, M., K. Langdon, and D. White, "Tactical Intelligence Data Fusion," (Tech report IEWD-RT-93-0026), Intelligence and Electronic Warfare Directorate, Warrenton, VA, September 1993.

Georgeff, M.P. and A.L. Lansky, "Procedural Knowledge", *Proc. IEEE*, Special Issue on Knowledge Representation, October 1986, pp. 1,383–1,398.

Goodman, I.R., "A General Theory for the Fusion of Data," *Proc. 1987 Tri-Service Data Fusion Symp.*, Laurel, MD, June 9–11, 1987, pp. 252–270.

Hafner, A.N., "Identification of Decision Related Information Elements: A Taxonometric Model of Information Schemas," *Proc. 1987 Tri-Service Data Fusion Symp.*, Laurel, MD, June 9–11, 1987, pp. 556–564.

Hall, D.L., and R.J. Linn, "Algorithm Selection for Data Fusion Systems," *Proc. 1987 Tri-Service Service Data Fusion Symp.*, Laurel, MD, June 1987, pp. 100–109.

Hall, D.L., and R.J. Linn, "A Taxonomy of Algorithms for Multi-Sensor Data Fusion," *Proc. 1990 Joint Service Data Fusion Symp.*, Laurel, MD, May 1990, pp. 593–610.

Hall, D.L., and J. Llinas, "A Survey of C3MIS Data Fusion Techniques," *Proc. Second Internat. Conf. on Command, Control, Communications and Management Information Systems*, Bournemouth, UK, March 1988.

Hanson, A.R.E., E.M. Riseman, and T.D. Williams, "Sensor and Information Fusion From Knowledge-based Constraints," *Proc. 1988 Tri-Service Data Fusion Symp.*, Laurel, MD, May 17–19, 1988, pp. 113–122.

Kadar, I., "Adaptive Predetection and Off-Board Track and Decision Level Fusion for Enhanced Surveillance," *Proc. 1994 IEEE Internat. Conf. on Multisensor Fusion and System Integration for Intelligent Systems*, Las Vegas, NV, October 2–5, 1994.

Mahler, R.P.S., "A Unified Foundation for Data Fusion," *Proc. 1994 Joint Services Data Fusion Symp.*, Laurel, MD, October 1994, pp. 153–173.

Wolfberg, A., "A Model of the Data Fusion Process," *Proc. 1989 Tri-Service Service Data Fusion Symp.*, Laurel, MD, May 16–18, 1989, pp. 467–493.

CHAPTER 3: SELECTED APPLICATIONS

Target Tracking

Bar-Shalom, Y., "Extension of Probabilistic Data Association Filter to Multi Target Environment," *Proc. Fifth Symp. of Nonlinear Estimation*, University of California, September 1, 1974.

Bar-Shalom, Y., "Survey Paper—Tracking Methods in a Multi-target Environment," *IEEE Trans. on Automation and Control*, Vol. AC-23, August 1978, pp. 618–626.

Bar-Shalom, Y., and E. Tse, "Tracking in a Cluttered Environment with Probabilistic Data Association," *Automatica*, Vol. 11, September 1975, pp. 451–460.

Bar-Shalom, Y. (ed.), *Multitarget-Multisensor Tracking: Applications and Advances*, Norwood, MA: Artech House, 1989.

Bar-Shalom, Y., and T.S. Fortmann, *Tracking and Data Association*, New York: Academic Press, 1988.

Blackman, S., *Multiple Target Tracking with Radar Applications*, Norwood, MA: Artech House, 1986.

Maybeck, P.S., "Kalman Filtering: Concepts, Design and Extensions," *Proc. Fifty-first MORS*, Air Force Institute of Technology, 1985.

Path Development and Planning

Alexander, R.S., and N.C. Rowe, "Path Planning by Optimal-Path-Map Construction for Homogeneous-Cost Two-Dimensional Regions," *Proc. IEEE Int. Conf. on Robotics and Automation*, Cincinnati, Ohio, May 13–18, Los Alamitos, CA: IEEE Computer Press, 1990, pp. 1,924–1,929.

Allen, J., J. Hendler, and A. Tate (eds.), *Readings in Planning*, Palo Alto, CA: Morgan Kaufmann, 1990.

Allen et al. (eds.), *Reasoning About Plans*, Palo Alto, CA: Morgan Kaufmann, 1990.

Bajcsy, R., and S. Kovacic, "Multiresolution Elastic Matching," *Computer Vision, Graphics, Image Processing*, Vol. 46, 1989, pp.1–21.

Barrera, R., and J. Vazquez-Gomez, "A Shortest Path Algorithm for Hierarchical Terrain Models," *Proc. Auto-Carto 9—Ninth Int. Symp. on Computer-Assisted Cartography*, Baltimore, MD, April 2–7, 1989, American Society of Photogrammetry and Remote Sensing and Congress of Surveying and Mapping, 1989, pp. 156–163.

Benton, J., and A. Brink, "Hierarchical Route Planner," *U.S. Army Science Conf*, Durham, NC, June 1990, pp. 87–97.

Benton, J., S.S. Inyengar, W. Deng, N. Breuer, and V.S. Subrahmanian, "Fine-Grained Data: New Computational Challenges for Route Planning," *AI J.* (submitted for publication).

Chen, Y., and P. Ramanan, "Euclidean Shortest Path in the Presence of Obstacles," *Networks*, Vol. 21, No. 3, New York: Wiley, 1991, pp. 257–265.

Fujimura, K., and H. Samet, "A Hierarchical Strategy for Path Planning Among Moving Obstacles," University of MD Center for Automation Research, CAR-TR-237, November 1986.

Georgeff, M., and A. Lansky, "Reactive Reasoning and Planning," *Proc. AAAI*, 1987.

Holmes, P.D., and E.R.A. Jungert, "Symbolic and Geometric Connectivity Graph Methods for Route Planning in Digitized Maps," *IEEE Trans. on Pattern Analysis and Machine Intelligence*, Vol. 14, No. 5, May 1992, pp. 549–565.

Holmes, P.D., and E.R.A. Jungert, "A Hierarchical Knowledge Structure for Heuristic Path Planning," *Proc. Second Int. Conf. on Intelligent Autonomous Systems*, Amsterdam, December 1989, pp. 230–240.

Khampampati, S., and L.S. Davis, "Multiresolution Path Planning for Mobile Robots," *IEEE J. Robotic Automation*, Vol. RA-2, No. 3, 1986, pp. 135–145.

King, J. H., D.W. Payton, K.E. Olin, D.Y. Tseng, "A Context Dependent Automatic Target Recognizer System," J. Gilmore (ed.), *Applications of Artificial Intelligence*, Vol. 485, Arlington, VA, May 3–4, 1984.

Lee, D.T., and F.P. Preparata, "Euclidean Shortest Path in the Presence of Rectilinear Barriers," *Networks*, Vol. 14, No. 3, New York: Wiley, Fall 1984, pp. 393–410.

Libby, V. (ed.), *Proc. SPIE Conf. on Data Structures and Target Classification*, Orlando, FL, Vol. 1470, 1991.

Metea, M.B., and J.J.-P. Tsai, "Route Planning for Autonomous Land Vehicles Using Hierarchical Terrain Representation," *Proc. IEEE Conf. on Robotics and Automation*, Rayleigh, NC, March 31–April 3, 1987, Los Alamitos, CA: IEEE Computer Society Press, 1987, pp. 1,947–1,952.

Meystel, A., A. Guez, and G. Hillel, "Minimum Time Path Planning for a Robot," *Proc. 1986 Int. Conf. on Robotics Automation*, San Francisco, CA, April 7–10, 1986, Vol. 3, pp. 1,678–1,687.

Minton, S., *Machine Learning Methods for Planning*, San Mateo, CA: Morgan Kaufmann, 1993.

Mitchell, J., D. Payton, and D. Keirsey, "Planning and Reasoning for Autonomous Vehicle Control," *Int. J. Intelligent Systems*, Ronald Yager (ed.), Vol. 2, No. 2, 1987, pp. 129–198.

Reid, D.B., "An Algorithm for Tracking Multiple Targets," *IEEE Trans. on Automatic Control*, Vol. AC-24, No. 6, December 1979, pp. 843–854.

Rowe, N.C., and R.S. Alexander, "Constructing Optimal-Path Maps for Path Planning Across Weighted Regions," Department of Computer Science, U.S. Naval Postgraduate School, Monterey, CA, 1992.

Sacerdoti, E., "Planning in a Hierarchy of Abstraction Spaces," *Artificial Intelligence*, Vol. 5, 1974, pp. 115–135.

Sheu, P.C.-Y., and Q. Xue, *Intelligent Robotic Planning Systems*, River Edge, NJ: World Scientific, 1993.

Sorenson, H.W., "Kalman Filtering and Techniques," *Advances in Control Systems*, Vol. 3, New York: Academic Press, 1966.

Stefik, M.J., "Planning with Constraints," *Artificial Intelligence*, Vol. 16, 1981, pp. 111–140.

Trovato, K., "General Method for Strategic Assessment and Planning Problems," *1993 Joint Services Data Fusion Symp.*, Laurel, MD, June 14–18, 1993, pp. 305–333.

Wilkins, D.E., "Extending the Classicial AI Planning Paradigm," *Readings in Planning*, J. Allen, J. Hendler, and A. Tate (eds.), Palo Alto, CA: Morgan Kaufmann, 1990.

Object Detection In Imagery

Advanced Decision Systems, *Advanced Digital Radar Imagery Exploitation System (ADRIES) Program Plan*, October 1986.

Rosenfeld, A., *Picture Processing by Computer*, New York: Academic Press, 1969.

Strat, T.M., *Natural Object Recognition*, New York: Springer-Verlag, 1992.

Grimson, W.E.L., *Object Recogntion by Computer: The Role of Geometrical Constraints*, Cambridge, MA: MIT Press, 1990.

Templating

Andrews, H.C., *Introduction to Mathematical Techniques in Pattern Recognition*, New York: Wiley-Interscience, 1973.

ARPA, *Proc. Speech and Natural Language Workshop*, San Mateo, CA: Morgan Kaufmann, 1992.

Friedman, D., *Detection of Signals by Template Matching*, Baltimore, MD: Johns Hopkins Press, 1969.

Gabriel, J.R., and M.H. Gabriel, "Data Representation and Matching for Events and Templates," *Proc 1988 Tri-Service Data Fusion Symp.*, Laurel, MD, May 1988, pp. 327–331.

Griesel, M.A., J.W. Gillis, R.R. Radbill, and N.L. Sizemore, "Cross-Correlation: Statistics, Templating, and Doctrine," U.S. Army Intelligence Center and School (NTIS AD-A155-624), Februrary 29, 1984.

Gulatti, S., J. Barhen, N. Toomarian, and M. Zak, "Automated Nodal Analysis for CRT Discrimination and Validation," *1993 Joint Service Data Fusion Symp.*, Laurel, MD, June 14–18, 1993, pp. 877–899.

Hall, D.L. and R.J. Linn, "Comments on the Use of Templating Techniques for Multi-Sensor Data Fusion," *Proc. 1989 Tri-Service Data Fusion Symp.*, Laurel, MD, May 16–18, 1989, pp. 345–353.

Headquarters, Department of the Army, *Intelligence Preparation of the Battlefield*, FM 34-130, May 1989.

Noble, D.F., "Template-Based Data Fusion for Situation Assessment," *Proc. 1987 Tri-Service Data Fusion Symp.*, June 1987, pp. 152–161.

Noble, D., "A Model to Support Development of Situation Assessment Decision Aids," *Decision Making in Action: Models and Methods*, G. Klein (ed.), Norwood, NJ: Ablex, 1993, pp. 287–305.

Noble, D., "Schema-based Knowledge Elicitation for Planning and Situation Assessment Aids," *IEEE Trans. on Systems, Man, and Cybernetics*, Vol. 19, 1989, pp. 473–482.

Pachowicz, P.W., "Semi-Autonomous Evolution of Object Models for Adaptive Object Recognition," *IEEE Trans. of Systems, Man, and Cybernetics*, Vol. 24, No. 8, August 1994, pp. 1,191–1,207.

Widrow, B., "The Rubber Mask Technique, Part I," *Pattern Recognition*, Vol. 5, 1973, pp. 175–211.

CHAPTER 4: CONVENTIONAL PROBLEM-SOLVING PARADIGMS

Pattern Recognition

Duda, R.O., and P.E. Hart, *Pattern Classification and Scene Analysis*, New York: Academic Press, 1973.

Fu, K.S., *Applications of Pattern Recognition*, Boca Raton, FL: CRC Press, 1982.

Fu, K.S., *Syntactic Methods of Pattern Recognition*, New York: Academic Press, 1974.

Fukunaga, K., *Introduction to Statistical Pattern Recognition*, New York: Academic Press, 1972.

Landy, M.S., and J.A. Moushon (eds.), *Computational Models of Visual Processing*, Cambridge, MA: MIT Press, 1989.

Pao, Y.-H., *Adaptive Pattern Recognition and Neural Networks*, New York: Addison-Wesley, 1989.

SPIE Proceedings on Advances in Image Compression and Automatic Target Recognition, Vol. 1099, Orlando, FL, March 30–31, 1989.

Probabilistic Reasoning

Aldenderfer, M.S., and R.K. Blashfield, "Cluster Analysis," Quantitative Applications in the Social Sciences Series, Paper 07-044, London: Sage Publications, 1984.

Brammer, K., and G. Stiffling, *Kalman-Bucy Filters*, Norwood, MA: Artech House, 1989.

Berger, J.O., *Statistical Decision Theory and Bayesian Analysis*, New York: Springer, 1985.

Chen, G. (ed.), *Approximate Kalman Filtering*, River Edge, NJ: World Scientific, 1993.

Dempster, A.P., "A Generalization of Bayesian Inference," *J. Roy. Statist. Soc. Ser. B*, Vol. 30, 1968.

Grefenstette, J. (ed.), *Proceedings of an International Conference on Genetic Algorithms and their Applications*, Carnegie Mellon University, Pittsburgh, PA, 1985.

Jaynes, E.T., "Probability Theory as Logic," *Maximum Entropy and Bayesian Methods*, P. F. Fougere (ed.), Boston, MA: Kluwer Academic, 1990, pp. 1–16.

Papoulis, A., *Probability, Random Variables and Stochastic Processes*, 2nd ed., New York: McGraw-Hill, 1985.

Patrick, E.A., and J.M. Fattu, *Artificial Intelligence with Statistical Pattern Recognition*, Englewood Cliffs, NJ: Prentice-Hall, 1986.

Pearl, J., *Probabilistic Reasoning in Intelligent Systems: Networks of Plausible Inference*, Palo Alto, CA: Morgan Kaufmann Publishers, 1988.

van Trees, H.L., *Detection, Estimation, and Modulation Theory: Part* I, New York: Wiley, 1968.

Fuzzy-Based Reasoning

Bezdek, J.C., and S.K. Pal, *Fuzzy Models for Pattern Recognition: Methods that Search for Structures in Data*, New York: IEEE Press, 1992.

Cox, E., *The Fuzzy Systems Handbook: A Practitioner's Guide to Building, Using, and Maintaining Fuzzy Systems,* San Diego, CA: Academic Press, 1994.

Klir, G.J., and T.A. Floger, *Fuzzy Sets, Uncertainty, and Information,* Englewood Cliffs, NJ: Prentice-Hall, 1988.

Kosko, B., *Neural Networks and Fuzzy Systems,* Englewood Cliffs, NJ: Prentice-Hall, 1992.

Goodman, I.R., and V.M. Bier, "A Reexamination of the Relationship Between Fuzzy Set Theory and Probability Theory," *Twenty-ninth Annual Bayesian Research Conf.*, University of Southern California, Los Angeles, CA, 1991.

Munakata, T., and J. Yashvant, "Fuzzy Systems: An Overview," *Comm. ACM,* Vol. 37, No. 3, March 1994, pp. 69–76.

Schmucker, K.J., *Fuzzy Sets, Natural Language Computations, and Risk Analysis,* Rockville, MD: Computer Science Press, 1984.

Sudkamp, T., "On Probability-Possibility Transformations," *Fuzzy Sets and Systems,* Vol. 51, 1992, pp. 73–81.

Takagi, H., "Fusion Technology of Fuzzy Theory and Neural Networks—Survey and Future Directions," *Proc. Int. Conf. on Fuzzy Logic and Neural Networks,* Iizuka, Japan, July 1990, pp. 13–26.

Terano, T., K. Asai, and M. Sugeno, *Fuzzy Systems Theory and Its Applications,* San Diego, CA: Academic Press, 1992.

Zadeh, L.A., "Fuzzy Sets as a Basis for a Theory of Possibility," *Fuzzy Sets and Systems,* Vol. 1, 1978, pp. 3–28.

Zadeh, L.A., and J. Kacprzyk (eds.), *Fuzzy Logic for the Management of Uncertainty,* New York: Wiley, 1992.

Zimmerman, H.J., *Fuzzy Set Theory and its Applications,* second edition, Dordrecht: Kluwer Academic Publishers, 1990.

Artificial Intelligence

Buchanan, B.G., and E.H. Shortliffe, *Rule-Based Expert Systems,* Reading, MA: Addison-Wesley Publishing Company, 1984.

Davis, E., *Representations of Commonsense Knowledge,* Palo Alto, CA: Morgan Kaufmann, 1990.

Davis, R., "Meta-Level Knowledge: Overview and Applications," *Int. Joint Conf. on Artificial Intelligence,* Vol. IJCAI-5, 1977, pp. 920–927.

Davis, R., B. Buchanan, and E. Shortliffe, "Production Rules as a Representation for a Knowledge-Based Consultation Program," *Artificial Intelligence,* Vol. 8, Amsterdam: Elsevier/North-Holland, 1977, pp. 15–45.

de Kleer, J., "An Assumption-Based TMS," *Artificial Intelligence,* Vol. 28, 1986, pp. 127–162.

Fikes, R., and T. Kehler, "The Role of Frame-Based Representation in Reasoning," *Comm. ACM,* Vol. 28, No. 9 (Special issue on architectures for knowledge-based systems), September 1985, pp. 904–920.

Graubard, S.R. (ed.), *The Artificial Intelligence Debate: False Starts, Real Foundations,* Cambridge, MA: MIT Press, 1988.

Hanson, S.J., R.L. Remmele, and R.L. Rivest (eds.), *Machine Learning: From Theory to Applications: Cooperative Research at Siemens and MIT,* New York: Springer, 1993.

Hayes, J.E., and D. Michie (eds.), *Intelligent Systems: The Unprecedented Opportunity,* New York: Halsted Press: A Division of John Wiley and Sons, 1984.

Hayes-Roth, F., and N. Jacobstein, "The State of Knowledge-Based Systems," *Comm. ACM,* Vol. 37, No. 3, March 1994, pp. 27–38.

Iolland, J.H., "Escaping Brittleness: The Possibilities of General-Purpose Machine Learning Algorithms Applied in Parallel Rule-Based Systems," *Machine Learning: An Artificial Intelligence Approach*, R.S. Michalki, J.G. Carbonell, and T.M. Mitchell (eds.), Vol. II, Los Altos, CA: Kaufmann, 1986.

ling, R.E., "A Paradigm for Reasoning by Analogy," *Artificial Intelligence*, Vol. 2, 1971, pp. 147–178.

Iinsky, M., "A Framework for Representing Knowledge," *The Psychology of Computer Vision*, New York: AcGraw-Hill, 1975, pp. 211–277.

Iinton, S., *Machine Learning Methods for Planning*, San Mateo, CA: Morgan Kaufmann, 1993.

Iorgenstern, L., and L.A. Stein, "Why Things Go Wrong: A Formal Theory of Causal Reasoning," *Readngs in Planning*, J. Allen, J. Hendler, A. Tate (eds.), Palo Alto, CA: Morgan Kaufmann, 1990.

Nii, H.P., and E.A. Feigenbaum, "Rule-Based Understanding of Signals," *Pattern-Directed Inference Systems*,).A. Waterman and F. Hayes-Roth (eds.), New York: Academic Press, 1978, pp. 483–501.

'uppe, F., *Systematic Introduction to Expert Systems: Knowledge Representations and Problem-Solving Methods*, New York: Springer-Verlag, 1993.

Vaterman, D.A., "Rule-Based Expert Systems, Machine Intelligence," *Infotech State-of-the-Art Report*, Inotech, Maidenhead, England, 1981.

Vaterman, D.A., and F. Hayes-Roth (eds.), *Pattern Directed Inference Systems*, New York: Academic Press, 978.

Vaterman, D. A., and F. Hayes-Roth (eds.), "Semantic Network Representations in Rule-Based Inference ystems," *Workshop on Pattern-Directed Inference Systems*, New York: Academic Press, 1978, pp. 203–221.

Veld, D.S., and J. de Kleer (eds.), *Readings in Qualitative Reasoning About Physical Systems*, Palo Alto, CA: Iorgan Kaufmann, 1990.

Uncertainty Management / Evidence Accumulation

yyub, B.M., M.M. Gupta, and L.N. Kanal (eds.), *Analysis and Managment of Uncertainty: Theory and Appliations*, New York: North-Holland, 1992.

eude, D.M., J.W. Martin, and J.C. Sipos, "Comparison of Bayesian and Dempster-Shafer Fusion," *Proc. 989 Tri-Service Data Fusion Symp.*, Laurel, MD, May 16–18, 1989, pp. 81–91.

ordon, J., and E.H. Shortliffe, "A Method for Managing Evidential Reasoning in a Hierarchical Hyothesis Space," *Artificial Intelligence*, Vol. 26, 1985, pp. 323–357.

Ioltzman, S., *Intelligent Decision Systems*, Reading, MA: Addison-Wesley, 1989.

lir, G.J., and T.A. Folger, *Fuzzy Sets, Uncertainty, and Information*, Englewood Cliffs, NJ: Prentice-Hall, 988.

yburg, H.E., Jr., "Bayesian and Non-Bayesian Evidential Updating," *Artificial Intelligence*, Vol. 31, 1987, p. 271–293.

rade, H., "A Computational Approach to Approximate and Plausible Reasoning with Applications to xpert Systems," *IEEE Trans. on Pattern Analysis and Machine Intelligence*, Vol. PAMI-7, 1985, pp. 260–283.

auch, H.E., "Probability Concepts for an Expert System Used for Data Fusion," *The AI Magazine*, Fall 984.

hachter, R.D., "Intelligent Probabilistic Inference," *Uncertainty in Artificial Intelligence*, L.N. Kanal and F. Lemmer (eds.), Amsterdam: North-Holland, 1986.

hafer, G.A., *A Mathematical Theory of Evidence*, Princeton, NJ: Princeton University Press, 1976.

hafer, G., and J. Pearl (eds.), *Readings in Uncertain Reasoning*, Palo Alto, CA: Morgan Kaufmann, 1990.

'inston, P.H., "Learning and Reasoning by Analogy," *Comm. ACM*, Vol. 23, No. 12, 1980, pp. 689–03.

Zadeh, L.A., "A Simple View of the Dempster-Shafer Theory of Evidence and Its Implication for the Rul of Combination," *AI Magazine*, Vol. 7, No. 2, June 1, 1986, pp. 85–90.

Zadeh, L.A., "A Theory of Approximate Reasoning," *Machine Intelligence*, J. Hayes, D. Michie, and L Mikulich (eds.), Vol. 9, Chichester, England: Ellis Horwood, 1979, pp. 149–194.

Bayes Nets

Chang, K.C., and R.M. Fung, "Node Aggregation for Distributed Inference in Bayesian Network," *Proc Eleventh IJCAI*, Detroit, Michigan, August 1989.

Cooper, G.F., "The Computational Complexity of Probabilistic Inference Using Bayesian Belief Ne works," *Artificial Intelligence*, Vol. 42, 1990, pp. 393–405.

Howard, R.A., and J.E. Matheson, "Influence Diagrams," *The Principles and Applications of Decision Analysi* R.A. Howard and J.E. Matheson (eds.), Vol. II, Strategic Decision Group, Menlo Park, CA, 1981.

Oliver, R M., and R.Q. Smith (eds.), *Influence Diagrams, Belief Networks and Decision Analysis*, New York Wiley, 1990.

Pearl, J., *Probabilistic Inference in Intelligent Systems: Networks of Plausible Inference*, San Mateo, CA: Morga Kaufmann, 1988.

Pearl, J., "Fusion, Propagation and Structuring in Belief Networks," *Artifical Intelligence*, Vol. 29, 1986, pp 241–288.

Neural Networks

Buchanan, J., and J. Pollack, "Introduction," *Advances In Connectionism and Neural Computation Theory*, J Barnden and J. Pollack (eds.), Vol. I, Norwood, NJ: Ablex Press, 1989.

Baum, E.B., and D. Haussler, "What Size Net Gives Valid Generalization?," *Neural Computation*, Vol. 1 No. 1, Spring 1989, pp. 151–160.

Geman, S., E. Bienenstock, and R. Doursat, "Neural Networks and the Bias/Variance Dilemma," *Neur Comp.*, Vol. 4, 1992, pp. 1–58.

Hecht-Nielsen, R., "Theory of the Backpropagation Neural Network," *Proc. Int. Joint Conf. on Neural Ne works*, Vol. 1, IEEE Press, 1989, pp. 593–611.

Hertz, J., A. Krough, and R.G. Palmer, *Introduction to the Theory of Neural Computation*, Reading, MA: A dison-Wesley, 1991.

Lippman, R.P., "An Introduction to Computing with Neural Nets," *IEEE ASSP-M Magazine*, Vol. 4, 198 pp. 4–22.

Sethi, I.K., and A.K. Jain (eds.), *Artificial Neural Networks and Statistical Pattern Recognition: Old and Ne Connections*, Amsterdam: North Holland, 1991.

Weiss, S.M., and I. Kapouleas, "An Empirical Comparison of Pattern Recognition, Neural Networks an Machine Learning Classification Methods," *Proc. Eleventh Int. Joint Conf. on Artificial Intelligence*, Detroi 1989, pp. 781–787.

Optimization

Ackley, D.H., *A Connectionist Machine for Genetic Hillclimbing*, Boston, MA: Kluwer Academic Publisher 1987.

Chui, C.K. (ed.), *Wavelets: A Tutorial in Theory and Applications*, Boston, MA: Academic Press, 1992.

Collins, N.E., R.W. Eglese, and B.L. Golden, "Simulated Annealing—An Annotated Bibliography," *Am J. Math. Management Sci.*, Vol. 8, 1988, pp. 209–307.

Cook, S.A., "An Overview of Computational Complexity," *Comm. ACM*, Vol. 26, No. 6, June 1983, p 400–408.

)avis, L. (ed.), *Genetic Algorithms and Simulated Annealing*, Los Altos, CA: Morgan Kaufmann, 1987.

'razier, M.W., and J. Benedetto (eds.), *Wavelets: Mathematics and Applications*, Boca Raton, FL: CRC Press, 994.

;ill, P.E., W. Murray, and M.H. Wright, *Practical Optimization*, New York: Academic Press, 1981.

;arp, R.M., "Reducibility Among Combinatoric Problems," *Complexity of Computer Computations*, R.E. /Iiller and J.W. Thatcher (eds.), New York: Plenum Press, 1972, pp. 85–104.

'an Laarhoven, P.J.M., and E.H.L. Aarts, *Simulated Annealing: Theory and Applications*, Utrecht, The Nethrlands: D. Riedel Publishing, 1987.

Villiams, H.P., *Model Solving in Mathematical Programming*, New York: Wiley, 1993.

General

;*ommunications of the ACM*, Vol. 37, No. 3 (Special issue surveys the state of knowledge-based systems, ızzy systems, speech technology, neural networks), March 1994.

yer, N., "CATA Algorithm Analysis Correlation Algorithms and Techniques Analysis," PAR Report 85-7 PAR, Technology Corporation, New Hartford, NY, August 30, 1985.

Iarris, L., and M. Jenkin (eds.), *Spatial Vision in Humans and Robots*, Cambridge, MA: Cambridge University Press, 1993.

Iayhurst, G., *Mathematical Programming Applications*, New York: Macmillan, 1987.

.owalski, R., "Algorithm = Logic + Control," *Comm. ACM*, Vol. 22, No. 7, 1979, pp. 424–436.

)ssenbruggen, P.J., *Fundamental Principles of Systems Analysis and Decision-Making*, New York: Wiley, 1994.

edgewick, R., *Algorithms*, 2nd Ed., Reading, MA: Addison-Wesley, 1988.

terling, L.S. (ed.), *Intelligent Systems: Concepts and Applications*, New York: Plenum Press, 1993.

/atson, S.R., and D.M. Beude, *Decison Synthesis: The Principles and Practice of Decision Analysis*, Cambridge, ngland: Cambridge University Press,1987.

Applications

ox, G., and S. Arkin, "From Laboratory to the Field: Practical Considerations in Large Scale Data Fuon System Development," *Proc. 1988 Tri-Service Data Fusion Symp.*, Laurel, MD, May 1988, pp. 277–295.

chroeder, F., and L. Wright, "Automatic Correlation of Multiple Intelligence Sources," *Proc. 1988 Tri-ervice Data Fusion Symp.*, May 1988, pp.425–432.

/right, F., "Correlation and Target Identification Process in the Battlefield Exploitation and Target Acuisition Program," *Proc. 1980 Tri-Service Conf. on Non-Cooperative Target Recognition*, 1980.

CHAPTER 6: DATABASE SYSTEMS

RDBMS

ate, C.J., *An Introduction to Database Systems*, Vol. 1, 5th Ed., Reading, MA: Addison-Wesley, 1990.

ate, C.J., *Relational Database: Selected Writings*, Reading, MA: Addison-Wesley, 1986.

Imasri, R., and S.B. Navathe, *Fundamentals of Database Systems*, Redwood City, CA: Benjamin/Cummings,)89.

RDBMS Extensions

Aref, W.G., and H. Samet, "Extending a DBMS with Spatial Operations," *Second Symp. on Advances in Spa tial Databases*, SSD'91, O. Gunther and H.J. Schek (eds.), 1991; Lecture Notes in Computer Science, Vol 525, Berlin: Springer-Verlag, 1991, pp. 299–318.

Batory, D.S., J.R. Barnett, J.F. Garza, K.P. Smith, K. Tsukuda, B.C. Twichell, and T.E. Wise, "GENESIS An Extensible Database Management System," *IEEE Trans. on Software Engineering*, Vol. 8, 1988, pp 1,711–1,730.

Codd, E.F., "Extending the Relational Database Model to Capture More Meaning," *ACM Trans. on Data base Systems*, Vol. 4, No. 4, December 1979, pp. 397–434.

Codd, E.F., "Fatal Flaws in SQL, Vol. 1," *Datamation*, August 15, 1988, pp. 45–48.

Heiler, S., "Extended Data Type Support in Distributed DBMS Products: A Technology Assessment an Forecast," Report No. TR-0170-12-91-165, GTE Laboratories Incorporated, December 31, 1991.

Ingram, K.J., and W.W. Phillips, "Geographic Information Processing Using SQL-Based Query Lan guage," *Proc. Eighth Int. Symp. on Automation in Cartography*, American Congress on Surveying and Map ping, March 1987.

Jeffrey, K. (ed.) *Expert Database Systems*, K. Jeffery (ed.), San Diego, CA: Academic Press, 1992.

Lorie, R., and A. Meier, "Using a Relational DBMS for Geographical Databases," *Geo-Processing*, Vol. 2 1984, pp. 243–257.

Lyngbaek, P., and V. Vianu, "Mapping a Semantic Database Model to the Relational Model," *Proc. ACM SIGMOD 1987 Annual Conf.*, New York: ACM Press, 1987.

Stonebraker, M., and G. Kemnitz, "The POSTGRES Next-Generation Database Management System, *Comm. ACM*, Vol. 34, No. 10, October 1991, pp. 78–92.

Tang, Y.T., "A Management System for an Integrated Database of Pictures and Alphanumerical Data, *Computer Graphics and Information Processing*, Vol.16, 1981, pp. 270–286.

Ullman, J.D., *Principles of Database Systems*, 2nd Ed., Rockville, MD: Computer Science Press, 1982.

Geographic Information Systems

Aangeenburg, R.T., "A Critique of GIS," *Geographical Information Systems: Principles and Applications*, D. Maguire, M.F. Goodchild, and D. Rhind (eds.), Essex: Longman Scientific and Technical, and New York Wiley, 1991, pp. 101–107.

Abel, D.J., "SIRO-DBMS: A Database Tool-Kit for Geographical Information Systems," *Int. J. Geograph Information Systems*, Vol. 3, No. 2 (April–June), 1989, pp. 103–116.

Aref, W.G., and H. Samet, "Extending a DBMS with Spatial Operations," *Advances in Spatial Databases*, O Gunther and H.-J. Schek (eds.), 1991, pp. 299–318.

Cook, B.G., "The Structural and Algorithmic Basis of a Geographic Data Base," *Proc. First Int. Advance Study on Topological Data Structures for Geographic Information Systems*, Dutton, G. (ed.), Harvard Papers o Geographic Information Systems, 1978.

Egenhofer, M.J., "Extending SQL for Graphical Display," *Cartography and Geographic Information System* Vol. 18. No. 4, 1991, pp. 230–245.

Frank, A.U. and I. Capari (eds.), "Spatial Information Theory: A Theoretical Basis for GIS," Europea Conference (COSIT '93), Elba Island, Italy, September 19–22, 1993, *Lecture Notes in Computer Science* 71 New York: Springer-Verlag, 1993.

Gambosi, G., M. Scholl, and H.-W. Six (eds.), *Geographic Database Management Systems*, New York: Sprin ger-Verlag, 1992, pp. 103–137.

Goodchild, M.F., "Towards an Enumeration and Classification of GIS Function," *Int. Geographic Information Systems (IGIS) Symp.: The Research Agenda,* R.T. Aangeenburg and Y.M. Schiffman (eds.), Vol. 2, Arlington, VA, November 16–18, 1988, pp. 67–77.

Goodchild, M.F., "A Spatial Analytical Perspective on Geographic Information Systems" *Int. J. Geogrpahic Information Systems,* Vol. 1, No. 4, October–December 1987, pp. 327–334.

Goodchild, M.F., and S. Gopal, *Accuracy of Spatial Databases,* New York: Taylor and Francis, 1989.

Guting, R.H., "GRAL: An Extensible Relational System for Geometric Applications," *Proc. Fifteenth Int. Conf. on Very Large Databases (VLDB),* P.M.G. Apers and G. Wiederhold (eds.), Amersterdam, August 1989, pp. 33–44.

Herring, J.R., "TIGRIS: Topologically Integrated Geographic Information System," *Proc. AUTOCARTO 8,* ASPRS/ACSM, Falls Church, VA, 1987, pp. 282–291.

Ingram, I.K., and W. Phillips, "Geographic Information Processing Using a SQL-Based Query Language," *Proc. AUTOCARTO 8,* ASPRS/ACSM, Falls Church, VA, 1987, pp. 326–335.

Jungert, E., "Extended Symbolic Projections Used as a Knowledge Structure for Spatial Reasoning," *Pattern Recognition 1989,* Heidelberg: Springer-Verlag, 1988, pp. 343–351.

Kasturi, R., R. Fernandez, M. L. Amlani, and W.-C. Feng, "Map Data Processing in Geographic Information Systems," *IEEE Computers,* December 1989, pp. 10–21.

Laurini, R., and D. Thompson, *Fundamentals of Spatial Information Systems,* New York: Academic Press, 1992.

Maguire, D.J., M.F. Goodchild, and D.W. Rhind, *Geographical Information Systems: Principles and Applications,* Vol. 1, New York: John Wiley and Sons, 1991.

Morehouse, S., "The Architecture of ARC/INFO," *Proc. AUTOCARTO 9,* ASPRS/ACSM, Falls Church, VA, 1989, pp. 266–277.

Ooi, B.C., *Efficient Query Processing for Geographic Information Systems,* Ph.D. thesis, Monash University, Victoria, Australia, 1988.

Orenstein, J.A., and F.A. Manola, "PROBE Spatial Data Modeling and Query Processing in an Image Database Application," *IEEE Trans. Software Engineering,* Vol. 14, No. 5, May 1988, pp. 611–629.

Peuquet, D.J., and D.F. Marble, *Introductory Readings in Geographic Information Systems,* London: Taylor and Francis, 1990.

Sacks-Davis, R., K.J. McDonell, and B.C. Ooi, "GEOQL—A Query Language for Geographic Information Systems," Technical Report 87/2, Melbourne, Victoria, Australia: Monash University, July 1987.

Samet, H., A. Rosenfeld, C.A. Shaffer, R.C. Nelson, Y.-G. Huang, and K. Fujimura, "Application of Hierarchical Data Structures to Geographical Information Systems: Phase IV," Technical Report CS-1578, University of Maryland, College Park, MD, December 1985.

Schek, H., and M. Scholl, "The Two Roles of Nested Relations in DASDBS Project," *Nested Relations and Complex Objects in Databases,* S. Abiteboul, P.C. Fischer, and H.-J. Sheck (eds.), Berlin: Springer-Verlag, 1989, pp. 50–68.

Shafer, C.A., H. Samet, and R.C. Nelson, "QUILT: A Geographic Information System Based on Quadtrees," *Int. J. Geographical Information Systems,* Vol. 4, No. 2, April–June 1990 , pp. 103–131.

Shapiro, L.G., and R.M. Haralick, "A Spatial Data Structure," *Geo-Processing,* Vol. 1, 1981, pp. 313—317.

Shapiro, L.G., and R.M. Haralick, *Computer and Robot Vision,* Vol. 1 and 2, Reading, MA: Addision-Wesley, 1993.

Star, J., and J.E. Estes, *Geographic Information Systems: An Introduction,* Englewood Cliffs, NJ: Prentice-Hall, 1990.

Understanding GIS: The ARC/INFO Method, New York: John Wiley and Sons, 1993.

USA-CERL, *Geographic Resources Analysis Support System (GRASS),* Version 3.0 User Manual, Champaign, IL: U.S Army Construction Engineering Research Laboratory, 1988.

van Oosterom, P., *Reactive Data Structures for Geographic Information Systems,* New York: Oxford University Press, 1993.

von Peter Mesenburg, H. (ed.), *Proceedings of the 16th International Cartographic Conference,* Cologne, Germany, May 3–9, 1993.

Vijlbrief, T., and P. van Oosterom, "The GEO++ system: an extensible GIS," *Proc. Fourth Int. Symp. on Spatial Data Handling,* Volume 1, Charleston, SC, August 1992, pp.40–50.

Waugh, T.C., and R.G. Healy, "The GEOVIEW Design: A Relational Data Base Approach to Geographical Data Handling," *Int. J. Geographic Information Systems,* Vol. 1, No. 2 , April–June 1987, pp. 101–118.

Object-Oriented Databases

Beech, D., and C. Ozbutun, "Object Databases as Generalizations of Relational Databases," *Proc. Object-Oriented Database Task Group Workshop,* October 23, 1990, Chateau Laurier Hotel, Ottowa, Canada, NISTIR 4488, National Institute of Standards and Technology, January 1991.

Bertino, E., and L. Martino, *Object-Oriented Database Systems: Concepts and Architectures,* Reading, MA: Addison-Wesley, 1993.

Cattell, R.G.G., *Object Data Management,* Reading MA: Addison-Wesley, 1991.

Chorafas, D., and H. Steinmann, *Object-Oriented Databases,* Englewood Cliffs, NJ: Prentice-Hall, 1993.

Dittrich, K.R., U. Dayal, and A.P. Buchmann (eds.), *On Object-Oriented Database Systems,* New York: Springer-Verlag, 1991.

Goldberg, A., and D. Robson, *Smalltalk-80: The Language and its Implementation,* Reading MA: Addison-Wesley, 1983.

Kim, W., *Introduction to Object-Oriented Databases,* Cambridge, MA: MIT Press, 1990.

Maier, D., "Development of an Object-Oriented DBMS," Technical Report No. CS/E-86-005, Oregon Graduate Center, Beaverton, OR, April 1986.

Peterson, R., "Object-Oriented Database Design," *AI Expert,* Vol. 2, No. 3, March 1987, pp. 76–31.

Penny, D., and J. Stein, "Class Modification in the GemStone Object-Oriented DBMS, *J. SIGPLAN,* Notices 22, 12, December, 1987, p. 111–117.

Unland, R., and G. Schlageter, "Object-Oriented Database Systems: State-of-the-Art and Research Problems," *Expert Database Systems,* K. Jeffery (ed.), San Diego, CA: Academic Press, 1992.

Won, K., *Introduction to Object-Oriented Databases,* Cambridge, MA: MIT Press, 1990.

Zdonick, S.B., and D. Maier, *Readings in Object-Oriented Databases,* San Mateo, CA: Morgan-Kaufmann, 1989.

CHAPTER 7: DATABASE DESIGN

Aho, A.V., J.E. Hopcroft, and J.D. Ullman, *Data Structures and Algorithms,* Reading, MA: Addison-Wesley, 1983.

Antony, R.T., "Representation Issues in the Design of a Spatial Database Management System," *Proc. U.S. Army Symp. on Artificial Intelligence Research for Exploitation of the Battlefield Environment*, El Paso, TX, November 15–16, 1988, pp. 79–100.

Aref, W.G., and H. Samet, "Optimization Strategies for Spatial Query Processing," *Proc. Seventeen Int. Conf. on Very Large Databases*, Barcelona, Spain, September 1991, pp. 81–90.

Brodie, M.L., J. Mylopoulos, and J.W. Schmidt (eds.), *On Conceptual Modelling: Perspectives from Artificial Intelligence, Databases, and Programming Languages*, New York: Springer-Verlag, 1984.

Brodie, M.L., and J. Mylopoulous, *On Knowledge Base Management Systems: Integrating Artificial Intelligence and Database Technologies*, New York: Springer-Verlag, 1986.

Chazelle, B., "A Functional Approach to Data Structures and Its Use in Multidimensional Searching," *SIAM J. Comput.*, Vol. 17, No. 3, June 1988, pp. 427–462.

Egenhofer, M.J., "Interaction with Geographic Information Systems via Spatial Queries," *J. Visual Languages Comput.*, Vol. 1, No. 4, December 1990, pp. 389–413.

Frakes, W.B., and R. Baeza-Yates (eds.), *Information Retrieval, Data Structures and Alkgorithms*, Englewood Cliffs, NJ: Prentice-Hall, 1992.

Frank, A.U., "Properties of Geographic Data: Requirements for Spatial Access Methods," *Advances in Spatial Databases—Second Symp.*, SSD'91, O. Gunther and H.J. Schek (eds.), 1991; Lecture Notes in Computer Science, Vol. 525, Berlin: Springer-Verlag, 1991, pp. 225–234.

Fry, J.P., "The Logical Record Access Approach to Database Design," *ACM Comput. Surveys*, Vol. 12, No. 2, June 1980, pp. 179–211.

Gotlieb, C.C., and L.R. Gotlieb, *Data Types and Data Structures*, Englewood Cliffs, NJ: Prentice-Hall, 1978.

Inmom, W.H., *Effective Data Base Design*, Englewood Cliffs, NJ: Prentice Hall, 1981.

Jarke, M., and J. Koch, "Query Optimization in Database Systems," *ACM Comput. Surveys*, Vol. 16, No. 2, June 1984, pp. 111–152.

King, T., *Dynamic Data Structure: Theory and Application*, San Diego, CA: Academic Press, 1992.

Kuipers, B., "Modeling Spatial Knowledge," *Cognitive Sci.*, Vol. 2, No. 2, 1978, pp. 129–153.

Laurini, R., and D. Thompson, *Fundamentals of Spatial Information Systems*, New York: Academic Press, 1992.

Litwin, W., and H.J. Schek (eds.), *Foundations of Data Organization and Algorithms*, Lecture Notes in Computer Science, Vol. 367, Berlin: Springer-Verlag, 1989.

Manola, F., and A. Pirotte, "An Approach to Multi-Model Database Systems," *Proc. Second Int. Conf. on Databases* (ICOD-2), S.M. Deen and P. Hammersley (eds.), New York: Wiley-Heyden Ltd., September 1983, pp. 53–75.

Peuquet, D.J., "Toward the Definition and Use of Complex Spatial Relations," *Proc. Third Int. Symp. on Spatial Data Handling*, Sydney, Australia, August 17–19, 1988, pp. 79–96.

Schiewietz, M., and H.P. Kriegel, "Query Processing of Spatial Objects: Complexity verses Redundancy," *Proc. Third Int. Symp. on Large Spatial Databases*, SSD'93, Singapore, June 1993.

Smith, H.F., *Data Structures: Form and Function*, New York: Harcount Brace Jovanovich Publishers, 1987.

Smith, J.M., and D.C.P. Smith, "Database Abstractions: Aggregation and Generalization," *ACM Trans. on Database Systems*, Vol. 2, June 1977, pp.105–133.

Standish, T.A., *Data Structure Techniques*, Reading, MA: Addison-Wesley, 1980.

Stonebraker, M., "Implementation of Integrity Constraints and Views on Query Modification," *Proc. 1975 ACM SIGMOD Int. Conf. on Management of Data*, 1975, pp. 65–78.

Tsichritzis, D.C., and F.H. Lochovsky, *Data Models*, Englewood Cliffs, NJ: Prentice-Hall, 1982.

Wiederhold, G., *Database Design*, New York: McGraw-Hill, 1977.

Wong, E., and K. Youssefi, "Decomposition—A Strategy for Query Processing," *ACM Trans. on Database Systems*, Vol. 1, No. 3, September 1976, pp. 223–241.

Yso, S.B. (ed.), *Principles of Database Design, Volume 1: Logical Organizations*, New York: Prentice-Hall, 1985.

CHAPTER 8: SEMANTIC/SPATIAL DATA STRUCTURES

Antony, R., "A Hybrid Spatial/Object-Oriented DBMS to Support Automated Spatial, Hierarchical and Temporal Reasoning," Chapter 3, *Advances in Spatial Reasoning*, S-S. Chen (ed.), Vol. 1, Norwood, NJ: Ablex, 1991, pp. 63–132.

Ballard, D., "Strip Trees: A Hierarchical Representation for Curves," *Comm. ACM*, Vol. 4, No. 5, May 1981, pp. 310–321.

Bauer, M.A., "Set Operations in Linear Quadtrees," *Computer Vision, Graphics, and Image Processing*, Vol 29, No. 2, February 1985, pp. 248–258.

Bentley, J.L., "Multidimensional Binary Search Trees Used for Associative Searching," *Comm. ACM*, Vol. 18, No. 9, September 1975, pp. 509–517.

Bentley, J.L., "Multi-Dimensional Divide-and-Conquer," *Comm. ACM*, Vol. 23, No. 4, April 1980, pp 214–229.

Bentley, J.L., and J.H. Friedman, "Data Structures for Range Searching," *Computing Surveys*, Vol. 11, No 4, December 1979, pp. 397–409.

Brunet, P., and D. Ayala, "Extended Octtree Representation of Free Form Surfaces," *Computer-Aided Geo metric Design*, Vol. 4, July 1987, pp. 141–154.

Buchmann et al. (eds.), "Design and Implementation of Large Spatial Databases," Lecture Notes in Computer Science, Vol. 409, New York: Springer-Verlag, 1990.

Burton, F.W., and J.G. Kollias, "Expected and Worst Case Storage Requirements for Quadtrees," *Pattern Recognition Lett.*, Vol. 3, No. 2, March 1985, pp. 131–135.

Chen, Z.T., and W.R. Tobler, "Quadtree Representations of Digital Terrain," *Proc. Auto-Carto London* September 1986, pp. 475–484.

Davis, E., *Representing and Acquiring Geographic Knowledge*, London: Pitman, 1986.

Dyer, C.R., "The Space Efficiency of Quadtrees," *Computer Graphics and Image Processing*, Vol. 19, No. 4 August 1982, pp. 335–348.

Elmes, G.A., "Data Structures of Quadtree-Addressed Entities on a Spatial Relational Database," *Proc. Int Symp. on Geographic Information Systems*, Vol. 2, Arlingrton, VA, November 1987, pp. 177–179.

Finkel, R.A., and J.L. Bentley, "Quad Trees—A Data Structure for Retrieval on Composite Keys," *Acta In formatica*, Vol. 4, 1974, pp. 1–9.

Fujimura, K., and T.L. Kunii, "A Hierarchical Space Indexing Method," *Proc. Computer Graphics '85*, To kyo, 1985, pp. 1–14.

Gadia, S.K., "Parametric Databases: Seamless Integration of Spatial, Temporal, Belief, and Ordinar Data," *SIGMOD Record*, Vol. 22, No. 1, March 1993, pp. 15–20.

Gao, P., and T. Smith, "Space Efficient Hierarchical Structures: Relatively Addressed Compact Quadtrees," *Proc. Int. Symp. on Geographic Information Systems*, Vol. 2, Arlington, VA, November 1987, pp. 405–414.

Garey, M.R., and D.S. Johnson, *Computers and Intractibility, A Guide to the Theory of NP-Completeness*, San Francisco, CA: W. H. Freeman and Company, 1979.

Gargantini, I., "An Effective Way to Represent Quadtrees," *Comm. ACM*, Vol. 25, No. 12, December 1982, pp. 905–910.

Gunther, O., "Efficient Structures for Geometric Data Management," Lecture Notes in Computer Science, Vol. 337, New York: Springer-Verlag, 1988.

Gunther, O., "The Design of the Cell Tree: An Object-Oriented Index Structure for Geometric Databases," *Proc. Fifth IEEE Int. Conf. on Data Engineering*, Los Angeles, CA, February 1989, pp. 598–605.

Guttman, A., "R-Trees: A Dynamic Index Structure for Spatial Searching," *Proc. ACM SIGMOD*, Vol. 13, Boston, MA, June 1984, pp. 47–57.

Hinrichs, K., and J. Nievergelt, "The Grid File: A Data Structure Designed to Support Proximity Queries on Spatial Objects," *Proc. Int. Workshop on Graphtheoretic Concepts in Computer Science*, M. Nagl, and J. Perl (eds.), Trauner Linz, Austria: Verlag, 1983, pp. 100–113.

Kleiner, A., and K.E. Brassel, "Hierarchical Grid Structures for Static Geographic Data Bases," *Proc. Auto-Carto London*, Vol. 1, September 1986, pp. 485–496.

Kriegel, H.P., P.H. Heep, S. Heep, M. Schiwietz, and R. Schneider, "An Access Method Based Query Processor for Spatial Database Systems," *Workshop Proc. Geographic Database Management Systems*, G. Gambosi, M. Scholl, and H.-W. Six (eds.), Capri, Italy, May 1991, Berlin: Springer-Verlag, pp. 194–211.

Mark, D.M., and J.P. Lauzon, "The Space Efficiency of Quadtrees: An Empirical Examination Including the Effects of Two-Dimensional Run-Encoding," *Geo-Processing*, Vol. 2, 1985, pp. 367–383.

Meer, P., C.A. Sher, and A. Rosenfeld, "The Chain Pyramid: Hierarchical Contour Processing," *IEEE Trans. on Pattern Analysis and Machine Intelligence*, Vol. 12, 1990, pp. 363–375.

Mei, G.G., and W. Liu, "Parallel Processing for Quadtree Problems," *Proc. IEEE Int. Conf. on Parallel Processing*, St. Charles, IL, August 1986, pp. 452–454.

Morganstern, M., "The Role of Constraints in Databases, Expert Systems, and Knowledge Representation," *Proc. First Int. Workshop on Expert Database Systems*, L. Kerschberg (ed.), Menlo Park, CA: Benjamin/Cummings Publishing, 1986, pp. 351–368.

Orenstein, J.A., "Spatial Query Processing in an Object-Oriented Database System," *Proc. SIGMOD Conference*, Washington, DC, May 1986, pp. 326–336.

Overmars, M.H., "The Design of Dynamic Data Structures," Lecture Notes in Computer Science, Vol. 156, New York: Springer-Verlag, 1983.

Peuquet, D.J., "A Hybrid Data Structure for Storage and Manipulation of Very Large Spatial Data Sets," *Computer Vision, Graphics, and Image Processing*, Vol. 24, No. 1, October 1983, pp. 14–27.

Robinson, J.T., "The K-D-B Tree: A Search Structure for Large Multidimensional Dynamic Indexes," *Proc. ACM SIGMOD Conf.*, Ann Arbor, MI, April 1991, pp. 10–18.

Samet, H., "A Top-Down Qadtree Traversal Algorithm," *IEEE Trans. Pattern Anal. and Machine Intelligence*, Vol. 7, No. 1, January 1985, pp. 94–98.

Samet, H., "The Quadtree and Related Hierarchical Data Structures," *Computing Surveys*, Vol. 16, No. 2, 1984, pp. 187–260.

Samet, H., "Region Representation: Quadtree to Raster Conversion," Report No. TR-768, University of Maryland Computer Science Department, Center for Automation Research, College Park, MD, June 1979.

Samet, H., and R.E. Webber, "A Comparison of Space Requirements of Multi-Dimensional Quadtree-Based File Structures," Report No. TR-1711, University of Maryland Computer Science Department, College Park, MD, September 1986.

Samet, H., and R.E. Webber, "On Encoding Boundaries with Quadtrees," *IEEE Trans. on Pattern Anal. and Machine Intelligence,* Vol. 6, No. 3, 1984.

Rishe, N., S. Navathe, and D. Tal (eds.), *Databases: Theory, Design, and Applications, PARBASE-90, Int. Conf. on Databases, Parallel Architectures and their Applications,* Miami Beach, FL, Los Alamitos, CA: IEEE Computer Society Press, 1990.

Shaffer, C.A., and H. Samet, "An In-Core Hierarchical Data Structure Organization for a Geographic Database," Report No. CS-TR-1886, University of Maryland, Center for Automation Research, College Park, MD, July 1987.

Tanimoto, S., and T. Pavlidis, "A Hierarchical Data Structure for Picture Processing," *Computer Graphics and Image Processing,* Vol. 4, No. 2, 1975, pp. 104–119.

van Oosterom, P., and J. van der Bos, "An Object-Oriented Approach to the Design of Geographic Information Systems," *Computers and Graphics,* Vol. 13, 1989, pp. 409–418.

CHAPTER 9: DATABASE EXTENSIONS

Database Design

Booch, G., *Object-Oriented Design with Applications,* Redwood City, CA: Benjamin/Cummings, 1991.

Brathwaite, K.S., *Object-Oriented Database Design: Concept and Applications,* San Diego, CA: Academic Press 1993.

Engenhofer, M.J., and A.U. Frank, "Towards a Spatial Query Language: User Interface Considerations,' *Proc. Fourteenth Int. Conf. on Very Large Databases,* F. Bachillon, and D. J. DeWhitt (eds.), Los Angeles, CA August 1988, pp. 124–133.

Gardarin, G., "Towards the Fifth Generation of Database Management Systems," *New Application of Databases,* G. Gardarin, and E. Gelenbe (eds.), San Diego, CA: Academic Press, 1984, pp. 3–15.

Ishikawa, H., *Object-Oriented Database System: Design and Implementation for Advanced Applications,* New York Springer-Verlag, 1993.

Mylopoulos, J., and M. Brodie, "Knowledge Bases and Databases: Current Trends and Future Directions," *Conf. on Information Systems and Artificial Intelligence: Integration Aspects,* D. Karagiannis (ed.), Lecture Notes in Computer Science, Vol. 474, New York: Springer-Verlag, 1991, pp. 153–180.

Query Optimization

Jarke, M., J. Bubenko, and K. Jeffery (eds.), *Advances in Database Technology—EDBT, Fourth Int. Conf. on Extending Database Technology,* Cambridge, United Kingdom, March 28–31, 1994.

Jarke, M., and J. Koch, "Query Optimization in Database Systems," *ACM Comput. Surveys,* Vol. 16, No. 2 June 1984, pp. 111–152.

Kim, W.D., D. Reiner, and D. Batory (eds.), *Query Processing in Database Systems,* New York: Springer-Verlag, February 1985.

Ooi, B.C., "Efficient Query Processing in Geographical Information Systems," Lecture Notes in Computer Science, Vol. 471, Heidelberg, Germany: Springer-Verlag, 1990.

Ooi, B.C., and R. Sacks-Davis, "Query Optimization in Extended DBMS," *Foundations of Data Organization and Algorithms,* W. Litwin and H.J. Shek (eds.), Lecture Notes in Computer Science, Vol. 367, Berlin: Springer-Verlag, 1989, pp. 48–63.

Wong, E., and K. Youssefi, "Decomposition—A Strategy for Query Processing," *ACM Trans. on Database Systems,* Vol. 1, No. 3, September 1976, pp. 223–241.

Expert Databases

Brodie, M.L.., "Future Intelligent Information Systems: AI and Database Technologies Working Together," *Proc. Second Int. Conference on Expert Database Systems,* L. Kerschberg (ed.), Menlo Park, CA: Benjamin/Cummings, 1989, pp. 623–641.

Chakravarthy, S., E. Anwar, and L. Maugis, "Active Database Architectures and the Design of Sentinel for Constraint Management," Rome Laboratory, RL-TR-93-190, Griffiss Air Force Base, New York, October 1993.

Jeffery, K. (ed.), *Expert Database Systems,* San Diego, CA: Academic Press, 1992.

Kerschberg, L. (ed.), *Proc. First Int. Workshop on Expert Database Systems,* Menlo Park, CA: Benjamin/Cummings, 1986.

Parsaye et al., *Intelligent Databases: Object-Oriented, Deductive Hypermedia Technologies,* New York: Wiley, 1989.

Stonebraker, M., and M. Hearst, "Future Trends in Expert Data Base Systems," *Proc.Second Int. Conf. on Expert Database Systems,* Menlo Park, CA: Benjamin/Cummings, 1988, pp. 3–20.

Ullman, J.D., *Principles of Database and Knowledge-Base Systems,* Vol. 1, Potomac, MD: Computer Science Press, 1988.

Database Mining/Trend Discovery

Glaeser, P.S. (ed.), *Data for Discovery, Proc.Twelfth Int. CODATA Conf.,* Columbus, OH, New York: Begell House, 1992.

Piatetsky-Shapiro, G., and W.J. Frawley (eds.), *Knowledge Discovery in Databases,* Menlo Park, CA: AAAI Press, 1991.

Distributed DB issues

Agrawal, R., and D. Bell, *Second Int. Symp. on Databases in Parallel and Distributed Systems,* Dublin, Ireland, Los Alamitos, CA: IEEE Computer Society Press, 1990.

Bertino, E., and M.T. Ozsu, "Distributed and Parallel Database Object Management," *Int. J. Distributed and Parallel Databases,* Vol. 2, No. 1, Boston, MA: Kluwer Academic Publishers, 1994.

Ceri, S., and G. Pelagatti, *Distributed Databases: Principles and Systems,* New York: McGraw-Hill, 1984.

Larson, J.A., and S. Rahimi, *Tutorial: Distributed Database Management,* Silver Spring, MD: IEEE Computer Society Press, 1985.

Multimedia

Chorafas, D., *Intelligent Multimedia Database: From Object Orientation and Fuzzy Engineering to Intentional Database Structures,* Englewood Cliffs, NJ: Prentice-Hall, 1994.

Computer, Vol. 24, No. 10, Special Issue on Multimedia Information Systems, IEEE Press, October 1991.

Cotton, B., and R. Oliver, *Understanding Hypermedia: From Multimedia to Virtual Reality*, London: Phaidon Press, 1993.

Gibbs, S., "Composite Multimedia and Active Objects," *Proc. OOPSLA'91 SIGPLAN Notices*, Vol. 26, No. 11, New York: ACM Press, November 1991.

Koegel Buford, J.F., *Multimedia Systems*, New York: ACM Press and Reading, MA: Addison-Wesley, 1994.

Kunii, T.L. (ed.), *Visual Database Systems*, Amsterdam: North-Holland, 1989.

Manola, F., "Object Data Language Facilities for Multimedia Data Types," Report No. TR-0169-12-91-165, GTE Laboratories Incorporated, Waltham, MA, December 31, 1991.

CHAPTER 11: MACHINE-BASED REASONING

Johnson-Laird, P.N., *The Computer and the Mind: An Introduction to Cognitve Science*, Cambridge, MA: Harvard Unversity Press, 1988.

Newell, A., and H.A. Simon, *Human Problem Solving*, Englewood Cliffs, NJ: Prentice Hall, 1972.

Rose, F., *Into the Heart of the Mind: An American Quest for Artificial Intelligence*, New York: Harper & Row, 1984.

Russell, S.J., *The Use of Knowledge in Analogy and Induction*, London: Pitman, 1989.

Schank, R.C., and R.P. Abelson, *Scripts, Plans, and Understanding: An Inquiry into Human Knowledge Structures*, Hillsdale, NJ: Lawrence Erlbaum Associates, distributed by Halsted Press, Division Wiley, 1977.

Willensky, R., *Planning and Understanding—A Computational Approach to Human Reasoning*, Reading, MA: Addison-Wesley, 1983.

CHAPTER 12: HIERARCHICAL, SPATIAL, AND TEMPORAL REASONING

Hierarchical Reasoning

Corkill, D.D., "Hierarchical Planning in a Distributed Environment," *International Joint Conference on Artificial Intelligence*, Tokyo, Japan, August 20–23, 1979, distributed by William Kaufman, Inc., Vol. IJCAI-6, 1979, pp.168–175.

Dean, T., R.J. Firby, and D. Miller, "Hierarchical Planning Involving Deadlines, Travel Time, and Resources," *Computational Intelligence*, Vol. 4, No. 4, 1988, pp. 381–398.

Erman, L.D., F. Hayes-Roth, V.R. Lesser, and D.R. Reddy, "The Hearsay-II Speech Understanding System: Integrating Knowledge to Resolve Uncertainty," *ACM Computing Surveys*, Vol. 12, No. 2, June 1980, pp. 213–253.

Graham, M.L., K.F. Gong, and A.G. Lindgren, "A Hierarchical Approach to Data Fusion for Tracking in an Underwater Environment," *Proc. 1987 Tri-Service Data Fusion Symp.*, Johns Hopkins University Applied Physics Lab, Laurel, MD, June 1987, pp. 576–585.

Kanal, L., and B. Chandrasekaran, "On Linguistic, Statistical and Mixed Models for Pattern Recognition," *Proc. Int. Conf. on Frontiers of Pattern Recognition*, S. Watanabe (ed.), New York: Academic Press, 1972, pp. 161–185.

Knoblock, C.A., *Generating Abstraction Hierarchies: An Automated Approach to Reducing Search in Planning*, Boston, MA: Kluwer Academic, 1993.

Miller, J.R., and J.H. Swaffield, "DACORR II: An Operational User, Hierarchical Knowledge-Based Tactical Decision Support System," *Proc. 1990 Joint Service Data Fusion Symp.*, Laurel, MD, 1990, pp. 511–518.

Sacerdoti, E.D., "Planning in a Hierarchy of Abstraction Spaces," *Artificial Intelligence*, Vol. 5, No. 2, 1974, pp. 115–135.

Spatial Reasoning

Al-Taha, K.K., R.T. Snodgrass, and M.D. Soo, "Bibliography of Spatiotemporal Databases," *SIGMOD Record*, Vol. 22, 1 March 1993, pp. 59–67.

Antony, R., "Spatial Reasoning Using an Object-Oriented Spatial DBMS," AAAI-sponsored Workshop on Spatial Reasoning and Multi-Sensor Fusion, St. Charles, IL, October 1987, pp. 42–51.

Antony, R., and P. Emmerman, "Spatial Reasoning and Knowledge Representation," *Geographic Information Systems in Government*, B. Opitz (ed.), Vol. 2, A. Deepak Publishing, 1986, pp. 795–814.

Chen, S.-S. (ed.), *Advances in Spatial Reasoning*, Vols. 1 and 2, Norwood, NJ: Ablex, 1990.

DeFloriani, L., G. Nagy, and E. Puppo, "Computing Line-of-Sight Network on a Terrain Model," *Proc. Fifth Int. Symp. on Spatial Data Handling*, International Geographic Union, Columbus, OH, 1992, pp. 672–681.

Deo, D., and C. Pang, "Shortest Path Algorithms: Taxonomy and Annotation," *Networks*, Vol. 14, 1984, pp. 275–323.

Gonzalez, R.C., and P. Wintz, *Digital Image Processing*, Reading, MA: Addison-Wesley, 1977.

Hayslip, I.C., and J.F. Gilmore, "A Multi-Level Knowledge Representation for Reasoning About Terrain," *U.S. Army Symp. on Artificial Intelligence Research for Exploitation of the Battlefield Environment*, El Paso, TX, November 15–16, 1988, pp. 123–137.

Hepner, G.F., T.L. Logan, and N.A. Bryan, "Spatial Query for Decision Support of Cross-Country Movement," *Photogrammetry, Engineering and Remote Sensing*, Vol. 54, No. 11, 1988, pp. 1,617–1,621.

Lozano-Perez, T., "Spatial Planning a Configuration Space Approach," *IEEE Trans. on Computers*, C-32, No. 2, 1983, pp. 108–120.

Magee, M., W. J. Wolfe, D. Mathis, and C. Weber, "Model-Based Spatial Reasoning for Hierarchically Organized Structured Objects," *Advances in Spatial Reasoning*, S.-S. Chen (ed.), Vol. 1, Norwood, NJ: Ablex, 1990, pp. 181–218.

Mitchell, J.S.B., "An Algorithmic Approach to Some Problems in Terrain Navigation," *Artificial Intelligence*, Vol. 37, 1988, pp. 171–201.

Pizano, A.P., A. Klinger, and A. Cardenas, "Specification of Spatial Integrity Constraints in Pictorial Databases," *Computer*, December 1989, pp. 59–71.

Preparata, F.P., and M.I. Shamos, *Computational Geometry: An Introduction*, New York: Springer-Verlag, 1985.

Samet, H., and M. Tamminen, "Computing Geometric Properties of Images Represented by Linear Quadtrees," *IEEE Trans. on Pattern Analysis and Machine Intelligence*, Vol. 7, No. 2, March 1985, pp. 229–240.

Slack, M.G., "Planning Paths Through a Spatial Hierarchy: Eliminating Stair-Stepping Effects," *SPIE Sensor Fusion: Spatial Reasoning and Scene Interpretation*, Vol. 3, 1988, pp. 350–357.

Wong, E.K., and K.S. Fu, "A Hierarchical Orthogonal Space Approach to Three-Dimensional Path Planning," *IEEE Trans. on Robotics Automation*, RA-2, No. 1, 1986, pp. 43–53.

Temporal Reasoning

Allen. J.F., "Towards a General Theory of Action and Time," *Artificial Intelligence,* Vol. 23, No. 2, July 1984, pp.123–154.

Bestougeff, H., and G. Ligozat, *Logical Tools for Temporal Knowledge Representation,* New York: Ellis Horwood, 1992.

Boddy, M., and T. Dean, "Solving Time-Dependent Planning Problems," *Proc. Eleventh Int. Joint Conf. on Artificial Intelligence,* Menlo Park, CA, 1989, pp. 979–984.

Dean, T., and M. Boddy, "An Analysis of Time-Dependent Planning," *Proc. Seventh National AAAI Conf. on Artificial Intelligence,* Menlo Park, CA, 1988, pp. 49–54.

de Kleer, J., "An Assumption-Based TMS," *Artificial Intelligence,* Vol. 28, 1986, pp. 127–162.

Doyle, J., "A Truth Maintenance System," *AI J.,* Vol. 12, 1979, pp 231–272.

Fall, T.C., "Evidential Reasoning with Temporal Aspects," *Fifth National Proc. AAAI,* Vol. 2, Philadelphia, PA, August 11–15, 1986, pp. 891–895.

Frank, A.U., I. Campari, and U. Formentini (eds.), *Proc. Int. Conf. GIS—From Space to Territory: Theories and Methods of Spatio-Temporal Reasoning in Geographic Space,* Pisa, Italy, New York: Springer-Verlag, September 1992.

McDermott, D.A., "A Temporal Logic for Reasoning About Processes and Plans," *Cognitive Science,* Vol. 6, 1982, pp. 101–155.

Meier, L., "Data Fusion Over Time Using Bayesian Reasoning," *1993 Joint Services Data Fusion Symp.*, Laurel, MD, 1993, pp. 919–935.

Noble, D., "Managing Temporal Uncertainty in Situation Assessment," *Proc. 1988 Tri-Service Data Fusion Symp.*, Laurel, MD, 1988, pp. 107–112.

Schank, R.C., "Scripts Plans and Knowledge," *Proc. Fourth Int. Joint Conf. on Artificial Intelligence,* Tbilisi, USSR, September 3–8, 1975, Artificial Intelligence Laboratory, Cambridge, MA, 1975, pp. 151–157.

Strosnider, J.K., and C.J. Paul, "A Structured View of Real-Time Problem-Solving," *AI Magazine,* Summer 1994, pp. 45–66.

Vere, S.A., "Planning in Time: Windows and Duration for Activities and Goals," *Readings in Planning,* J. Allen, J. Hendler, and A. Tate (eds.), Palo Alto, CA: Morgan Kaufmann, 1990.

About the Author

Richard Antony has been involved in data fusion research since the early 1980s. He has been an Army member of the Joint Directors of Laboratories Data Fusion Group since 1987. His previous experience included large-scale digital simulations, fuze end-game modeling, systems analysis, and target classification. He received his BS in Electrical Engineering in 1968 and his MS in Communications Theory in 1972 from the University of Maryland.

Mr. Antony lives in the foothills of the Blue Ridge Mountains with his wife and two children. In addition to his interests in growing and pruning quadtrees, the author is a part-time nurseryman, assisting his nursery-manager wife in the evenings and on weekends with the family's wholesale azalea nursery.

Index

The Artech House Radar Library

David K. Barton, Series Editor

Active Radar Electronic Countermeasures, Edward J. Chrzanowski

Adaptive Signal Processing for Radar, Ramon Nitzberg

Airborne Early Warning System Concepts, Maurice Long, *et al.*

Airborne Pulsed Doppler Radar, Guy V. Morris

AIRCOVER: Airborne Radar Vertical Coverage Calculation Software and User's Manual, William A. Skillman

Analog Automatic Control Loops in Radar and EW, Richard S. Hughes

Aspects of Modern Radar, Eli Brookner *et al.*

Aspects of Radar Signal Processing, Bernard Lewis, Frank Kretschmer, and Wesley Shelton

Bistatic Radar, Nicholas J. Willis

CARPET (Computer-Aided Radar Performance Evaluation Tool): Radar Performance Analysis Software and User's Manual, developed at the TNO Physics and Electronic Laboratory by Albert G. Huizing and Arne Theil

Coherent Radar Performance Estimation, James A. Scheer and James L. Kurtz, editors

Cryptology Yesterday, Today, and Tomorrow, Cipher Deavours *et al.*

Cryptology: Machines, History, and Methods, Cipher Deavours *et al.*

DETPROB: Probability of Detection Calculation Software and User's Manual, William A. Skillman

Designing Automatic Control Systems, W. L. Rubin

Detectability of Spread-Spectrum Signals, Robin A. Dillard and George M. Dillard

Electronic Countermeasure Systems Design, Richard Wiegand

Electronic Homing Systems, M. V. Maksimov and G. I. Gorgonov

Electronic Intelligence: The Analysis of Radar Signals, Second Edition, Richard G. Wiley

Electronic Intelligence: The Interception of Radar Signals, Richard Wiley

Electronic Warfare Receiving Systems, Dennis D. Vaccaro

Engineer's Refractive Effects Prediction Systems—EREPS, Herbert Hitney

EREPS: Engineer's Refractive Effects Prediction System Software and User's Manual, developed by NOSC

Estimation and Tracking: Principles, Techniques, and Software, Yaakov Bar-Shalom and Xiao-Rong-Li

High Resolution Radar, Donald R. Wehner

High Resolution Radar Cross-Section Imaging, Dean Mensa

Interference Suppression Techniques for Microwave Antennas and Transmitters, Ernest R. Freeman

Introduction to Electronic Defense Systems, Fillippo Neri

Introduction to Electronic Warfare, D. Curtis Schleher

Introduction to Sensor Systems, S. A. Hovanessian

IONOPROP: Ionospheric Propagation Assessment Software and Documentation, Herbert Hitney

Kalman-Bucy Filters, Karl Brammer

Laser Radar Systems, Albert V. Jelalian

Lidar in Turbulent Atmosphere, V. A. Banakh and V. L. Mironov

Logarithmic Amplification, Richard Smith Hughes

Machine Cryptography and Modern Cryptanalysis, Cipher A. Deavours and Louis Kruh

Millimeter-Wave Radar Clutter, Nicholas Currie, Robert Hayes, and Robert Trebits

Modern Radar System Analysis, David K. Barton

Modern Radar System Analysis Software and User's Manual, David K. Barton and William F. Barton

Monopulse Radar, A. I. Leonov and K. I. Fomichev

MTI and Pulsed Doppler Radar, D. Curtis Schleher

Multifunction Array Radar Design, Dale R. Billetter

Multifuctional Array Radar: System Design and Analysis, Sergio Sabatini and Marco Tarantino

Multisensor Data Fusion, Edward L. Waltz and James Llinas

Multiple-Target Tracking with Radar Applications, Samuel S. Blackman

Multitarget-Multisensor Tracking: Advanced Applications, Volumes I and II, Yaakov Bar-Shalom, editor

Over-The-Horizon Radar, A. A. Kolosov *et al.*

Photonic Aspects of Modern Radar, Henry Zmuda and Edward N. Toughlian

Principles and Applications of Millimeter-Wave Radar, Charles E. Brown and Nicholas C. Currie, editors

Principles of Data Fusion Automation, Richard T. Antony

Principles of Traveling Wave Tubes, A. S. Gilmour

Pulse Train Analysis Using Personal Computers, Richard G. Wiley and Michael B. Szymanski

Radar and the Atmopshere, Alfred J. Bogush, Jr.

Radar Anti-Jamming Techniques, M. V. Maksimov *et al.*

Radar Cross Section Analysis and Control, A. K. Bhattacharyya and D. L. Sengupta

Radar Cross Section, Second Edition, Eugene F. Knott *et al.*

Radar Electronic Countermeasures System Design, Richard J. Wiegand

Radar Engineer's Sourcebook, William Morchin

Radar Evaluation Handbook, David K. Barton *et al.*

Radar Meteorology, Henri Sauvageot

Radar Polarimetry for Geoscience Applicatons, Fawwaz Ulaby

Radar Propagation at Low Altitudes, M. L. Meeks

Radar Reflectivity Measurement: Techniques and Applications, Nicholas C. Currie, editor

Radar Signals: An Introduction to Theory and Application, Charles E. Cook and Marvin Bernfeld

Radar System Design and Analysis, S. A. Hovanessian

Radar Technology, Eli Brookner, editor

Radar Vulnerability to Jamming, Robert N. Lothes, Michael B. Szymanski, and Richard G. Wiley

RGCALC: Radar Range Detection Software and User's Manual, John E. Fieldng and Gary D. Reynolds

Russian-English and English-Russian Dictionary of Radar and Electronics, Sergey A. Leonov and William F. Barton

SACALC: Signal Analysis Software and User's Guide, William T. Hardy

Signal Detection and Estimation, Mourad Barkat

Small-Aperture Radio Direction Finding, Herndon Jenkins

Solid-State Radar Transmitters, Edward D. Ostroff *et al.*

Space-Based Radar Handbook, Leopold J. Cantafio, editor

Spaceborne Weather Radar, Robert M. Meneghini and Toshiaki Kozu

Statistical Signal Characterization, Herbert Hirsch

Statistical Signal Characterization: Algorithms and Analysis Programs Software and User's Manual, Herbert Hirsch

Surface-Based Air Defense System Analysis, Robert Macfadzean

Surface-Based Air Defense System Analysis Software and User's Manual, Robert Macfadzean and James M. Johnson

Synthetic-Aperture Radar and Electronic Warfare, Walter W. Goj

VCCALC: Vertical Coverage Plotting Software and User's Manual, John E. Fielding and Gary D. Reynolds

For further information on these and other Artech House titles, contact:

Artech House
685 Canton Street
Norwood, MA 02062
617-769-9750
Fax: 617-769-6334
Telex: 951-659
email: artech@world.std.com

Artech House
Portland House, Stag Place
London SW1E 5XA England
+44 (0) 171-973-8077
Fax: +44 (0) 171-630-0166
Telex: 951-659
bookco@artech.demon.co.uk